AF411368

Microstructure and Mechanical Properties of Titanium Alloys

Microstructure and Mechanical Properties of Titanium Alloys

Editor

Artur Shugurov

MDPI • Basel • Beijing • Wuhan • Barcelona • Belgrade • Manchester • Tokyo • Cluj • Tianjin

Editor
Artur Shugurov
Russian Academy of Sciences
Russia

Editorial Office
MDPI
St. Alban-Anlage 66
4052 Basel, Switzerland

This is a reprint of articles from the Special Issue published online in the open access journal *Metals* (ISSN 2075-4701) (available at: https://www.mdpi.com/journal/metals/special_issues/ microstructure_mechanical_titanium).

For citation purposes, cite each article independently as indicated on the article page online and as indicated below:

LastName, A.A.; LastName, B.B.; LastName, C.C. Article Title. *Journal Name* **Year**, *Volume Number*, Page Range.

ISBN 978-3-0365-2283-8 (Hbk)
ISBN 978-3-0365-2284-5 (PDF)

Contents

About the Editor

Prof. Artur Shugurov was born in 1967. He graduated (1992) from Physical Faculty of Tomsk State University, Tomsk, Russia. And, finished (2000) post-graduate study in Institute of Strength Physics and Materials Science, Siberian Branch of Russian Academy of Science, Tomsk, Russia. Prof. Artur Shugurov was awarded with PhD (2021), and Doctor of Science degree (2016). Since 2021, he has becomed a leading researcher in Institute of Strength Physics and Materials Science, Siberian Branch of Russian Academy of Science, Tomsk, Russia. Until now, published more than 150 scientific works on Physics and Materials Science.

Editorial

Microstructure and Mechanical Properties of Titanium Alloys

Artur Shugurov *

Institute of Strength Physics and Materials Science, Siberian Branch, Russian Academy of Sciences, 634055 Tomsk, Russia

Citation: Shugurov, A. Microstructure and Mechanical Properties of Titanium Alloys. *Metals* **2021**, *11*, 1617. https://doi.org/10.3390/met11101617

Received: 28 September 2021
Accepted: 11 October 2021
Published: 12 October 2021

1. Introduction and Scope

Titanium and its alloys are widely used engineering materials within the aerospace, automotive, energy, and chemical industries. Their unique combination of high strength-to-weight ratio, strong resistance to creep, excellent corrosion resistance, and low heat conductivity makes them suitable for many applications. A large variety of microstructures, including lamellar, martensitic, equiaxed globular, and bimodal (duplex) microstructures, can be obtained in titanium alloys depending on the thermomechanical processing routes. Despite a large amount of work in the investigation of microstructure evolution and mechanical properties of titanium alloys, detailed studies of the effect of their microstructure on the mechanical behaviour are still necessary because of ever-increasing demands for structural materials to optimize their properties for different applications, by varying processing parameters and resulting microstructures.

This Special Issue focuses on various aspects of microstructure evolution in titanium alloy samples obtained using traditional and additive technologies and subjected to different processing techniques, as well as on the relation between their microstructure and mechanical behaviour. The presented original articles cover the areas of preparation and experimental characterization of titanium alloys, as well as computer simulation of their mechanical behaviour under different loading conditions.

2. Contributions

Seventeen papers of high scientific quality have been published in the present Special Issue of *Metals*, covering the fields of thermomechanical processing, mechanical behavior, surface modification, electrochemical micromachining, and bonding mechanisms of different titanium alloys. The contents of the published manuscripts are briefly summarized below.

Despite a large variety of developed thermomechanical processing routes, the issue of thermomechanical treatment of titanium alloys is still very topical because of the emergence of new titanium and titanium-based alloys and the need to further improve the performance of existing alloys and therefore modify their microstructure. This is confirmed by the fact that eight papers consider microstructure evolution and the mechanical properties of Ti-based alloys subjected to different thermomechanical treatments. The investigation of the two-pass thermal compression behaviour of a near-β Ti-55511 alloy revealed the characteristic features of its dynamic recovery and recrystallization [1]. The evolution of inelastic and plastic strains in a $Ti_{49.3}Ni_{50.7}$ alloy subjected to torsional deformation was studied in [2]. It was found that in the temperature range studied, the maximum inelastic strain that could be obtained in the coarse-grained samples subjected to torsion could be completely returned due to the superelasticity effect of 4–7%. The effect of true strains on the microstructure and mechanical properties of a $Ti_{49.8}Ni_{50.2}$ alloy subjected to isothermal abc pressing was analysed in [3]. It was shown that the grain structure of the alloy was refined by increasing the true strain, which increased its yield stress and strain-hardening coefficient. Defects of the crystal structure formed in $Ti_{49.4}Ni_{50.6}$ and $Ti_{50}Ni_{47.3}Fe_{2.7}$ alloys after equal-channel angular pressing were studied by positron lifetime spectroscopy [4]. The results obtained indicate that the formation of ultrafine-grained structures in a $Ti_{50}Ni_{47.3}Fe_{2.7}$ alloy does not change the martensite transformation sequence B2 $\leftrightarrow$ R $\leftrightarrow$ B19′ but reduces the start temperature of the R $\leftrightarrow$ B19′ transition by 18 K, and narrows the B19′ $\rightarrow$ R temperature interval by 10 K. The microstructural evolution of a titanium-based Ti-3.3Al-5Mo-5V alloy subjected to hot pressing and radial forging was investigated [5]. The radial forging of the alloy was found to lead to structural refinement and changes in grain morphology from predominately lamellar to uniform globular, as well as to variations in the phase composition of the alloy and the suppression of powder formation. The microstructure and mechanical properties of a newly developed Fe-microalloyed Ti–6Al–4V titanium alloy subjected to different heat treatments were studied by Liu et al. [6]. Optimal mechanical performance of the alloy was shown to be achieved via β-annealing at 1005 °C for 70 min, followed by air cooling to room temperature and aging at 722 °C for 2 h with subsequent air cooling to room temperature. Kashin et al. reported the results of the experimental study of the microstructure and martensite transformations of $Ti_{49.8}Ni_{50.2}$ alloy after *abc* pressing at 573 K [7]. They showed that increasing the true strain resulted in grain–subgrain refinement on different scales and proposed possible mechanisms for this effect. The structure and multistage martensite transformation in nanocrystalline Ti-50.9Ni alloy was studied [8]. The research revealed that the presence of different types of internal interfaces in the nanostructure contributed to the heterogeneous distribution of coherent Ti_3Ni_4 nanoparticles in the volume of the B2 matrix, which was associated with the precipitation of particles in the region of low-angle sub-boundaries and the suppression of the Ti_3Ni_4 precipitation in nanograins with high-angle boundaries.

Two papers considered the mechanical behaviour of titanium alloys under different loadings. The first manuscript presents the results of systematic investigations of the cyclic behaviour of a Ti-3Al-8V-6Cr-4Mo-4Zr alloy with three different microstructures [9]. The cyclic stress response was shown to be highly related to the applied strain amplitude and precipitated phase. The second paper analysed the deformation behaviour of Ti-6Al-4V alloy samples with lamellar and bimodal microstructures subjected to scratch testing experimentally and using molecular dynamics simulation [10]. It was found that the scratch depth in the sample with a bimodal microstructure was twice as shallow as that measured in the sample with a lamellar microstructure. This effect was attributed to the greater hardness of the sample with a bimodal microstructure and the larger amount of elastic recovery of scratch grooves in this sample. Based on the results of molecular dynamics simulation, a mechanism was proposed, which associates the recovery of the scratch grooves with the inhomogeneous vanadium distribution in the β-areas.

Studies of the surface modification of titanium alloys also contributed a substantial part to the issue. Smyslova et al. investigated near-surface layer microstructure of Ti-6Al-4V alloy samples subjected to plasma electrolytic polishing with subsequent high-energy nitrogen ion implantation [11]. It was found that the initial structural state of the Ti-6Al-4V alloy substrate had a significant effect on the transformation of the dislocation substructure during the treatment. Experimental and theoretical studies of the surface roughening and the microstructure refinement in the surface layer of commercially pure titanium during ultrasonic impact treatment were performed in [12]. It was shown that the surface plastic strains of the titanium sample proceeded according to the plastic ploughing mechanism, which was accompanied by dislocation sliding, twinning, and the transformations of the microstructure and phase composition. The role of the electronic subsystem in the development of the strain-induced phase transformations during ultrasonic impact treatment was discussed. The titanium alloy surface was modified with copper ions to improve ceramic coatings' strength properties, adhesion, and thermal cycling resistance [13]. A multilevel micro- and nanoporous nanocrystalline structure was shown to form in the surface layer of the titanium alloy samples, which increased the adhesion and thermal cyclic resistance of the overlying Si-Al-N coating. The physical mechanism, reasons, and conditions of nanocrystal formation in an amorphous NiTi metal film stimulated by infrasonic action are formulated [14]. The transformation of the amorphous film into the nanostructured one was explained by the accumulation of the potential energy of inelastic deformation to a critical value equal to the latent heat of the transformation.

Other papers are related to an analysis of the influence of sodium-chloride-based electrolytes on machining a Ti-6Al-4V alloy [15], the effect of the microstructure of a titanium alloy with a trimodal microstructure on plastic deformation and crack growth mechanisms [16], and the adhesion properties of the TiAl/TiO2 interface estimated in dependence on interfacial layer composition and contact configuration using the projector-augmented wave method [17].

3. Conclusions and Outlook

The contributions included in the present Special Issue of *Metals* cover a wide range of research on titanium alloys, representing a well-balanced combination of theoretical and practical efforts. The papers provide a comprehensive overview of recent progress in designing and experimentally characterizing titanium alloys and investigating their microstructure evolution and mechanical behaviour. Hopefully, the present studies will be interesting for a wide range of readers and stimulate further investigations of titanium alloys' microstructure and mechanical properties.

As guest editor, I am very happy to report the success of this Special Issue and expect that the papers will be useful to scientists and engineers working in the development of new materials and improvement of their performance. I am sincerely grateful to the authors for their contributions and the reviewers for their significant efforts in providing high-quality publications. Sincere thanks to editors and editorial assistants of *Metals* for their continuous support during the preparation of this volume. In particular, I would like to warmly acknowledge Ms. Sammi Meng for her valuable assistance.

Funding: This research received no external funding.

Conflicts of Interest: The authors declare no conflict of interest.

References

1. Wang, H.; Ge, J.; Zhang, X.; Chen, C.; Zhou, K. Investigation of the Dynamic Recovery and Recrystallization of Near-β Titanium Alloy Ti-55511 during Two-Pass Hot Compression. *Metals* **2021**, *11*, 359. [CrossRef]
2. Zhapova, D.; Grishkov, V.; Lotkov, A.; Timkin, V.; Gusarenko, A.; Rodionov, I. Behavior of Inelastic and Plastic Strains in Coarse-Grained Ti49.3Ni50.7(at%) Alloy Deformed in B2 States. *Metals* **2021**, *11*, 741. [CrossRef]
3. Kashin, O.; Krukovskii, K.; Lotkov, A.; Grishkov, V. Effect of True Strains in Isothermal abc Pressing on Mechanical Properties of Ti49.8Ni50.2 Alloy. *Metals* **2020**, *10*, 1313. [CrossRef]
4. Lotkov, A.; Baturin, A.; Kopylov, V.; Grishkov, V.; Laptev, R. Structural Defects in TiNi-Based Alloys after Warm ECAP. *Metals* **2020**, *10*, 1154. [CrossRef]
5. Zuev, L.B.; Shlyakhova, G.V.; Barannikova, S.A. Effect of Radial Forging on the Microstructure and Mechanical Properties of Ti-Based Alloys. *Metals* **2020**, *10*, 1488. [CrossRef]
6. Liu, Y.; Chen, F.; Xu, G.; Cui, Y.; Chang, H. Correlation between Microstructure and Mechanical Properties of Heat-Treated Ti–6Al–4V with Fe Alloying. *Metals* **2020**, *10*, 854. [CrossRef]
7. Kashin, O.; Lotkov, A.I.; Grishkov, V.; Krukovskii, K.; Zhapova, D.; Mironov, Y.; Girsova, N.; Kashina, O.; Barmina, E. Effect of abc Pressing at 573 K on the Microstructure and Martensite Transformation Temperatures in Ti49.8Ni50.2 (at%). *Metals* **2021**, *11*, 1145. [CrossRef]
8. Poletika, T.M.; Girsova, S.L.; Lotkov, A.I.; Kudryachov, A.N.; Girsova, N.V. Structure and Multistage Martensite Transformation in Nanocrystalline Ti-50.9Ni Alloy. *Metals* **2021**, *11*, 1262. [CrossRef]
9. Zhang, S.; Zhang, H.; Hao, J.; Liu, J.; Sun, J.; Chen, L. Cyclic Stress Response Behavior of near β Titanium Alloy and Deformation Mechanism Associated with Precipitated Phase. *Metals* **2020**, *10*, 1482. [CrossRef]
10. Shugurov, A.R.; Panin, A.V.; Dmitriev, A.I.; Nikonov, A.Y. Recovery of Scratch Grooves in Ti-6Al-4V Alloy Caused by Reversible Phase Transformations. *Metals* **2020**, *10*, 1332. [CrossRef]
11. Smyslova, M.K.; Valiev, R.R.; Smyslov, A.M.; Modina, I.M.; Sitdikov, V.D.; Semenova, I.P. Microstructural Features and Surface Hardening of Ultrafine-Grained Ti-6Al-4V Alloy through Plasma Electrolytic Polishing and Nitrogen Ion Implantation. *Metals* **2021**, *11*, 696. [CrossRef]
12. Panin, A.; Dmitriev, A.; Nikonov, A.; Kazachenok, M.; Perevalova, O.; Sklyarova, E. Transformations of the Microstructure and Phase Compositions of Titanium Alloys during Ultrasonic Impact Treatment. Part I. Commercially Pure Titanium. *Metals* **2021**, *11*, 562. [CrossRef]

13. Fedorischeva, M.; Kalashnikov, M.; Bozhko, I.; Perevalova, O.; Sergeev, V. Effect of Surface Modification of a Titanium Alloy by Copper Ions on the Structure and Properties of the Substrate-Coating Composition. *Metals* **2020**, *10*, 1591. [CrossRef]
14. Slyadnikov, E.E. Infrasonic Nanocrystal Formation in Amorphous NiTi Film: Physical Mechanism, Reasons and Conditions. *Metals* **2021**, *11*, 1390. [CrossRef]
15. Thangamani, G.; Thangaraj, M.; Moiduddin, K.; Mian, S.H.; Alkhalefah, H.; Umer, U. Performance Analysis of Electrochemical Micro Machining of Titanium (Ti-6Al-4V) Alloy under Different Electrolytes Concentrations. *Metals* **2021**, *11*, 247. [CrossRef]
16. Tan, C.; Fan, Y.; Sun, Q.; Zhang, G. Improvement of the Crack Propagation Resistance in an α + β Titanium Alloy with a Trimodal Microstructure. *Metals* **2020**, *10*, 1058. [CrossRef]
17. Bakulin, A.V.; Schmauder, S.; Kulkov, S.S.; Kulkova, S.E.; Hocker, S. First Principles Study of Bonding Mechanisms at the TiAl/TiO2 Interface. *Metals* **2020**, *10*, 1298. [CrossRef]

Article

Investigation of the Dynamic Recovery and Recrystallization of Near-β Titanium Alloy Ti-55511 during Two-Pass Hot Compression

Hande Wang, Jinyang Ge, Xiaoyong Zhang *, Chao Chen * and Kechao Zhou

State Key Laboratory of Powder Metallurgy, Central South University, Lu Mountain South Road, Changsha 410083, China; 183312092@csu.edu.cn (H.W.); gjy416065@126.com (J.G.); zhoukechao@csu.edu.cn (K.Z.)
* Correspondence: zhangxiaoyong@csu.edu.cn (X.Z.); pkhqchenchao@126.com (C.C.)

Abstract: The two-pass thermal compression behavior of near-β Ti-55511 alloy was investigated. The first-pass restoration mechanisms changed from dynamic recrystallization (DRX) to dynamic recovery (DRV) as the first-pass deformation temperature increased from 700 °C to 850 °C. The occurrence of recrystallization reduced the dislocation density, resulting in a slower grain growth rate in the subsequent process. Because of the static recrystallization (SRX) and β grain growth, the β grain size increased and the morphology became less uniform during the subsequent β holding process, which also changed the restoration mechanism during second-pass compression. The level of continuous dynamic recrystallization (CDRX) and discontinuous dynamic recrystallization (DDRX) become weaker during second-pass deformation. The changes in the restoration mechanism and the microstructures slightly increased the peak stress during the second-pass deformation.

Keywords: near-β titanium alloy; two-pass hot compression; dynamic recovery (DRV); continuous dynamic recrystallization (CDRX); discontinuous dynamic recrystallization (DDRX)

Citation: Wang, H.; Ge, J.; Zhang, X.; Chen, C.; Zhou, K. Investigation of the Dynamic Recovery and Recrystallization of Near-β Titanium Alloy Ti-55511 during Two-Pass Hot Compression. *Metals* **2021**, *11*, 359. https://doi.org/10.3390/met11020359

Academic Editor: Artur Shugurov

Received: 1 February 2021
Accepted: 17 February 2021
Published: 20 February 2021

Publisher's Note: MDPI stays neutral with regard to jurisdictional claims in published maps and institutional affiliations.

1. Introduction

Near-β titanium Ti-55511 is a high-strength lightweight alloy with good fatigue performance that is widely used in the aerospace field [1–3]. The mechanical properties of the Ti-55511 alloy are closely related to its microstructure [4,5]. In general, titanium alloys with a fine microstructure have better mechanical properties than those with coarse microstructures [6]. Forging is a widely used industrial grain refinement method [7]. Deformation parameters, such as temperature, strain rate, and strain, play important roles during thermal forging [8–11]. For example, reducing the deformation temperature in the α + β region can effectively reduce the β grain size [10], but this is not conducive to grain refinement in the β region [12]. A relatively low strain rate and large strain are favorable for the recrystallization of β grains, which promotes β grain refinement [13]; however, it is very difficult to process titanium alloys by single-pass large plastic deformation due to the large deformation resistance [8]. Multi-pass deformation was developed to efficiently process titanium alloys. Zhan et al. [14] found that grain refinement occurred by multi-pass deformation, and the contents of <5 μm β grains increased significantly upon increasing the number of compression passes from 1 to 5 at 800 °C. Zherebtsov et al. [10] found that the deformation temperature during subsequent passes could be reduced appropriately because the workability was improved after the previous pass deformation. The grain size was reduced by decreasing the deformation temperature, and ultrafine grains with an average size of 150 nm were obtained at a deformation temperature of 475 °C. To obtain fine and homogeneous β grains, a "high-low-high" process was proposed which involved primary working in the β region, followed by working in the α + β region, and finally β working again [15].

Dynamic recovery (DRV) and dynamic recrystallization (DRX) are the main mechanisms that influence the grain refinement during multi-pass deformation, and they are very

5

sensitive to the deformation temperature [13,16]. Warchomicka et al. [13] found that DRV is the dominant restoration mechanism at the beginning of deformation. Subsequently, geometric dynamic recrystallization (GDRX) occurs below the β transition temperature, while continuous dynamic recrystallization (CDRX) occurs above the β transition temperature. The emergence of DRV and DRX influences the microstructure deformation. Ning et al. [17] found that the periodic competition between DRX and DRV resulted in the incomplete growth of recrystallized grains. Different initial microstructures also produce differences in the restoration behavior. For example, Wu et al. [18] found that, since the recrystallization rate of the deformed microstructure is significantly higher than that of the undeformed microstructure, β grains were refined during subsequent deformation. Dehghan-Manshadi et al. [19] found that the recrystallization mechanism changed from DDRX to CDRX upon decreasing the initial grain size.

The fundamental factors affecting recrystallization are nucleation and growth rate, which are related to the dislocation density [20]. The α phase plays an important role in the movement and entanglement of dislocations because it has limited slip systems during deformation [21]. As a result, dislocations are more likely to accumulate around α precipitates, which then promote β grain nucleation and inhibit β grain growth [22,23]. Moreover, recrystallization can be improved by changing the α morphology before deformation [11,21]. The typical initial microstructures, lamellar and equiaxed, have different effects on the dislocation motion. Wei et al. [24] found that dislocations more easily accumulated around equiaxed α grains and then formed high-angle grain boundaries after hot-rolling with a strain of 80% at 750 °C. As a result, the size of the β grain with an initial equiaxed α microstructure was finer than that with an initial lamellar α microstructure. On the other hand, more slip systems were activated by increasing the number of deformation passes in which the slip mode changed from base slip to pyramid slip [25]. Multi-pass deformation leads to a higher dislocation density and makes it easier to form finer grains in the β matrix [26].

Previous research on the recrystallization mechanism has mainly focused on single-pass deformation. Multi-pass forging is the main processing method of Ti-55511 ingots, and there is limited research on the interactions between deformation mechanisms during multi-pass deformation. In this study, four types of two-pass hot compression methods were designed to investigate the influence of the first-pass deformation restoration behavior on subsequent deformation. This work for Ti-55511 provides essential references for process parameter optimizing and microstructure controlling under multi-pass forging conditions.

2. Materials and Methods

Ti-55511 ingot with the composition (ωt%) of 5.16 Al, 4.92 Mo, 4.96 V, 1.10 Cr, 0.98 Fe, and Ti (balance) was provided by Xiangtou Goldsky Titanium Industry (Changde, Hunan, China). The β-transition temperature was found to be approximately 875 °C using the metallographic method. Figure 1 shows the original microstructure composed of large β grains with an average size of 300 μm. Inside β grains, there are α duplex microstructures including (1) an equiaxed α grain with a diameter of 5–8 μm and (2) a lamellar α grain with a length of 3–5 μm.

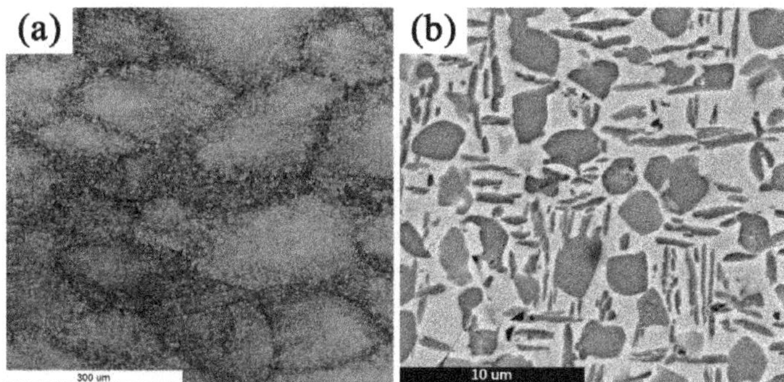

Figure 1. (**a**) Metallographic image and (**b**) SEM image of the original microstructure of Ti-55511.

Samples of $\Phi 10 \times 15$ mm sizes were line-cut from an ingot and then mechanically polished. Thermal compression was conducted on a Gleeble 3800 thermo-mechanical simulator (Data sciences international, INC, New Brighton, MN, USA) at a strain rate of 0.1 s^{-1}, including three stages as shown in Figure 2: (1) compression to a true strain of 0.36 at 700–850 °C, followed by water quenching (Stage-D1); (2) heating to 950 °C and holding for 3 min (Stage-H); (3) continuing to compress at a true strain of 0.56 and then water quenching (Stage-D2). During compression, tantalum wafers, which play the role of a lubricant, were placed between the samples and die to maintain uniform deformation. Thermocouples were welded to samples to measure the processing temperatures.

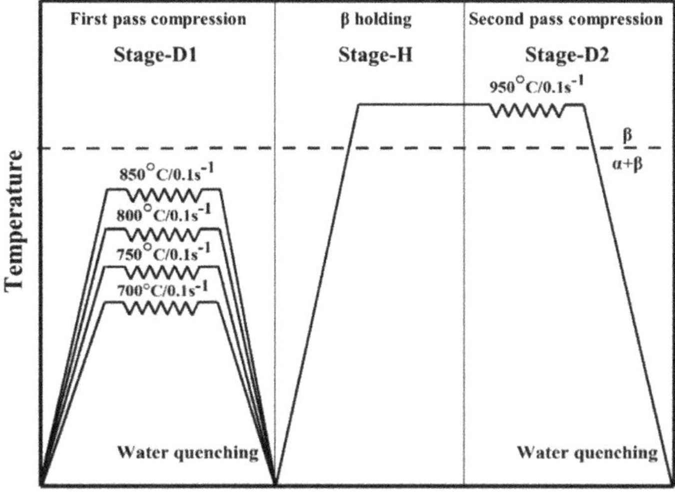

Figure 2. Schematic diagram of the thermal compression process.

The hot-compressed samples were cut along the compression direction using electro-discharge machining. Then, they were mechanically ground by SiC paper and finally polished by a twin-jet electropolisher at a −30 °C surrounding temperature and 30 V operating voltage. Specimens were jet-thinned by Kroll's reagent, which was composed of 300 mL methanol, 175 mL N-butyl alcohol, and 25 mL perchlorate acid.

The microstructure of the compressed samples was characterized by scanning electron microscope (SEM) (FEI Czech Republic s.r.o., Brno, Czech Republic) and electron backscattered diffraction (EBSD) (FEI Czech Republic s.r.o., Brno, Czech Republic) using 0.2 μm and 0.6 μm step sizes and a 20 kV operating voltage. Information about the misorientation and dislocation density was obtained by data analysis using the OIM software (EDAX). All the samples were obtained from the center of each compression sample.

3. Results and Discussion

3.1. Flow Behavior

Figure 3a depicts the stress–strain curves of the first-pass deformation at a strain rate of $0.1\ \text{s}^{-1}$. The stress quickly reaches the peak in the early stage of the first-pass deformation. The deformation temperature has a great influence on the flow behavior, and the peak stress decreases upon increasing the deformation temperature. A higher deformation temperature leads to a lower α-phase content and more operative slip systems [27,28]. After the peak stress, the curve decreases significantly at a strain of 0.12 in the 700 °C first compression sample; however, the curves at other deformation temperatures quickly reach a steady state after passing the peak stress.

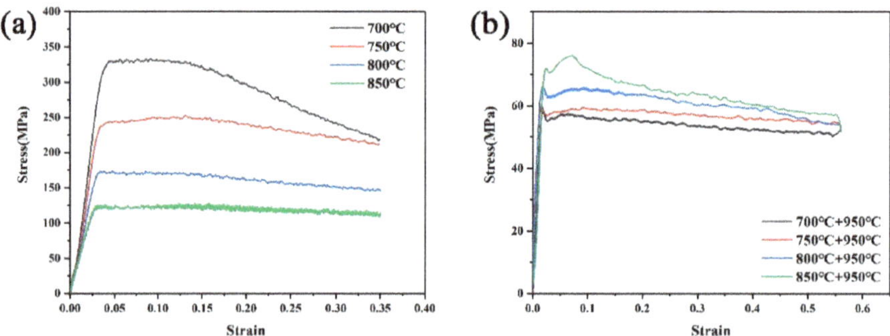

Figure 3. Stress-strain curves of duplex Ti-55511 during (**a**) first-pass and (**b**) second-pass thermal compression.

Figure 3b shows the stress–strain curves of the second-pass compression. The peak stress of the second-pass deformation increases upon increasing the deformation temperature during first-pass deformation. Obvious discontinuous yielding also appears due to recrystallization, which is closely related to the generation of movable dislocations [19,29]. After the discontinuous yield phenomenon, limited work-hardening is observed, and then flow stress enters the steady stage. The flow oscillation phenomenon is still obvious in this stage, which indicates the promoting of DRX after the first-pass compression [16].

3.2. Microstructure Characteristics after First-Pass Compression (Stage-D1)

Figure 4a–d show the microstructure characteristics after first-pass compression. In Figure 4a, the fine lamellar α phase is uniformly distributed among the equiaxed α phase. The average lamellar α phase is 3–5 μm long, and the average equiaxed α phase radius is about 5 μm. When the deformation temperature reaches 750 °C, the lamellar α phase disappears and only the equiaxed α phase exists in the matrix. The sample deformed at 750 °C contains less α phase. Upon further increasing the deformation temperature, the content of α phase continues to decrease, and the equiaxed α phase becomes smaller.

To reveal the change in the β matrix, Figure 5 shows the inverse pole figures (IPF) at different deformation temperatures. High-angle grain boundaries (HAGBs; misorientation angle >15°) and low-angle grain boundaries (LAGBs; 2°< misorientation angle <15°) are represented by black and white lines, respectively. Figure 5a shows that a large number of HAGBs appear at the α/β interface due to the occurrence of DRX, which explains the sudden drop of stress in the strain–stress curve mentioned earlier (Figure 4a); however, the recrystallization is still in the early stage, and the recrystallized grain is only a few microns in size. Figure 5b shows that at 750 °C, the LAGBs are uniformly distributed in the matrix and the LAGB fraction reaches 0.99, which means that the dominant restoration mechanism changes to DRV at 750 °C. However, in some regions (circled areas in Figure 5b) a small number of recrystallized grains form in the α grooves. Upon increasing the deformation temperature, the content of LAGBs in the β matrix continuously decreases, and no recrystallized grains are observed in the β matrix. The deformation process is mainly controlled by the DRV mechanism, and the level of DRV gradually decreases.

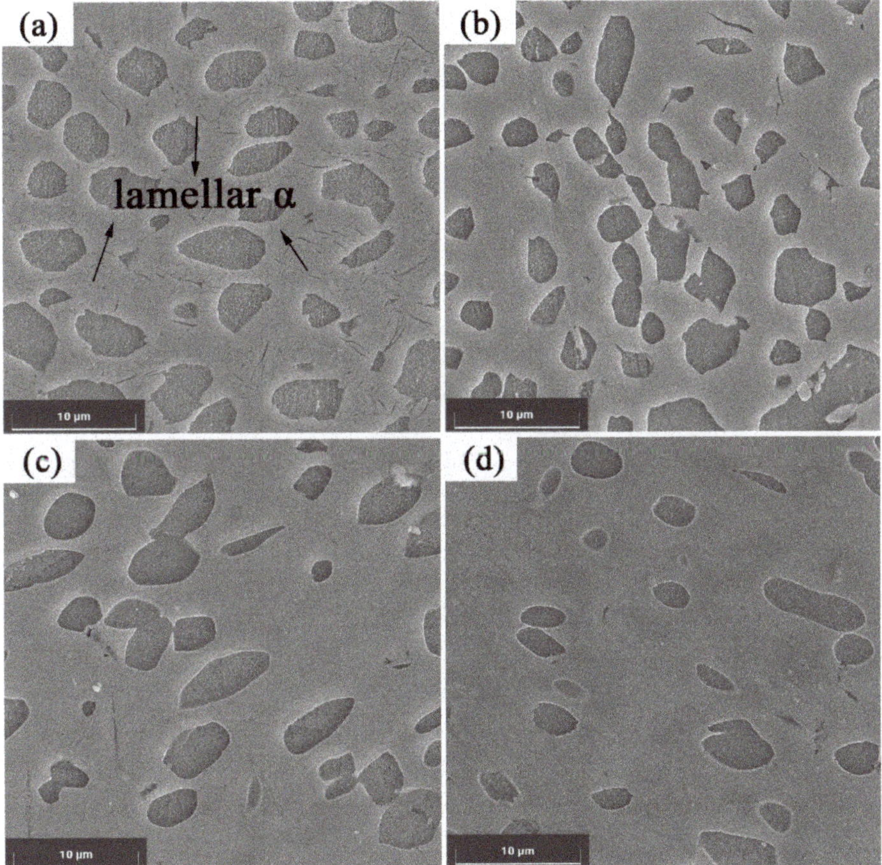

Figure 4. SEM figures after first-pass compression at (**a**) 700 °C, (**b**) 750 °C, (**c**) 800 °C, (**d**) 850 °C.

DRX and DRV are related to the dislocation density. To reveal the dislocation distribution, kernel average misorientation (KAM) maps were obtained, as shown in Figure 6. A higher KAM value indicates a higher dislocation content and more severe deformation. In Figure 6a, the KAM value at 700 °C is lower than at other temperatures because the recrystallization effectively eliminates the residual dislocations in deformed grains [30,31]. The different dislocation densities affect the growth rate of recrystallized grains during subsequent holding.

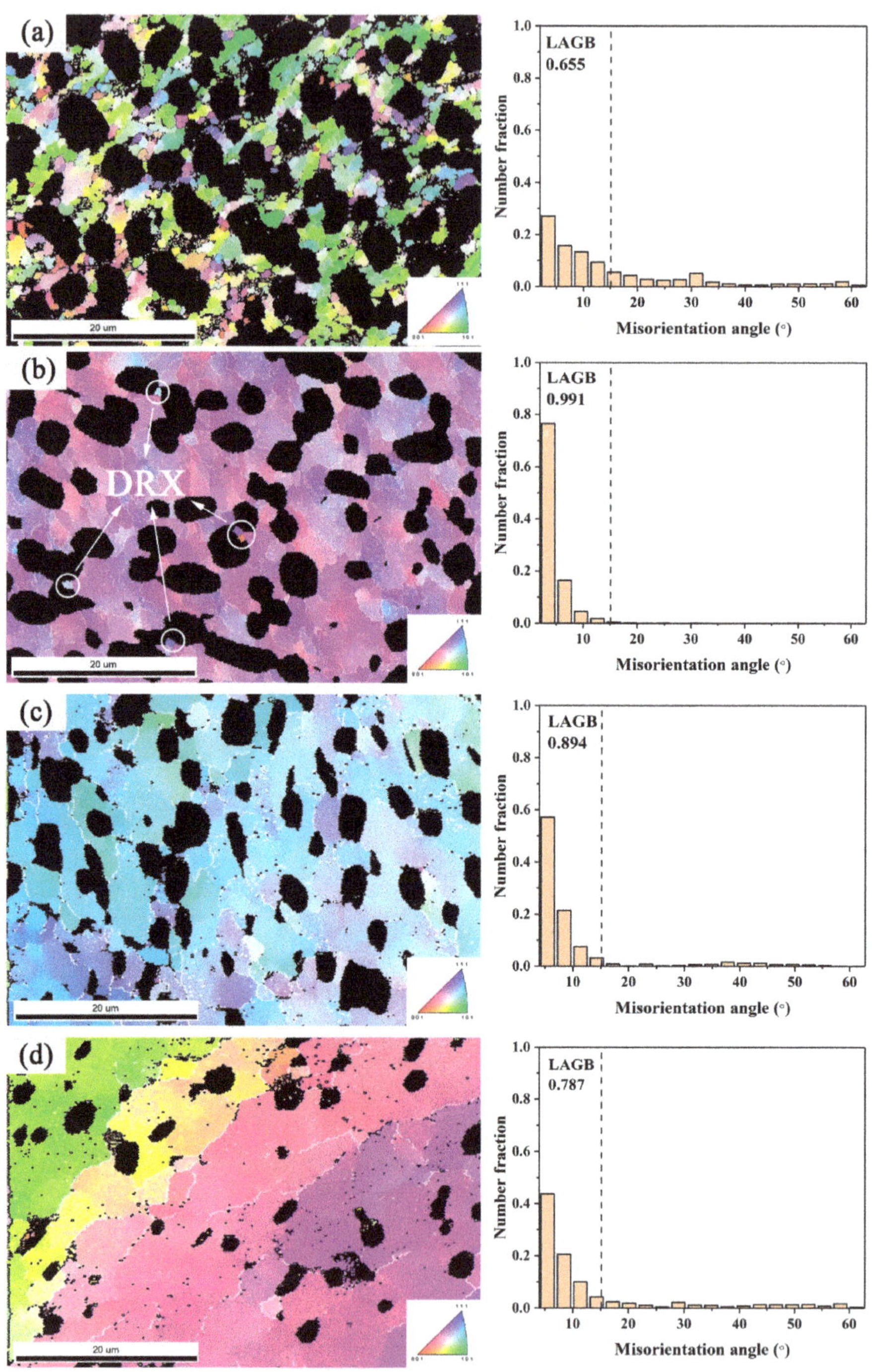

Figure 5. Inverse pole figures (IPF) figures after first-pass compression at (**a**) 700 °C, (**b**) 750 °C, (**c**) 800 °C, (**d**) 850 °C.

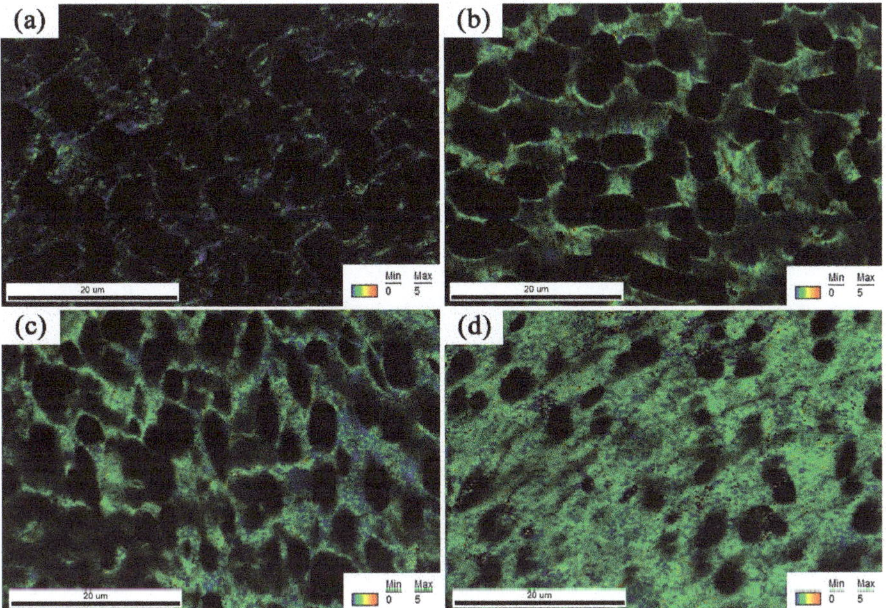

Figure 6. Kernel average misorientation (KAM) figures of samples after first-pass compression at (**a**) 700 °C, (**b**) 750 °C, (**c**) 800 °C, (**d**) 850 °C.

3.3. Microstructure Characteristics after Holding at 950 °C (Stage-H)

Figure 7a shows the sample microstructure after being deformed at 700 °C and then holding at 950 °C for 3 min. The β grain boundaries show zigzag lines, and the β grain size is significantly finer than that of other samples. This is because the β grain size is related to the recrystallization nucleation rate and the grain boundary migration rate [32]. A high nucleation rate and a small grain boundary migration rate are conducive to grain refinement. Meanwhile, the rate of grain boundary migration (v) and the effective pressure (P_{Eff}) on the grain boundary satisfy equation:

$$v = mP_{Eff}, \tag{1}$$

where effective pressure (P_{Eff}) is the sum of driving pressure (P_D) and retarding pressure (P_R). Driving pressure (P_D) is related to dislocation density:

$$P_D = (\rho_2 - \rho_1)\tau, \tag{2}$$

where τ is the energy of per unit length dislocation and ρ is the dislocation density. A large amount of recrystallized nucleation has been observed in the sample deformed at 700 °C, and the generation of β recrystallized grains reduces the dislocation density in the β matrix. This results in the β grain size in Figure 7a being much smaller than that of other samples. The microstructure of the sample after deforming at 750 °C and holding at 950 °C for 3 min is shown in Figure 7b. The average crystal grain size greatly increases to 128 μm and the grain boundaries become straight. Compared with the sample deformed at 700 °C, the average dislocation density is higher, which leads to the rapid growth of the recrystallized grains. The circled small grains are observed at the junction of boundaries where dislocation plugging drives the generation of SRX. In addition, some sub-crystals still exist in the matrix (arrows in Figure 7) because the strain of the first-pass deformation is only 0.36, which provides insufficient deformation storage energy to convert all LAGBs to HAGBs.

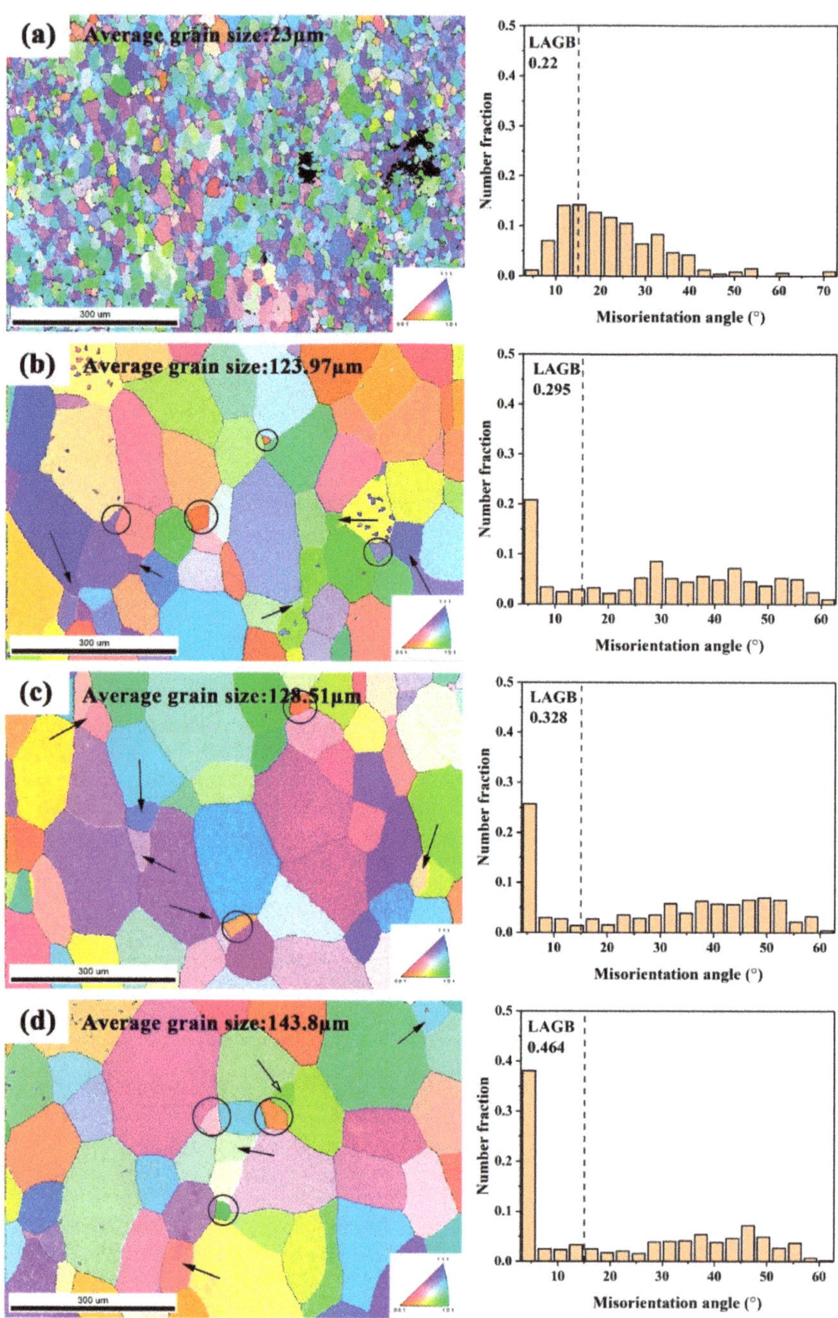

Figure 7. Electron backscattered diffraction (EBSD) figures after (**a**) 700 °C, (**b**) 750 °C, (**c**) 800 °C, (**d**) 850 °C first-pass compressing and then holding at 950 °C for 3 min.

As the first-pass deformation temperature constantly increases, the average grain size after holding at 950 °C increases slightly. Almost no recrystallization is observed after the first-pass deformation at 750 °C, and the small grains observed in Figure 7c,d are all the SRX grains that formed during the holding stage. The LAGBs content increases as the first-pass deformation temperature increases. These different microstructures change the recrystallization behavior during subsequent deformation.

3.4. Microstructural Characteristics after Second-Pass Compression (Stage-D2)

Figure 8 shows the microstructures at the beginning of the secondary deformation (before peak strain). The growth of the β grains is very rapid during hot working, and the grains grow to large sizes at the beginning of the deformation. In the sample first compressed at 700 °C and then second-pass deformed at 950 °C, CDRX is the main restoration mechanism in the early stage of deformation (square area in Figure 8a). This phenomenon has also been observed in 304 austenitic stainless steel [19] when the initial grain size decreased from 35 μm to 8 μm. Since CDRX requires less energy than DDRX, it occurs sooner. A small amount of DDRX occurs at grain boundaries (shown in the circled area). The occurrence of these two kinds of recrystallization significantly increases the LAGB fraction compared with the previous stage.

As for the samples subjected to first-pass compression at other temperatures, CDRX in the matrix almost disappears and a small amount of DDRX appears at the grain boundaries (circled in Figure 8d). The β grain morphology does not change much after deformation. Meanwhile, the LAGB content is significantly reduced compared with the previous stage. Generally speaking, the LAGB content should increase upon increasing the strain, and a slight increase in the LAGB content should be observed after the early deformation stage. This is because some of the residual LAGBs are preferentially transformed into HAGBs during the early stage of second-pass deformation.

The grain morphology and KAM maps after the second-pass deformation are shown in Figure 9. The morphology and grain size of the β matrix change significantly after deformation. Perpendicular to the direction of compression, β grains are elongated and grain boundaries are serrated. Discontinuous dynamic recrystallized grains are observed at the junctions of grain boundaries, and they form by bulging out from grain boundaries as non-deformed new grains, as circled in Figure 9. The recrystallized grain size is independent of the grain size before deformation [31]. After first-pass compression at a lower temperature, the volume fraction of the recrystallized grains is higher due to the grain refinement during the β holding stage. As a high-energy region, grain boundaries promote recrystallization nucleation and fine grains have more grain boundaries than coarse grains [20].

In the sample that was first compressed at 700 °C and then secondarily deformed at 950 °C (Figure 9a), dislocations preferentially accumulate at grain boundaries during second-pass deformation, and grain growth occurs simultaneously. Then, the LAGBs formed by dislocations entanglement stay in their original locations, which leads to LAGBs becoming uniformly distributed within the matrix. The KAM map shows the uniform dislocation density distribution. In the samples first compressed at 750–850 °C then secondarily deformed at 950 °C (Figure 9b–d), the density of the dislocations decreases and is non-uniformly distributed.

Meanwhile, many substructures form in β grains after the second compression. LAGBs are mainly concentrated at locations with a high dislocation density, such as β grain boundaries. During the early stage of second deformation, partial deformation energy is consumed, and the number of LAGBs generated by the remaining deformation energy decreases; therefore, after second-pass deformation, the number of LAGBs increases but the LAGB content decreases obviously with the increase in the first deformation temperature. CDRX requires less energy than DDRX, and it is mainly observed in fine grains; thus, the LAGB fraction of the samples subjected to first-pass compression at 700 °C is the highest, and the LAGB content decreases upon increasing the first-pass compression temperature.

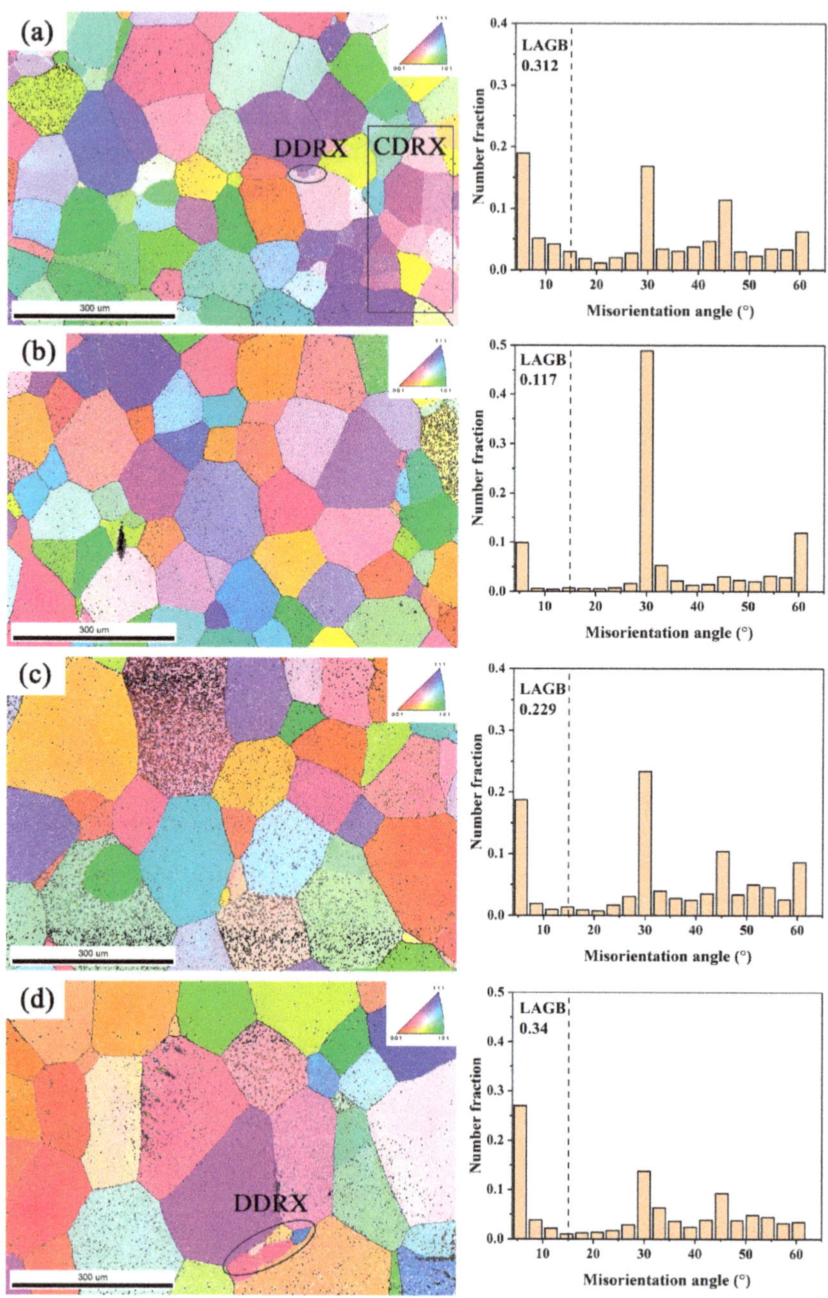

Figure 8. EBSD figures of samples after secondary-pass compression at 950 °C with 0.02 strain with the first-pass deformation temperature of (**a**) 700 °C, (**b**) 750 °C, (**c**) 800 °C, (**d**) 850 °C.

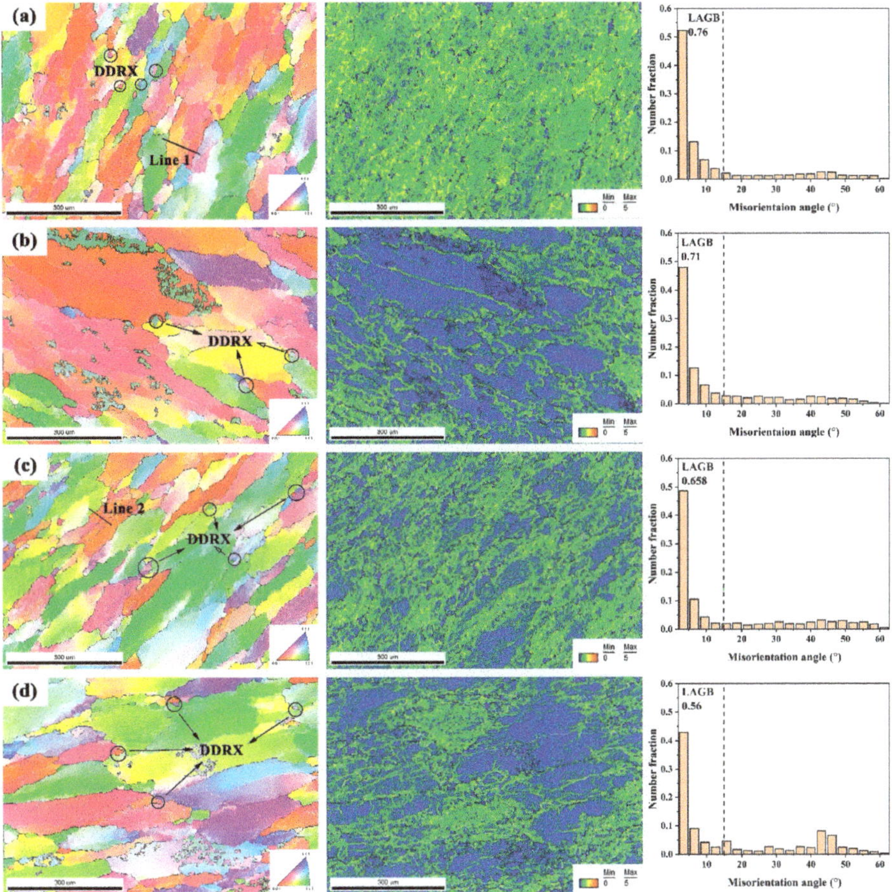

Figure 9. EBSD and KAM figures of samples after second compression at 950 °C with 0.56 strain and a first-pass deformation temperature of (**a**) 700 °C, (**b**) 750 °C, (**c**) 800 °C, (**d**) 850 °C.

There are two deformation modes of CDRX in the β region after deformation (line 1 and line 2 in Figure 9a,c). The first is the formation of new subgrains that are relatively uniformly distributed in the matrix. HAGBs formed after recrystallization cannot form in the case of insufficient strain, which leads to the formation of substructures and a higher dislocation density in the matrix [33]. Continuous dynamic recrystallization has no obvious nucleation and growth process but is accomplished by the continuous rotation of sub-crystals at grain boundaries [34]. This is shown in Figure 9 L1, and the orientation change is shown in Figure 10. According to the grain orientation difference, this region is divided into five parts. The point-to-point misorientation from region B to region D is a low angle of <15°. Since the dislocation density is homogeneously distributed within the grains, subgrains are formed in the original grain at the same time and have similar sizes and similar misorientations. This leads to only a small change in the point-to-original misorientation between regions B, C, and D. These subgrains are mainly distributed parallel to the compression axis.

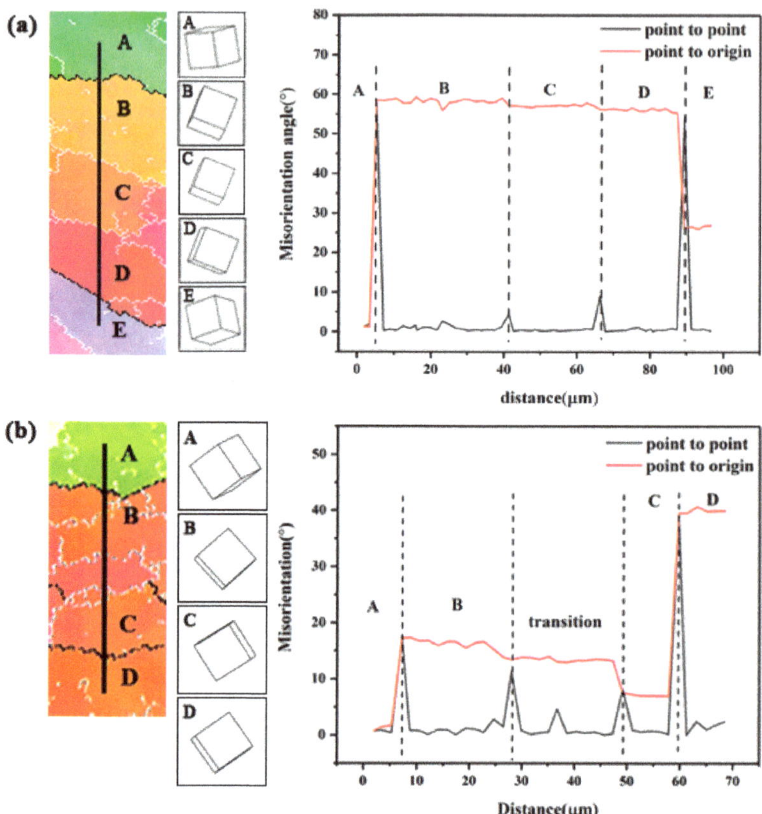

Figure 10. Orientation change along (**a**) line 1 and (**b**) line 2.

The second deformation mode is the continuous transformation from LAGBs to HAGBs, which occurs non-uniformly in the material. After deforming at 950 °C with 0.56 strain, some grain boundaries are observed in the matrix. These grain boundaries that are perpendicular to the deformation direction divide the original grains into several substructures to form grain-scale deformation bands [15,35]. These characteristics are markedly different from those of the substructures formed in the first mode. Point-to-original misorientation decreases in a gradient, and there is an obvious transformation region between regions B and C, in which misorientations undergo an obvious change. The EBSD figures in L2 indicate that this is a continuous process, and the boundaries between the transformation regions B and C are composed of alternating LAGBs and HAGBs. Non-uniform dislocation in the sample is an important reason for the formation of deformation bands, which are more easily observed in the coarse initial grains. Since deformation bands are mainly distributed at grain boundaries with severe deformation, more substructures are formed by this mode than the first mode.

4. Conclusions

In this study, the deformation behavior and microstructure of Ti-55511 alloy during two-pass thermal compression were investigated. The conclusions can be summarized as follows:

(1) The peak stress during first-pass deformation decreases upon increasing the deformation temperature. The peak stress during second-pass deformation increases upon increasing the first-pass deformation temperature.

(2) During first-pass compression, the dominant restoration mechanisms are DRX and DRV at 700 °C with lamellar α phase in the matrix. At higher deformation temperatures, lamellar α phase disappears and deformation is dominated by the recovery mechanism. Meanwhile, the level of recovery decreases at higher deformation temperatures.

(3) During the holding stage at 950 °C for 3 min, the average β grain size decreases upon decreasing the first-pass deformation temperature. When the first-pass deformation temperature is 700 °C, the dislocation density decreases due to recrystallization, the grain growth rate becomes slower, and the grain refinement is obvious.

(4) Two restoration mechanisms co-exist during second-pass deformation: CDRX and DDRX. At the beginning of the deformation, residual LAGBs formed during first-pass deformation preferentially transform to HAGBs. Then, CDRX is observed in initial fine grains but it is not observed in the initial coarse grains. Meanwhile, a small amount of DDRX is observed at grain boundaries.

(5) The CDRX behavior during second-pass deformation is affected by the grain size. The formation of substructures is related to the grain size before deformation. The substructures in the initial fine grains form at the same time and have the same misorientation angle as the original grains. The substructures in the initial coarse grains are formed by deformation bands and gradually transform from a non-uniform deformation region. The misorientation angle displays a gradient to the original grains, which is closely related to the dislocation density distribution.

Author Contributions: Data curation, H.W. and J.G.; Writing—original draft preparation, H.W.; Writing—review and editing, H.W. and X.Z.; Supervision, K.Z.; Project Administration, K.Z. and C.C.; Funding Acquisition, X.Z. and K.Z. All authors have read and agreed to the published version of the manuscript.

Funding: The authors are grateful to thanks to funding support from the National Natural Science Foundation of China (No. 51871242), the National Key R&D Program of China (2018YFB0704100) and Scientific and Technological Innovation Projects of Hunan Province, China (No. 2017GK2292).

Data Availability Statement: All the raw data supporting the conclusion of this paper were provided by the authors.

Acknowledgments: The authors would like to acknowledge financial support from the National Natural Science Foundation of China (No. 51871242); Scientific and Technological Innovation Projects of Hunan Province, China (No. 2017GK2292); and the National Key R&D Program of China (2018YFB0704100). Besides this, the authors would like to thank Jinyang Ge for assistance with the experiments.

Conflicts of Interest: The authors declare that they have no known competing financial interests or personal relationships that could have appeared to influence the work reported in this paper.

References

1. Li, C.; Zhang, X.Y.; Zhou, K.C.; Peng, C.Q. Relationship between lamellar α evolution and flow behavior during isothermal deformation of Ti-5Al-5Mo-5V-1Cr-1Fe near B titanium alloy. *Mater. Sci. Eng. A* **2012**, *558*, 668–674. [CrossRef]

2. Tsybanev, G.V.; Ageev, M.A.; Titarenko, R.V. Analysis of the specific features of loading of the elements of landing gears of aircrafts aimed at the evaluation of the load-bearing capacity of the structure. *Strength Mater.* **2008**, *40*, 458–462. [CrossRef]

3. Banerjee, D.; Williams, J.C. Perspectives on titanium science and technology. *Acta Mater.* **2013**, *61*, 844–879. [CrossRef]

4. Zhou, W.; Ge, P.; Zhao, Y.; Wu, H.; Li, Q.; Chen, J. Response of heat-treatment on A near-β titanium alloy. *Xiyou Jinshu Cailiao Yu Gongcheng/Rare Met. Mater. Eng.* **2010**, *39*, 723–726.

5. Ravichandran, K.S. Fracture mode transitions during fatigue crack growth in Ti-6Al-4V alloy. *Scr. Metall. Mater.* **1990**, *24*, 1275–1280. [CrossRef]

6. Estrin, Y.; Vinogradov, A. Extreme grain refinement by severe plastic deformation: A wealth of challenging science. *Acta Mater.* **2013**, *61*, 782–817. [CrossRef]

7. Wang, B.; Wang, X.; Li, J. Formation and Microstructure of Ultrafine-Grained Titanium Processed by Multi-Directional Forging. *J. Mater. Eng. Perform.* **2016**, *25*, 2521–2527. [CrossRef]
8. Fan, X.G.; Zhang, Y.; Zheng, H.J.; Zhang, Z.Q.; Gao, P.F.; Zhan, M. Pre-processing related recrystallization behavior in β annealing of a near-β Ti-5Al-5Mo-5V-3Cr-1Zr titanium alloy. *Mater. Charact.* **2018**, *137*, 151–161. [CrossRef]
9. Mironov, S.; Murzinova, M.; Zherebtsov, S.; Salishchev, G.A.; Semiatin, S.L. Microstructure evolution during warm working of Ti–6Al–4V with a colony-α microstructure. *Acta Mater.* **2009**, *57*, 2470–2481. [CrossRef]
10. Zherebtsov, S.; Kudryavtsev, E.; Kostjuchenko, S.; Malysheva, S.; Salishchev, G. Strength and ductility-related properties of ultrafine grained two-phase titanium alloy produced by warm multiaxial forging. *Mater. Sci. Eng. A* **2012**, *536*, 190–196. [CrossRef]
11. Bézi, Z.; Krállics, G.; El-Tahawy, M.; Pekker, P.; Gubicza, J. Processing of ultrafine-grained titanium with high strength and good ductility by a combination of multiple forging and rolling. *Mater. Sci. Eng. A* **2017**, *688*, 210–217. [CrossRef]
12. Jia, J.; Zhang, K.; Lu, Z. Dynamic recrystallization kinetics of a powder metallurgy Ti-22Al-25Nb alloy during hot compression. *Mater. Sci. Eng. A* **2014**, *607*, 630–639. [CrossRef]
13. Warchomicka, F.; Poletti, C.; Stockinger, M. Study of the hot deformation behavior in Ti–5Al–5Mo–5V–3Cr–1Zr. *Mater. Sci. Eng. A* **2011**, *528*, 8277–8285. [CrossRef]
14. Hu, Z.; Zhou, X.; Nie, X.A.; Zhao, S.; Liu, H.; Yi, D.; Zhang, X. Finer subgrain microstructure induced by multi-pass compression in α+β phase region in a near-β Ti-5Al-5Mo-5V-1Cr-1Fe alloy. *J. Alloys Compd.* **2019**, *788*, 136–147. [CrossRef]
15. Fan, X.G.; Zhang, Y.; Gao, P.F.; Lei, Z.N.; Zhan, M. Deformation behavior and microstructure evolution during hot working of a coarse-grained Ti-5Al-5Mo-5V-3Cr-1Zr titanium alloy in beta phase field. *Mater. Sci. Eng. A* **2017**, *694*, 24–32. [CrossRef]
16. Wang, K.; Li, M.Q. Flow behavior and deformation mechanism in the isothermal compression of the TC8 titanium alloy. *Mater. Sci. Eng. A* **2014**, *600*, 122–128. [CrossRef]
17. Ning, Y.Q.; Luo, X.; Liang, H.Q.; Guo, H.Z.; Zhang, J.L.; Tan, K. Competition between dynamic recovery and recrystallization during hot deformation for TC18 titanium alloy. *Mater. Sci. Eng. A* **2015**, *635*, 77–85. [CrossRef]
18. Wu, W.X.; Jin, L.; Dong, J.; Zhang, Z.Y.; Ding, W.J. Effect of initial microstructure on the dynamic recrystallization behavior of Mg-Gd-Y-Zr alloy. *Mater. Sci. Eng. A* **2012**, *556*, 519–525. [CrossRef]
19. Dehghan-Manshadi, A.; Hodgson, P.D. Dependency of recrystallization mechanism to the initial grain size. *Metall. Mater. Trans. A Phys. Metall. Mater. Sci.* **2008**, *39*, 2830–2840. [CrossRef]
20. Lütjering, G.; Williams, J.C. *Titanium*, 2nd ed.; Springer: Berlin, Germany, 2007; pp. 72–79.
21. Dikovits, M.; Poletti, C.; Warchomicka, F. Deformation mechanisms in the near-β titanium alloy Ti-55531. *Metall. Mater. Trans. A Phys. Metall. Mater. Sci.* **2014**, *45*. [CrossRef]
22. Balachandran, S.; Kumar, S.; Banerjee, D. On recrystallization of the α and β phases in titanium alloys. *Acta Mater.* **2017**, *131*, 423–434. [CrossRef]
23. Wagner, F.; Bozzolo, N.; Van Landuyt, O.; Grosdidier, T. Evolution of recrystallisation texture and microstructure in low alloyed titanium sheets. *Acta Mater.* **2002**, *50*, 1245–1259. [CrossRef]
24. Chen, W.; Lv, Y.; Zhang, X.; Chen, C.; Lin, Y.C.; Zhou, K. Comparing the evolution and deformation mechanisms of lamellar and equiaxed microstructures in near β-Ti alloys during hot deformation. *Mater. Sci. Eng. A* **2019**, *758*, 71–78. [CrossRef]
25. Tang, B.; Xiang, L.; Yan, Z.; Kou, H.; Li, J. Effect of strain distribution on the evolution of α phase and texture for dual-phase titanium alloy during multi-pass forging process. *Mater. Chem. Phys.* **2019**, *228*, 318–324. [CrossRef]
26. Li, L.; Luo, J.; Yan, J.J.; Li, M.Q. Dynamic globalization and restoration mechanism of Ti–5Al–2Sn–2Zr–4Mo–4Cr alloy during isothermal compression. *J. Alloys Compd.* **2015**, *622*, 174–183. [CrossRef]
27. Li, L.X.; Lou, Y.; Yang, L.B.; Peng, D.S.; Rao, K.P. Flow stress behavior and deformation characteristics of Ti-3Al-5V-5Mo compressed at elevated temperatures. *Mater. Des.* **2002**, *23*, 451–457. [CrossRef]
28. Jia, W.; Zeng, W.; Zhou, Y.; Liu, J.; Wang, Q. High-temperature deformation behavior of Ti60 titanium alloy. *Mater. Sci. Eng. A* **2011**, *528*, 4068–4074. [CrossRef]
29. Huang, Y.; Humphreys, F.J. Transient dynamic recrystallization in an aluminum alloy subjected to large reductions in strain rate. *Acta Mater.* **1997**, *45*, 4491–4503. [CrossRef]
30. Cram, D.G.; Zurob, H.S.; Brechet, Y.J.M.; Hutchinson, C.R. Modelling discontinuous dynamic recrystallization using a physically based model for nucleation. *Mater. Sci. Forum* **2012**, *715–716*, 492–497. [CrossRef]
31. Huang, K.; Logé, R.E. A review of dynamic recrystallization phenomena in metallic materials. *Mater. Des.* **2016**, *111*, 548–574. [CrossRef]
32. Vandermeer, R.A.; Jensen, D.J.; Woldt, E. Grain boundary mobility during recrystallization of copper. *Metall. Mater. Trans. A Phys. Metall. Mater. Sci.* **1997**, *28*, 749–754. [CrossRef]
33. Xiang, L.; Tang, B.; Xue, X.; Kou, H.; Li, J. Characteristics of the dynamic recrystallization behavior of Ti-45Al-8.5Nb-0.2W-0.2B-0.3Y alloy during high temperature deformation. *Metals* **2017**, *7*, 261. [CrossRef]
34. Tikhonova, M.; Kaibyshev, R.; Fang, X.; Wang, W.; Belyakov, A. Grain boundary assembles developed in an austenitic stainless steel during large strain warm working. *Mater. Charact.* **2012**, *70*, 14–20. [CrossRef]
35. Kuhlmann-Wilsdorf, D. Overview No. 131: "regular" deformation bands (DBs) and the leds hypothesis. *Acta Mater.* **1999**, *47*. [CrossRef]

Article

Behavior of Inelastic and Plastic Strains in Coarse-Grained Ti$_{49.3}$Ni$_{50.7}$(at%) Alloy Deformed in B2 States

Dorzhima Zhapova *, Victor Grishkov, Aleksandr Lotkov, Victor Timkin, Angelina Gusarenko and Ivan Rodionov

Institute of Strength Physics and Materials Science of the Siberian Branch of the Russian Academy of Science, 634055 Tomsk, Russia; grish@ispms.ru (V.G.); lotkov@ispms.ru (A.L.); timk@ispms.ru (V.T.); angel.ru09@mail.ru (A.G.); rodionov231@mail.ru (I.R.)
* Correspondence: dorzh@ispms.tsc.ru; Tel.: +7-382-228-69-82

Abstract: The regularities of the change in inelastic strain in coarse-grained samples of the Ti$_{49.3}$Ni$_{50.7}$ (at%) alloy are studied when the samples are given torsional strain in the state of the high-temperature B2 phase. During cooling and heating, the investigated samples underwent the B2–B19$'$ martensite transformation (MT); the temperature of the end of the reverse MT was A_f = 273 K. It was found that at the temperature of isothermal cycles "loading-unloading" A_f + 8 K, when the specimen is assigned a strain of 4%, the effect of superelasticity is observed. With an increase in the torsional strain, the shape memory effect is clearly manifested. It is assumed that the stabilization of the B19$'$ phase in unloaded samples is due to the appearance of dislocations during deformation due to high internal stresses at the interphase boundaries of the B2 phase and the martensite phase during MT. The appearance of dislocations during the loading of samples near the temperatures of forward and reverse MT can also be facilitated by the "softening" of the elastic moduli of the alloy in this temperature range. At a test temperature above A_f + 26 K, the superelasticity effect dominates in the studied samples.

Keywords: TiNi-based alloys; superelasticity; shape memory effect; inelastic strain; plastic strain

Citation: Zhapova, D.; Grishkov, V.; Lotkov, A.; Timkin, V.; Gusarenko, A.; Rodionov, I. Behavior of Inelastic and Plastic Strains in Coarse-Grained Ti$_{49.3}$Ni$_{50.7}$(at%) Alloy Deformed in B2 States. *Metals* **2021**, *11*, 741. https://doi.org/10.3390/met11050741

Academic Editor: Maciej Motyka

Received: 30 March 2021
Accepted: 28 April 2021
Published: 29 April 2021

Publisher's Note: MDPI stays neutral with regard to jurisdictional claims in published maps and institutional affiliations.

1. Introduction

The significant interest in smart TiNi materials owes to the unique combination of their functional, strength, and plastic properties beneficial for engineering and medicine [1–12]. Their superelasticity (SE) and shape memory effect (SME) are provided by thermoelastic martensite transformations (MT) from a cubic B2 phase to a rhombohedral R or a monoclinic B19$'$ phase. When cooled and heated free of load or at low internal stress, such materials remain macroscopically invariant as their thermoelastic transformations result in a polyvariant system of self-accommodated martensite domains [4], but when exposed to external or oriented internal stresses, they display superelasticity and shape memory. Shape memory allows a material to accumulate reversible inelastic strains at $T_d < M_f$ (where T_d is the deformation temperature and M_f is the martensite finish temperature on cooling) and to recover them at above A_f (which is the austenite finish temperature on heating). Superelasticity allows a material to change its macroscopic shape or linear dimensions through direct B2 $\rightarrow$ B19$'$ or B2 $\rightarrow$ R $\rightarrow$ B19$'$ transformations under external stress (tension, torsion, bending) at $T > A_f$ and to take it back through inelastic strain recovery under subsequent isothermal unloading (because no martensite free of load can exist at this temperature). In both cases, the reversible inelastic strain can reach 6–8%. The inelastic strain, whether from SE or SME, depends largely on the crystallography limit of the recoverable strain whose value in TiNi-based alloys measures, on average, 11% [13–15].

However, reversible inelastic strains 1.5–2 times higher than the above limit have been found in TiNi-based alloys [16–23] after bending [15–18,20], torsion [20–23], and tension [19], with the total strain comprising an inelastic and a plastic component. For example, a reversible inelastic strain of up to 18% is attainable in Ti$_{49.3}$Ni$_{50.7}$ (at%) after torsion at $T < M_f$ against a high (12%) plastic strain [23]. Its recovery is stepwise: via SE under unloading at T_d and via SME under further heating above the finish temperature of $B19' \rightarrow B2$ MT.

In one of the cited studies [16], $Ti_{49.3}Ni_{50.7}$ (at%) specimens differing in structure were bent to a total strain of 15% and were then kept constrained at 310 K (to prevent shape recovery), cooled to 77 K with keeping at this temperature for 30 s, and unloaded with further heating to room temperature. In nanocrystalline $Ti_{49.3}Ni_{50.7}$ (at%) (grain size 30–70 nm), the total strain was recovered completely so that 7.5% fell on SE and 7.5% on SME. In $Ti_{49.3}Ni_{50.7}$ (at%) with a mixed structure (nanocrystalline, subgrain), the strain recovery via SME was 14.8% against 0.2% of plastic strain, and in its specimens with microcrystalline structure (maximum grain size 10 μm), the SME value was 10% against up to 5% of plastic strain and zero SE.

In another study [17], $Ti_{50}Ni_{50}$ (at%) specimens annealed after cold drawing were bent at 5 K above A_f with further keeping at this temperature for 30 s, cooling in their constrained state to 273 K, and keeping at this temperature for 30 s. Thereafter, estimates of their strain recovery were taken under unloading (SE, including small elastic strain) and under further heating to 373 K (SME). The plastic strain is the residual one at 373 K. In $Ti_{50}Ni_{50}$ (at%) with a polygonized B2 substructure (annealing at 623 K for 1 h, subgrain size $\leq$ 200 nm), the strain recovery via SE and SME at a total strain of 18% measured 10.3% and 7.6%, respectively, against 0.1% of plastic strain; the total inelastic strain was 17.9%. In $Ti_{50}Ni_{50}$ (at%) with a polygonized B2 substructure comprising individual recrystallized grains of up to 3 μm (annealing at 723 K for 1 h), the strain recovery via SE and SME at the same total strain measured 7.2% and 8%, respectively, against 2.8% of plastic strain; the total inelastic strain reached 15.2%.

Thus, it is still unclear what conditions can bring the reversible inelastic strain in TiNi materials to above their theoretical limit. Clarifying this issue needs additional studies of TiNi alloys differing in chemical composition, structure (grain-subgrain size), and phase state. Here, we analyze the behavior of inelastic and plastic strains in coarse-grained $Ti_{49.3}Ni_{50.7}$ (at%) deformed at $T_d > A_f$.

2. Materials and Methods

The test material was $Ti_{49.3}Ni_{50.7}$ (at%) supplied as hot-swaged bars of diameter 30 mm (Matek-Sma Ltd., Moscow, Russia). The bars were spark cut into specimens (cross-sectional area ~1 mm^2, gage length ~10 mm), rinsed in ethanol, grinded with an abrasive and diamond paste, and electrolytically polished with plane stainless steel electrodes in a cold $CH_3COOH/HClO_4$ solution (75/25 vol%) at a voltage of 12–17 V for 10–15 s. For further structural analysis, they were chemically etched in a $HNO_3/HF/H_2O$ mixture (14/4/82 vol%) for 15 s.

Their structure and phase state were analyzed on a DRON-7 diffractometer in Co-K_α radiation (Bourevestnik JSC, Saint-Petersburg, Russia) and on an AXIOVERT-200 MAT optical microscope (Carl Zeiss AG, Oberkochen, Germany). According to the analysis, the specimens at room temperature were in the state of a high-temperature B2 phase (CsCl superstructure) with less than 5 vol% of $Ti_4Ni_2(O,N,C,H)_x$. The average grain size was 53 ± 11 μm. The deformation-induced surface microrelief after torsion were studied by scanning electron microscopy LEO EVO 50XVP (Carl Zeiss AG, Oberkochen, Germany). These studies were conducted using the equipment of Nanotech shared Use Center of ISPMS SB RAS.

The temperatures and sequence of martensite transformations were determined by temperature resistometry ($\rho(T)$ measurements). On cooling and heating, the specimens experienced martensite transformations $B2 \leftrightarrow B19'$. The start and finish temperatures of $B2 \rightarrow B19'$ MT are $M_S = 252$ K, $M_f = 223$ K. The start and finish temperatures of $B19' \rightarrow B2$ MT are $A_S = 258$ K, $A_f = 273$ K.

The inelastic and plastic strains in the material were studied on an inverted torsion pendulum with an operating temperature of 573–120 K. The τ–γ dependences in isothermal loading–unloading cycles, and the inelastic strain recovery on further heating to 500 K (227 K above A_f) of unloaded specimens were obtained. In each cycle, the total strain γ_t was successively increased up to fracture. The temperature of loading–unloading cycles was 281, 299, 309, 315, and 339 K, i.e., T_d was above A_f and specimens were in the B2 state.

All components of the total strain γ_t were determined. The total strain is the sum of inelastic and plastic strains: $\gamma_t = \gamma_{SID} + \gamma_{pl}$. The total inelastic strain is $\gamma_{SID} = \gamma_{SE} + \gamma_{SME}$, where the summands stand for inelastic strains recovered via superelasticity, γ_{SE}, under isothermal unloading (including a small Hook strain of ~1.5%) and via shape memory, γ_{SME}, under further heating to complete shape recovery (via B19$'$ $\rightarrow$ B2 transformation). In more detail, the total inelastic strain has the form:

$$\gamma_{SID} = (\gamma_t - \gamma_r) + (\gamma_r - \gamma_{pl}) = \gamma_{SE} + \gamma_{SME} \tag{1}$$

where γ_r is the residual strain after isothermal unloading (295 K), and γ_t, γ_r, and γ_{pl} are equal to arctgS_t, arctgS_r, and arctgS_{pl}, with $S_t = (r\varphi_t)/l$; $S_r = (r\varphi_r)/l$; and $S_{pl} = (r\varphi_{pl})/l$; having r and l for the specimen cross-section radius and gage length and φ_t, φ_r, and φ_{pl} for the torsion angles in radians after loading, unloading, and heating to 500 K, respectively. The plastic strain γ_{pl} is equal to the residual strain at 500 K. The γ_t, γ_r, γ_{pl}, γ_{SE} and γ_{SME} strains are presented in Figure 1. The measurement error for the quantities depended on the total strain, and at $\gamma_t = 36.6\%$, it was $\Delta\gamma_t = 0.3\%$, $\Delta\gamma_{SE} = 0.4\%$, $\Delta\gamma_{SME} = 0.4\%$, and $\Delta\gamma_{pl} = 0.2\%$.

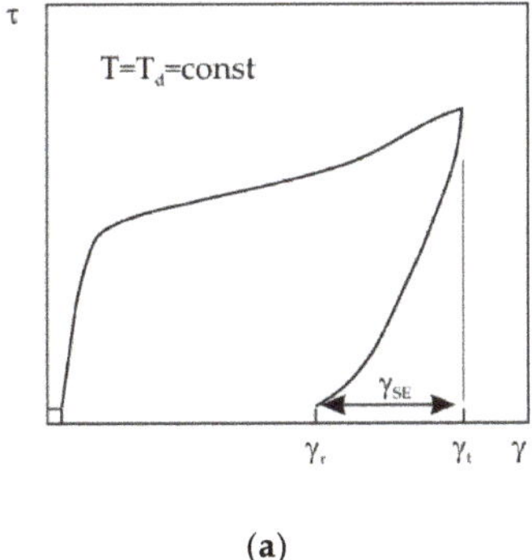

(a)

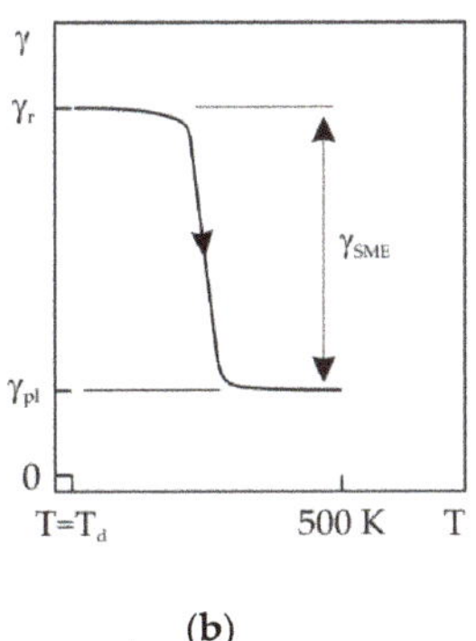

(b)

Figure 1. The experimental strains determined in isothermal τ–γ cycle (**a**) and on subsequent heating of unloaded specimen (**b**).

3. Results

Figure 2 shows the τ–γ dependences of Ti$_{49.3}$Ni$_{50.7}$ (at%) in loading-unloading cycles at 281, 299, 309, and 339 K. As can be seen, the dependences at 281 and 299 K reveal four deformation stages. Stage I is quasi-elastic and its strain increases linearly as the applied stress is increased. At $\tau = \tau_m$ (martensite shear stress), stage I passes into stage II, at which the strain increment is large while the external stress build-up is comparatively small. Stage II represents a so-called pseudo-yield plateau which ends, in our case, at about $\gamma_t = 8\%$. Obviously, at 281 and 299 K (Figure 2, curves 1, 2), the pseudo-yield plateau results from the generation of B19$'$ martensite under loading [4,13]. Stage II is followed by stage III, at which γ_t increases almost linearly from 8% to 18–20% as τ is increased from ~450 to 800 MPa. Stage III represents strain hardening, which passes into active plastic flow at stage IV. From Figure 2, it is seen that when deformed at 309 K (curve 3) and 339 K (curve 4), the material shows no pseudo-yield plateau and its quasi-elastic stage passes directly into parabolic flow.

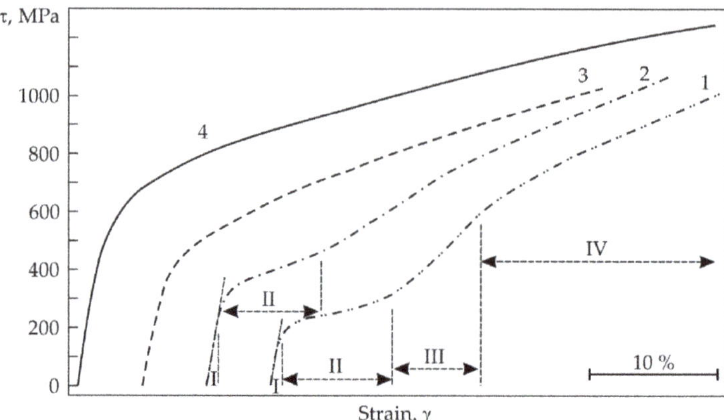

Figure 2. Stress–strain dependence of $Ti_{49.3}Ni_{50.7}$ (at%) specimens deformed at 281 K (1), 299 K (2), 309 K (3), and 339 K (4). The start strain for each τ–γ dependence corresponds $\gamma = 0$.

Figures 3 and 4 show the τ–γ dependences of $Ti_{49.3}Ni_{50.7}\cdot$(at%) in loading–unloading cycles with successively increasing γ_t and its inelastic strain recovery on further heating at different test temperatures. From Figure 3a, it is seen that as the number of loading–unloading cycles with $T_d = 281$ K is increased, the martensite shear stress τ_m grows and the extent of the pseudo-yield plateau first increases, then decreases (after the fifth cycle). It should be noted that the martensite shear stress τ_m increases from 180 MPa in the first loading–unloading cycle to 380 MPa in the seventh one. After the first loading–unloading cycle with $\gamma_t = 4.1\%$, the strain recovery via superelasticity (under unloading) was 3.9%, and via shape memory (at heating), it was 0.2%. Although T_d was higher than A_f by about 8 K, the SE value in the next isothermal τ–γ cycles was low compared to γ_t. Even after the second cycle with $\gamma_t = 9.9\%$, the SE value was $\gamma_{SE} = 3.6\%$ against $\gamma_{SME} = 6.3\%$ (Figure 3a), and no plastic component was detected within the measurement error. After the third cycle with $\gamma_t = 14.4\%$, the SE and SME values were $\gamma_{SE} = 3.0\%$ and $\gamma_{SME} = 11.0\%$ against $\gamma_{pl} = 0.4\%$ (Figure 3b). As the number of loading–unloading–heating cycles was further increased, γ_{SE} increased monotonically. The γ_{SME} value first increased and then decreased, and γ_{pl} became larger (Figure 3b).

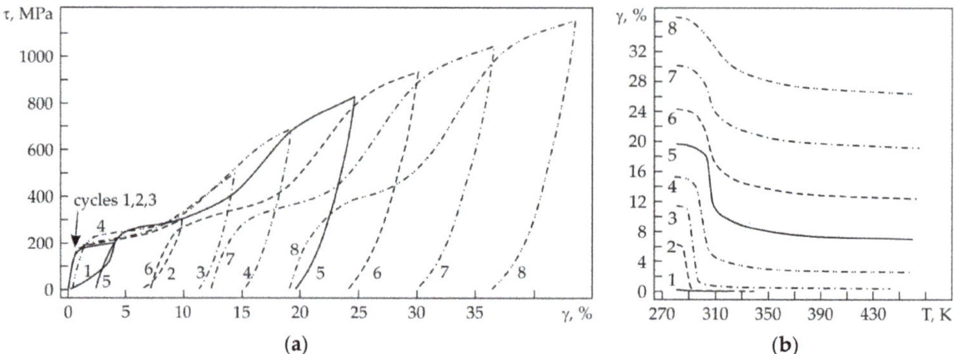

Figure 3. Accumulation and recovery of inelastic strain in loading–unloading cycles at 281 K (**a**) and its recovery on further heating (**b**).

From Figure 4a it is seen that at $T_d = 299$ K (26 K above A_f), the first two loading–unloading cycles with increasing γ_t result in flag-shaped τ–γ dependences characteristic of superelasticity so that the total strain is fully recovered after unloading. It should be noted that all loading–unloading cycles at this temperature are dominated by the SE effect. After the first cycle with $\gamma_t = 3.7\%$, $\gamma_{SE} = 3.6\%$, and after the second cycle with $\gamma_t = 7.3\%$, its value is $\gamma_{SE} = 6.9\%$, i.e., almost the whole strain is recovered under unloading. After the third cycle, γ_{SE} is equal to about 10.6%, but some strain remains unrecovered under unloading. As the number of loading–unloading cycles as well as the total strain is further increased, γ_{SE} varies little. It should be noted that at $T_d = 299$ K, compared to $T_d = 281$ K, the SE effect is much more pronounced and the SME value γ_{SME} on heating is small (Figure 4b). It should also be noted that after the third loading–unloading cycle with up to $\gamma_t = 12.3\%$, the strain is mostly inelastic and almost completely recovered via SE and SME under unloading and heating, respectively. However, in the next cycles, the plastic strain component steeply increases. Noteworthy also is that the pseudo-yield plateau at $T_d = 299$ K is poorly detectable in all loading–unloading cycles (Figure 4a).

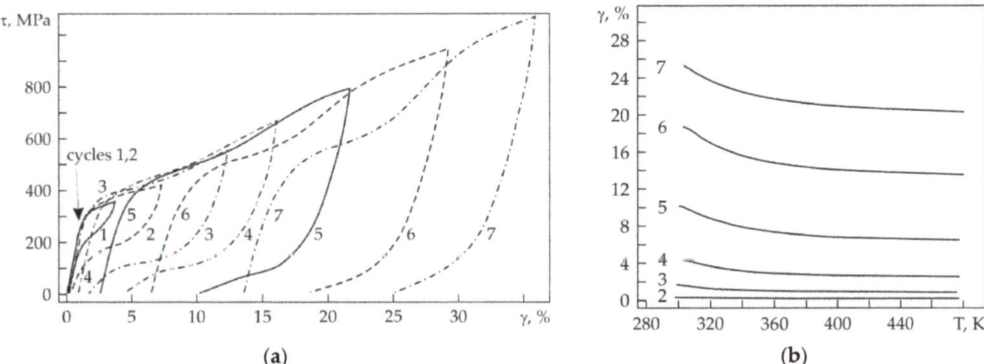

Figure 4. Accumulation and recovery of inelastic strain in loading–unloading cycles at 299 K (**a**) and its recovery on further heating (**b**).

Figure 5 shows the γ_t dependences of γ_{SID}, γ_{SE}, and γ_{SME} at different temperatures of isothermal $\tau-\gamma$ cycles. It is seen that at 281 K, γ_{SME} reaches 12.7% (Figure 5a). At the same time, the temperature interval of shape recovery is narrow on heating, spanning from 285 to 310 K (Figure 3b). Note that at all temperatures of isothermal loading–unloading cycles, except for $T_d = 281$ K, the γ_t dependences of γ_{SID}, γ_{SE}, and γ_{SME} are qualitatively similar: as γ_t is increased, these quantities first reach a maximum and then decrease. The γ_t strains at which γ_{SID}, γ_{SE}, and γ_{SME} reach their maximum values differ but fall on the stage of active plastic strain accumulation (Figure 5).

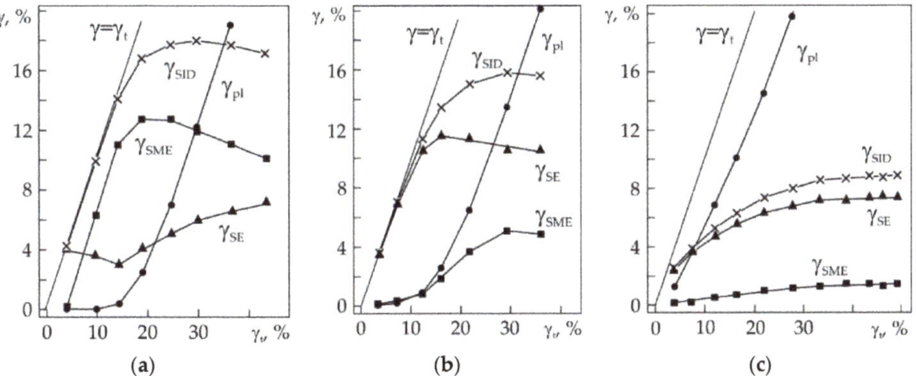

Figure 5. Dependences of γ_{SE} (▲), γ_{SME} (■), γ_{SID} (×), and γ_{pl} (●) on γ_t after isothermal loading-unloading at 281 K (**a**), 299 K (**b**), 339 K (**c**) and subsequent heating of unloaded specimens.

4. Discussion

Certainly, at first glance, it is rather strange that the total strain measuring about a mere 4.1% at 281 K (8 K above A_f) results in an SME of 0.2% (Figure 3b). With such a small total strain at 281 K, one would expect its complete recovery via SE after unloading [24]. Moreover, after the second loading–unloading cycle with $\gamma_t = 9.9\%$, the SE value decreases to 3.6% against 3.9% after the first cycle, whereas the SME value increases from 0.2% in the first cycle to 6.3% in the second one (Figure 3b). Although no plastic strain is found in the second loading–unloading cycle, the rise of SME after the first and second cycles at 8 K above A_f is most likely due to the generation of dislocations at the interfaces of B2 and B19′ phases under internal stress, which results from their lattice misfit [25]. It has been known that as the temperature of TiNi-based alloys is decreased and brought to about M_S, their elastic constants C' and C_{44} decrease greatly [26–28], and the decrease in the elastic constants should decrease the martensite shear stress in this temperature range. It should be noted that at $T_d > A_f$ in the temperature range of decreased elastic constants, the recovery of inelastic strains via SE at comparatively low external stresses can be incomplete due to the rise of small plastic strains lying within 0.3%. The stabilization of martensite B19′ caused by the development of plastic microdeformations as a result of the appearance of dislocations at small, specified strains in binary alloys with ~50 and 50.8 at% Ni was observed in [29–31]. The generation of lattice defects and, primarily, dislocations during plastic deformation and martensite transformations impedes the recovery of inelastic strains via SE and can even block this process. On heating after unloading, the recovery is contributed to by SME. Thus, it is reasonable to suggest that the rise of SME in the test material after the first and second loading–unloading cycles at 281 K, which is 8 K higher than A_f, is due to dislocation generation at comparatively low stresses. When the temperature is increased to 299 K, the strain recovery after the second cycle with $\gamma_t = 7.2\%$ measures 6.9% via SE and only 0.1% via SME against 0.2% of plastic strain (Figures 4 and 5b). In addition, the analysis of experimental data shows that increasing T_d from 309 to 339 K leads to a decrease in γ_{SE} from 6% to 4% in isothermal cycles τ–γ with $\gamma_t \approx 7\%$. At the same time, the residual deformation increases from 1% to 3% after isothermal unloading of samples. Hence, the temperature interval of manifestation of superelasticity (4–7% and practically without residual deformation) is very narrow and localized from 281 K (A_f + 8 K) to 299 K (A_f + 26 K).

The fact that in loading–unloading cycles at $T_d \geq 309$ K, the SE and the SME value decreases and the plastic component steeply increases, is explained by the deformation temperature approaching M_d, at which no B2 $\rightarrow$ B19′ transformation is possible in the material under applied stress. The higher the temperature of isothermal loading–unloading cycles, the lower the total inelastic strain γ_{SID} and the higher the plastic component γ_{pl} (Figure 5). In this work, we did not investigate the nature of the development of reversible and plastic strains during torsion of the samples (nonhomogeneous or homogeneous [32]). At the same time, it was shown in [31] that the plastic strain develops by dislocation sliding and twinning in an alloy of close composition (50.8 at% Ni) with an average grain size of 500 nm.

Plastic deformation accumulated during torsion causes the appearance of deformation microrelief on the original polished surface of the samples. Figure 6 shows the specimen surface microrelief after torsion at 299 K. Dependences of inelastic deformations γ_{SE}, γ_{SME}, γ_{SID} and plastic deformation γ_{pl} on a total γ_{pl} were presented in Figure 6. The surface microrelief image was obtained after the final cycle τ–γ with $\gamma_t = 38\%$, heating the unloaded sample to 500 K and subsequent cooling to 299 K. The sample after cooling had a B2 phase structure. Plastic deformation $\gamma_{pl} \approx 20\%$, Figure 5c. Figure 6 clearly shows large deformation fragments with an intense microband structure within them. The average size of the large fragments, estimated from three similar images, is 46 $\pm$ 5 µm. This value correlates well with the average size of the original grains (53 $\pm$ 11 µm). Similar deformation fragments with sizes equal to those of the original grains arise as a result of mutual reversals and displacements of neighboring grains relative to each other during tensile plastic deformation of various materials not undergoing martensitic transformations: for example, aluminum-based alloy (Al 1100-0 [33], Al-Mg [34], AA-6022 [35]), AISI 1010 [36] steel, and others. At the same time, the formation of a fine microband structure within the fragments is due to the development of intra-grain plastic deformation processes by the mechanisms of dislocation sliding [29,31].

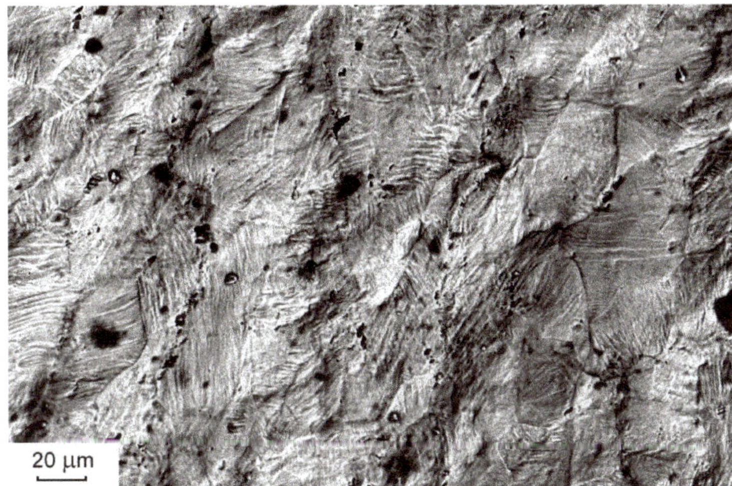

Figure 6. Deformation-induced surface microrelief after torsion at 299 K of coarse-grained specimen of $Ti_{49.3}Ni_{50.7}$ (at%) alloy (details in text). Scanning electronic microscopy.

The maximum total reversible inelastic strain in torsion is γ_{SID} = 18% and is attained at 281 K (8 K higher than A_f) with γ_{pl}· = 12% against $\gamma_t \approx$ 30%. Interestingly, the same result is reported for $Ti_{49.3}Ni_{50.7}$ (at%) after torsion at T < M_f [23]: γ_{SID} = 18% with γ_{pl} = 12%. We think that this coincidence is not by chance but results from a certain common mechanism.

In conclusion, it should be noted that a direct comparison of inelastic and plastic strains obtained under torsion with similar strains obtained under compression, tension, or bending, and with the crystallography limit of martensitic strains at $B19' \leftrightarrow B2$ MT, which is determined by compression-tension of the initial phase crystal lattice, is impossible. This is due to the difference in the deformation modes and the determination of the corresponding strains. In the first approximation, for a correct comparison of the total inelastic strain and plastic strain obtained in torsion with the results of studies of similar strains obtained in tension and bending and the crystallography limit of martensitic strain at $B2 \leftrightarrow B19'$ MT, equal to ~10% for the polycrystalline $Ti_{49.3}Ni_{50.7}$ (at%) alloy [14], it was the concept of equivalent true strain at different loading modes, used earlier in [20]. True tensile strain $e_1 = \ln(1 + \varepsilon)$, where ε is the relative elongation. True torsional strain $e_2 = S/\sqrt{3}$, where S = tgγ is the accumulated shear strain. Tensile strains ε_t, ε_r, ε_{pl}, corresponding to torsional strains γ_t, γ_r, γ_{pl}, can be obtained from the relation $e_1 = e_2$. Using these values of ε_t, ε_r, and ε_{pl}, the inelastic strains recovered by the realization of SE, ε_{SE}, and SME, ε_{SME}, and the total inelastic strain ε_{SID} can be determined. The tensile strains corresponding to torsion strains after loading at 281K (γ_{SID}· = 18%, γ_{pl} = 12%, γ_t· = 30%) are: ε_t = 19.6%, ε_{SID}· = 12.4%, ε_{pl}· = 7.2%. To take into account that both γ_{SID} and ε_{SID} include Hooke's elastic strain, it can be concluded that the maximum reversible inelastic strain γ_{SID} during torsion approximately corresponds to the known crystallography limit of martensitic strain at $B2 \leftrightarrow B19'$ MT in our coarse-grained alloy samples with 50.7 at% Ni. This result corresponds to the data of [16,31]. When bending samples of alloy $Ti_{49.3}Ni_{50.7}$ (at%) with microcrystalline structure (grain size $\leq$ 10 μm), the maximum inelastic strain equal also to 10% was obtained [16]. It was shown in [31] that the maximum total recoverable inelastic strain ε_{SID} in σ-ε cycles during tension and the subsequent heating of unloaded binary alloy samples with 50.9 at% Ni and an average grain size of 500 nm is 13%, which is close to the maximum total inelastic strain in our coarse-grained alloy with 50.7 at% Ni. Thus, abnormally high inelastic strains (~18% [23] and ~15% [16] obtained under the bending of ultrafine-grained TiNi-based binary alloys) are not observed in the coarse-grained samples of binary alloy with 50.7 at% Ni.

5. Conclusions

It was found that the maximum value of the total inelastic torsional strain (γ_{SID}) of coarse-grained samples of $Ti_{49.3}Ni_{50.7}$ (at%) alloy, which was achieved in this work, is 18% and is observed with developed plastic deformation of about 12% (a total specified deformation $\gamma_t \approx 30\%$). It was shown that the maximum value of total inelastic strain γ_{SID}. = 18% obtained during torsion corresponds to the crystallographic limit of martensitic strain at B2 $\leftrightarrow$ B19$'$ MT.

It was shown that very narrow temperature range of a bright manifestation of the superelasticity effect (4–7% and practically without the residual strain) in coarse-grained specimens of $Ti_{49.3}Ni_{50.7}$ (at%) alloy is from 281 K (A_f + 8 K) to 299 K (A_f + 26 K).

It is shown that in the temperature range from A_f + 8 K to A_f + 26 K, the maximum inelastic strain that can be obtained in coarse-grained samples of alloy $Ti_{49.3}Ni_{50.7}$ (at%) during torsion and can be returned completely as a superelasticity effect is 4–7%. When the external stress is increased in order to achieve a larger value of the superelasticity effect, the yield strength of the samples is exceeded. This leads to plastic strain of samples and hinders the return of inelastic strain in the form of superelasticity effect. This is the reason for the appearance and increase in SME when the γ_t total strain increases under torsion of samples at temperatures above A_f.

Author Contributions: Conceptualization, D.Z., V.G., and A.L.; investigations, D.Z., I.R., A.G., and V.T.; writing–original draft preparation, D.Z.; writing–review and editing, D.Z. and A.L.; project administration, D.Z. and V.G.; funding acquisition, D.Z. All authors have read and agreed to the published version of the manuscript.

Funding: This research was funded by Government research assignment for ISPMS SB RAS (project FWRW-2021-0004) and a grant of the President of the Russian Federation No. MK-1057.2020.8.

Data Availability Statement: Not applicable.

Conflicts of Interest: The authors declare no conflict of interest.

References

1. Melton, K.N. Ni-Ti Based Shape Memory Alloys. In *Engineering Aspects of Shape Memory Alloys*; Duerig, T.W., Melton, K.N., Stöckel, D., Wayman, C.M., Eds.; Elsevier: Amsterdam, The Netherlands, 1990; pp. 21–35. [CrossRef]
2. Tabanli, R.; Simha, N.; Berg, B. Mean stress effects on fatigue of NiTi. *Mater. Sci. Eng. A* **1999**, *273–275*, 644–648. [CrossRef]
3. Jarrige, I.; Holliger, P.; Nguyen, T.; Ip, J.; Jonnard, P. From diffusion processes to adherence properties in NiTi microactuators. *Microelectron. Eng.* **2003**, *70*, 251–254. [CrossRef]
4. Otsuka, K.; Ren, X. Physical metallurgy of Ti–Ni-based shape memory alloys. *Prog. Mater. Sci.* **2005**, *50*, 511–678. [CrossRef]
5. Jani, J.M.; Leary, M.; Subic, A.; Gibson, M.A. A review of shape memory alloy research, applications and opportunities. *Mater. Des.* **2014**, *56*, 1078–1113. [CrossRef]
6. Kang, G.; Song, D. Review on structural fatigue of NiTi shape memory alloys: Pure mechanical and thermo-mechanical ones. *Theor. Appl. Mech. Lett.* **2015**, *5*, 245–254. [CrossRef]
7. AbuZaiter, A.; Nafea, M.; Ali, M.S.M. Development of a shape-memory-alloy micromanipulator based on integrated bimorph microactuators. *Mechatronics* **2016**, *38*, 16–28. [CrossRef]
8. Ryklina, E.; Korotitskiy, A.; Khmelevskaya, I.; Prokoshkin, S.; Polyakova, K.; Kolobova, A.; Soutorine, M.; Chernov, A. Control of phase transformations and microstructure for optimum realization of one-way and two-way shape memory effects in removable surgical clips. *Mater. Des.* **2017**, *136*, 174–184. [CrossRef]
9. Cardone, D.; Angiuli, R.; Gesualdi, G. Application of Shape Memory Alloys in Historical Constructions. *Int. J. Arch. Heritage* **2019**, *13*, 390–401. [CrossRef]
10. Geetha, M.; Dhanalakshmi, K. Structural Design and Realization of Electromechanical Logic Elements Using Shape Memory Alloy Wire Actuator. *Phys. Mesomech.* **2020**, *23*, 446–456. [CrossRef]

11. Racek, J.; Duchoň, J.; Vronka, M.; Cieslar, M. TEM observation of twins in surface grains of superelastic NiTi wire after cyclic loading. *Mater. Sci. Eng. A* **2020**, *782*, 139271. [CrossRef]
12. Wang, J.; Pan, Z.; Carpenter, K.; Han, J.; Wang, Z.; Li, H. Comparative study on crystallographic orientation, precipitation, phase transformation and mechanical response of Ni-rich NiTi alloy fabricated by WAAM at elevated substrate heating temperatures. *Mater. Sci. Eng. A* **2021**, *800*, 140307. [CrossRef]
13. Brailovski, V.; Prokoshkin, S.; Terriault, P.; Trochu, F. *Shape Memory Alloys: Fundamental, Modelling and Applications*; École de Technologie Supéreure: Quebec, QC, Canada, 2003.
14. Prokoshkin, S.D.; Korotitskiy, A.V.; Brailovski, V.; Inaekyan, K.E.; Dubinskiy, S.M. Crystal lattice of martensite and the reserve of recoverable strain of thermally and thermomechanically treated Ti-Ni shape-memory alloys. *Phys. Met. Met.* **2011**, *112*, 170–187. [CrossRef]
15. Sehitoglu, H.; Hamilton, R.; Canadinc, D.; Zhang, X.Y.; Gall, K.; Karaman, I.; Chumlyakov, Y.; Maier, H.J. Detwinning in NiTi alloys. *Met. Mater. Trans. A* **2003**, *34*, 5–13. [CrossRef]
16. Ryklina, E.P.; Prokoshkin, S.D.; Chernavina, A.A. Peculiarities of implementation of abnormally high shape memory effects in thermomechanically treated Ti-Ni alloys. *Inorg. Mater. Appl. Res.* **2013**, *4*, 348–355. [CrossRef]
17. Ryklina, E.P.; Prokoshkin, S.D.; Kreitsberg, A.Y. The feasibility of abnormally high shape memory effects of Ti-50.0Ni (at%) in different austenite states. *Izv. Ross. Akad. Nauk. Seriya Fiz.* **2013**, *77*, 1644–1652.
18. Ryklina, E.P.; Polyakova, K.A.; Prokoshkin, S.D. Role of Nickel Content in One-Way and Two-Way Shape Recovery in Binary Ti-Ni Alloys. *Metals* **2021**, *11*, 119. [CrossRef]
19. Chumlyakov, Y.I.; Kireeva, I.V.; Panchenko, E.Y.; Timofeeva, E.E.; Pobedennaya, Z.V.; Chusov, S.V.; Karaman, I.; Maier, H.; Cesari, E.; Kirillov, V.A. High-temperature superelasticity in CoNiGa, CoNiAl, NiFeGa, and TiNi monocrystals. *Russ. Phys. J.* **2008**, *51*, 1016–1036. [CrossRef]
20. Grishkov, V.N.; Lotkov, A.I.; Baturin, A.A.; Timkin, V.N.; Zhapova, D.Y. Comparative analysis of inelastic strain recovery and plastic deformation in a Ti49.1Ni50.9 (at %) alloy under torsion and bending. *AIP Conf. Proc.* **2015**, *1683*, 020067. [CrossRef]
21. Lotkov, A.; Zhapova, D.; Grishkov, V.; Cherniavsky, A.; Timkin, V. Effect of warm rolling on the martensite transformation temperatures, shape memory effect, and superelasticity in Ti49.2Ni50.8 alloy. *AIP Conf. Proc.* **2016**, *1783*, 020137. [CrossRef]
22. Lotkov, A.; Grishkov, V.; Zhapova, D.; Timkin, V.; Baturin, A.; Kashin, O. Superelasticity and shape memory effect after warm abc-pressing of TiNi-based alloy. *Mater. Today Proc.* **2017**, *4*, 4814–4818. [CrossRef]
23. Lotkov, A.I.; Grishkov, V.N.; Zhapova, D.Y.; Gusarenko, A.A.; Timkin, V. Impact of Plastic Straining in the Martensitic State on the Development of the Superelasticity and Shape Memory Effects in Titanium-Nickelide-Based Alloys. *Technol. Phys. Lett.* **2018**, *44*, 995–998. [CrossRef]
24. Miyazaki, S.; Otsuka, K.; Suzuki, Y. Transformation pseudoelasticity and deformation behavior in a Ti-50.6at%Ni alloy. *Scr. Met.* **1981**, *15*, 287–292. [CrossRef]
25. Pelton, A.; Huang, G.; Moine, P.; Sinclair, R. Effects of thermal cycling on microstructure and properties in Nitinol. *Mater. Sci. Eng. A* **2012**, *532*, 130–138. [CrossRef]
26. Kuznetsov, A.V.; Muslov, S.A.; Lotkov, A.I.; Khachin, V.N.; Grishkov, V.N.; Pushin, V.G. Elastic constants of TiNi near martensitic transformations. *Izv. Vuzov. Fiz.* **1987**, *7*, 98–99.
27. Muslov, S.A.; Kuznetsov, A.V.; Khachin, V.N.; Lotkov, A.I.; Pushin, V.G.; Grishkov, V.N. Anomalies of elastic constants of Ti50Ni48Fe2 single crystals near martensitic transformations. *Izv. Vuzov. Fiz.* **1987**, *8*, 104–105.
28. Sedlák, P.; Janovská, M.; Bodnárová, L.; Heczko, O.; Seiner, H. Softening of Shear Elastic Coefficients in Shape Memory Alloys Near the Martensitic Transition: A Study by Laser-Based Resonant Ultrasound Spectroscopy. *Metals* **2020**, *10*, 1383. [CrossRef]
29. Liu, Y.; Favier, D. Stabilisation of martensite due to shear deformation via variant reorientation in polycrystalline NiTi. *Acta Mater.* **2000**, *48*, 3489–3499. [CrossRef]
30. Predki, W.; Klönne, M.; Knopik, A. Cyclic torsional loading of pseudoelastic NiTi shape memory alloys: Damping and fatigue failure. *Mater. Sci. Eng. A* **2006**, *417*, 182–189. [CrossRef]
31. Chen, Y.; Molnárová, O.; Tyc, O.; Kadeřávek, L.; Heller, L.; Šittner, P. Recoverability of large strains and deformation twinning in martensite during tensile deformation of NiTi shape memory alloy polycrystals. *Acta Mater.* **2019**, *180*, 243–259. [CrossRef]
32. Sun, Q.-P.; Li, Z.-Q. Phase transformation in superelastic NiTi polycrystalline micro-tubes under tension and torsion—From localization to homogeneous deformation. *Int. J. Solids Struct.* **2002**, *39*, 3797–3809. [CrossRef]
33. Dai, Y.Z.; Chiang, F.P. On the Mechanism of Plastic Deformation Induced Surface Roughness. *J. Eng. Mater. Technol.* **1992**, *114*, 432–438. [CrossRef]
34. Stoudt, M.R.; Ricker, R.E. The relationship between grain size and the surface roughening behavior of Al-Mg alloys. *Met. Mater. Trans. A* **2002**, *33*, 2883–2889. [CrossRef]
35. Stoudt, M.; Levine, L.; Creuziger, A.; Hubbard, J. The fundamental relationships between grain orientation, deformation-induced surface roughness and strain localization in an aluminum alloy. *Mater. Sci. Eng. A* **2011**, *530*, 107–116. [CrossRef]
36. Stoudt, M.; Hubbard, J. Analysis of deformation-induced surface morphologies in steel sheet. *Acta Mater.* **2005**, *53*, 4293–4304. [CrossRef]

Article

Effect of True Strains in Isothermal *abc* Pressing on Mechanical Properties of Ti$_{49.8}$Ni$_{50.2}$ Alloy

Oleg Kashin *, Konstantin Krukovskii, Aleksandr Lotkov and Victor Grishkov

Institute of Strength Physics and Materials Science of the Siberian Branch of the Russian Academy of Sciences, 634055 Tomsk, Russia; kvk@ispms.tsc.ru (K.K.); lotkov@ispms.ru (A.L.); grish@ispms.tsc.ru (V.G.)
* Correspondence: okashin@ispms.ru; Tel.: +7-3822-286-920

Received: 14 August 2020; Accepted: 28 September 2020; Published: 30 September 2020

Abstract: The paper analyzes the microstructure and mechanical properties of Ti$_{49.8}$Ni$_{50.2}$ alloy (at.%) under uniaxial tension at room temperature after isothermal *abc* pressing to true strains $e = 0.29 - 8.44$ at $T = 723$ K. The analysis shows that as the true strain e is increased, the grain–subgrain structure of the alloy is gradually refined. This leads to an increase in its yield stress σ_y and strain hardening coefficient $\theta = d\sigma/d\varepsilon$ at linear stage III of its tensile stress–strain curve according to the Hall–Petch relation. However, the ultimate tensile strength remains invariant to such refinement. The possible mechanism is proposed to explain why the ultimate tensile strength can remain invariant to the average grains size (d_{av}). It is assumed that the sharp increase of the ultimate tensile strength σ_{UTS} begins when (d_{av}) is less than the critical average grain size $(d_{av})_{cr}$. In our opinion, for the investigated alloy $(d_{av})_{cr} \approx 0.5$ µm. In our study, the attained average grain size is larger the critical one. The main idea of the mechanism is next. In alloys with an average grain size (d_{av}) less than the critical one, a higher external stress is required for the nucleation and propagation of the main crack.

Keywords: TiNi; isothermal *abc* pressing; mechanical properties; average grain–subgrain size

---

1. Introduction

The capability of TiNi-based alloys for thermoelastic martensite transformations, which lie behind their superelasticity and shape memory, makes them beneficial for use in engineering and medicine [1–4]. In practical terms, it is often required that such alloys be mechanically strong. One of the ways of increasing the yield stress and ultimate strength of these alloys is to refine their grain structure down to submicro- and nanoscales with no change of chemical composition. Such refinement is possible by conventional forming (hot and cold rolling, extrusion, rotary forging) and by severe plastic deformation (equal channel angular pressing [5–7], high pressure torsion [5,8,9], *abc* pressing [10–16]), of which the former can provide a nanocrystalline and even an amorphous structure in the alloys while the latter with its constrained conditions allows more efficient transformations of their grain–subgrain and dislocation structures. However, to take full advantage of TiNi-based alloys, we should understand the mechanisms of their grain–subgrain transformation by severe plastic deformation (SPD) and the SPD effect on their functional and mechanical properties. After SPD transformation, TiNi-based alloys normally show a certain increase in their yield stress and ultimate tensile strength, but for attaining the record-breaking strength characteristics at the least cost, we should know what SPD modes (strain temperature, total true strain) can impart a desired grain structure to one or another type of this material.

Despite abundant research data [5–31], it is still unclear what threshold holds for the average grain size below which the ultimate tensile strength of TiNi-based alloys shows an efficient increase and how this increase is contributed by grain size distributions. In many of the works cited above, it is shown that a grain structure in which regions with different average grain sizes (d_{av}) (coarse-grained, submicrocrystalline, nanocrystalline) coexist after SPD of TiNi-based alloys. The different volume fraction of such areas, as well as the uniformity of the distribution of such areas over the volume of the material can significantly affect its mechanical properties. The study of this influence is one of the tasks of this work. Some TiNi-based alloys still need a study to clarify their strain characteristics in relation with average grain sizes. All the foregoing is true for *abc* pressed TiNi-based alloys, among which only two have been examined in terms of the above issues: non-aging $Ti_{49.9}Ni_{50.1}$ [10] and $Ti_{49.8}Ni_{50.2}$ [11–16]. Hereinafter the alloy composition is given in at.%.

In particular, $Ti_{49.8}Ni_{50.2}$ with an initial grain size of 20–70 µm was *abc* pressed in steps, each at a true strain $e \approx 3.6$ but at a different temperature in order of its gradual decrease from 873 to 573 K [11], such that each pressing event could be considered as transforming a different grain–subgrain structure. When *abc* pressed at 873 and 773 K, the alloy remained coarse-grained (5–10 µm), and when the temperature was decreased to below 673 K, its bulk was represented by submicrocrystalline grains sized to 100–700 nm with a small amount of nanograins sized to 20–100 nm. In the alloy *abc* pressed in the same way at $e = 7.7$ and 573 K, the volume fraction of its nanostructural elements of size 20–50 nm was about 30% [12,13].

At $e = 8.44$, was reached after *abc* pressing at 723 K [14–16], the alloy was dominated by a submicrocrystalline structure with a noticeable nanograin fraction, residing mainly around coarse precipitation of the Ti_2Ni type, and at this temperature (723 K), intensive recovery processes occurred.

In $Ti_{49.9}Ni_{50.1}$ *abc* pressed at 423 K and 673 K [10], tangled dislocations whose density increased with increasing the true strain were found after *abc* pressing at 423 K but no quantitative estimates of the dislocation density were presented. No estimates were also taken of the average size and volume fraction of grains sized to ~20 nm found in the alloy microstructure after *abc* pressing at 423 K. It was reported that the *abc* pressing temperature 673 K provided a dynamic recovery in the alloy by rearrangement and lowering of high density dislocations, the formation of dislocation-free recrystallized grains with a random grain size distribution, whose pattern was not given.

In $Ti_{49.8}Ni_{50.2}$ *abc* pressed to grain–subgrain sizes of 2–40 µm at 873 K, the true ultimate tensile strength under uniaxial tension at room temperature measured 1560 MPa [13] and decreased to 1450 MPa when the alloy was further pressed to grain–subgrain sizes of 100–700 nm at 723 K. It was concluded that the ultimate tensile strength of the alloy was little influenced by the grain–subgrain size varying from coarse-crystalline to submicrocrystalline [13].

In $Ti_{49.9}Ni_{50.1}$ *abc* pressed at 673 K, the true ultimate tensile strength σ_{UTS} increased from 1320 to 1490 MPa as the strain e was increased from 1.5 to 6 [10]. Decreasing the *abc* pressing temperature to 423 K increased the true ultimate tensile strength σ_{UTS} to 1550 MPa at $e \approx 2.2$ and to 1700 MPa at $e \approx 9.0$. It was concluded that the high strength and the small elongation to fracture in the *abc* pressed material was governed by strain hardening through the formation of a tangled substructure with a high dislocation density and that the lower values of σ_{UTS} after *abc* pressing at 673 K was due to recovery processes [10].

Thus, the available experimental data give only an idea of the main trends of structural transformations in *abc* pressed non-aging TiNi alloys with a near-equiatomic composition but the relation between their structure and mechanical properties remains unclear.

In this context, our aim here is in providing systematic experimental data from which to judge the effect of isothermal *abc* pressing at 723 K on the grain size in $Ti_{49.8}Ni_{50.2}$ at different true strains e and the effect of its grain–subgrain structure on its yield stress, ultimate tensile strength, and deformation behavior at room temperature. As shown in [11–16], at room temperature an alloy of this composition contains a high temperature phase B2, an intermediate martensitic R-phase with a rhombohedral crystal lattice, and a martensitic B19′ phase with a monoclinic crystal lattice.

Typically, the $Ti_{49.8}Ni_{50.2}$ alloy is considered as non-aging one. The probability formation of Ti_3Ni_4 phase precipitates in this alloy during thermomechanical treatments is small. We suggest that the volume fraction of these precipitates is small and will not affect the properties of the alloy. The temperature 723 K, being much higher than M_d (M_d is the maximum temperature at which a martensitic transformation of B2(R)→B19′ induced by an external applied stress is possible), provides *abc* pressing in the B2 phase of the alloy, its recovery processes, and a substantial decrease in its internal stresses.

Note that studying the structure and properties of TiNi-based alloys after *abc* pressing not only extends our knowledge of the capabilities of SPD processes but it can also identify proper technological modes of manufacturing various TiNi products, the more so as *abc* pressing is applicable to large-sized billets.

2. Materials and Methods

The test material was $Ti_{49.8}Ni_{50.2}$ alloy (hereinafter the alloy composition is given in at.%) supplied as bars of diameter 20 mm (MATEK-SMA Ltd., Moscow, Russia). The bars were cut into cylindrical segments of length 25 mm which were then forged in three mutually perpendicular directions at 1123 K to obtain cubic specimens sized to about $20 \times 20 \times 20$ mm^3. The cubic specimens, hereinafter referred to as initial, were *abc* pressed in a die at 723 K (0.46 T_{mel}). A hydraulic press MIS-6000 K (MISIS, Moskow, Russia) with a force of 600 tons was used to perform deformation during *abc* pressing. At this temperature, the B2 phase of the alloy has a low plastic strain resistance such that the die after *abc* pressing remains intact. As has been shown [16,21,22], equal channel angular pressing at 723 K provides an ultrafine-grained structure in $Ti_{49.8}Ni_{50.2}$ and a substantial increase in its mechanical properties. Applying *abc* pressing at 723 K to $Ti_{49.8}Ni_{50.2}$ allows us to compare the different SPD methods in terms of the effect on the grain–subgrain structure of the alloy and on its yield stress and ultimate tensile strength.

The initial specimens placed in dies were kept in a furnace at 723 K for 10 min and were then hydraulically pressed with a rate of 0.16–0.18 s^{-1}. Each pressing cycle was a sequence of three mutually perpendicular compression events. Once compressed in one direction, the specimens were removed from the dies and were again placed in them without cooling so that the next compression event would be perpendicular to the previous one. At the end of each compression event, the specimen temperature decreased by no more than 10 K. Before each event, the dies with the specimens were heated to 723 K. The true strain in the specimens singly pressed in one direction was from $e \approx 0.15$ to $e \approx 0.30$. Its values 0.29, 0.62, 1.82, 4.15, 6.44, and 8.44 were attained by repeating the cycle of *abc* pressing.

For tensile tests, the *abc* pressed specimens were shaped as a dumbbell with a gage area of $7 \times 1 \times 1$ mm^3 on an electrical discharge machine (Del'ta-test Ltd., Moscow, Russia), hand polished with silicon carbide abrasives gradually decreasing in roughness to 1200, and electrolytically polished in a mixture of acetic (Reahim Ltd., Moscow, Russia) and perchloric acids (Reahim Ltd., Moscow, Russia) with a ratio of 70:30. The tensile tests were performed at room temperature on a Walter + Bai AG LFM-125 testing machine (Walter + Bai AG, Löhningen, Switzerland) with Dionpro software (Version 7, Walter + Bai AG, Löhningen, Switzerland). The rate of movement of the mobile gripper of testing machine was constant and was 0.007 mm/s. For a sample with a working base length of 7 mm, this corresponds to an initial relative strain rate of 10^{-3} s^{-1}. For each true strain after *abc* pressing, four to six specimens were tested.

The temperature of martensite transformations were determined by an experimental setup (ISPMS SB RAS, Tomsk, City, Russia). The measurement method is described in details in [15].

The microstructure of the alloy was examined by light optical microscopy (Zeiss Axiovert-200M, (Carl Zeiss AG, Oberkochen, Germany) and by scanning and transmission electron microscopy (Zeiss LEO EVO 50 XVP (Carl Zeiss AG, Oberkochen, Germany) and JEM-2100 (JEOL Ltd., Tokyo, Japan) provided by Tomsk Regional Center for Collective Use TSC SB RAS).

The specimens for optical and SEM microstructure studying were cut using an electric discharge machine along the plane passing though the specimen center and parallel to any two arbitrary opposite edges of the specimen. Such cross section provided analysis of the specimen microstructure in its overall bulk. After cutting, the specimen surfaces were mechanically polished using a Saphir 350

(Audit Diagnostics, Bussiness & Technology Park, Carrigtwohill. Co. Cork, Ireland) grinder, and then they were chemically etched and electrolytically polished.

Transmission electron microscopy TEM microstructure studying was performed using thin foils. Thin plates with a thickness of 0.2 mm were cut from the blanks obtained after *abc* pressing in the selected section using an electric discharge machine. Finally, the specimen thinned electrolytically.

3. Results

3.1. Martensite Transformation Temperatures and Phase States

Figure 1 illustrates the martensitic transformations temperatures after *abc* pressing with various true strain. Here M_S and M_f are the start and finish temperatures of the direct martensitic transformation B2(R)→B19′ at cooling, respectively; A_S and A_f are the start and finish temperatures of the reverse martensitic transformation at heating, respectively; T_R is the start temperature of the martensitic transformation B2→R at cooling. The dotted line on Figure 1 shows the tensile test temperature T_{test}.

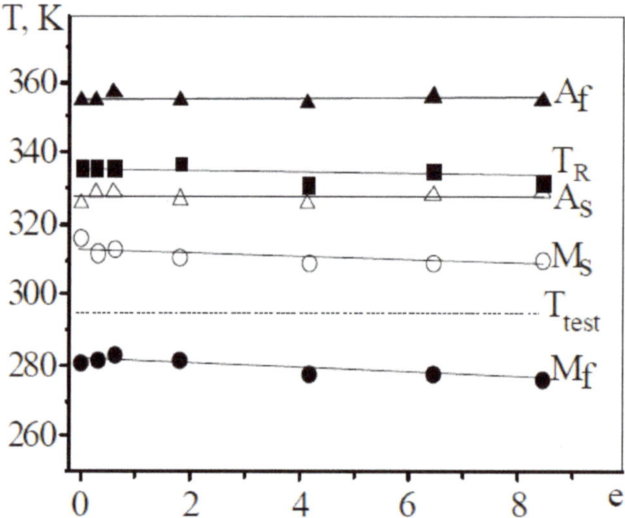

Figure 1. Martensite transformation temperatures versus true strain in Ti$_{49.8}$Ni$_{50.2}$ after *abc* pressing.

As can be seen from Figure 1, the temperatures of direct (at cooling) and reverse (at heating) martensite transformations in Ti$_{49.8}$Ni$_{50.2}$ are almost independent of its true strain after *abc* pressing (as in Ref. [15]). Its direct martensite transformation, irrespective of the true strain, proceeds as B2→R→B19′. It is also seen that the state of all specimens at room temperature corresponds to the interval between the martensite start and finish temperatures M_S and M_f.

3.2. Microstructure

Figures 2 and 3 present the microstructures for Ti$_{49.8}$Ni$_{50.2}$ before and after *abc* pressing to *e* = 8.44. The data exemplify that the bulk of all specimens is occupied by B19′ martensite (Figures 2c and 3b) with a small (5–10 vol.%) amount of R (Figure 2f) and B2 phases (Figures 2f and 3b) which fail to transform to B19′ on cooling to room temperature.

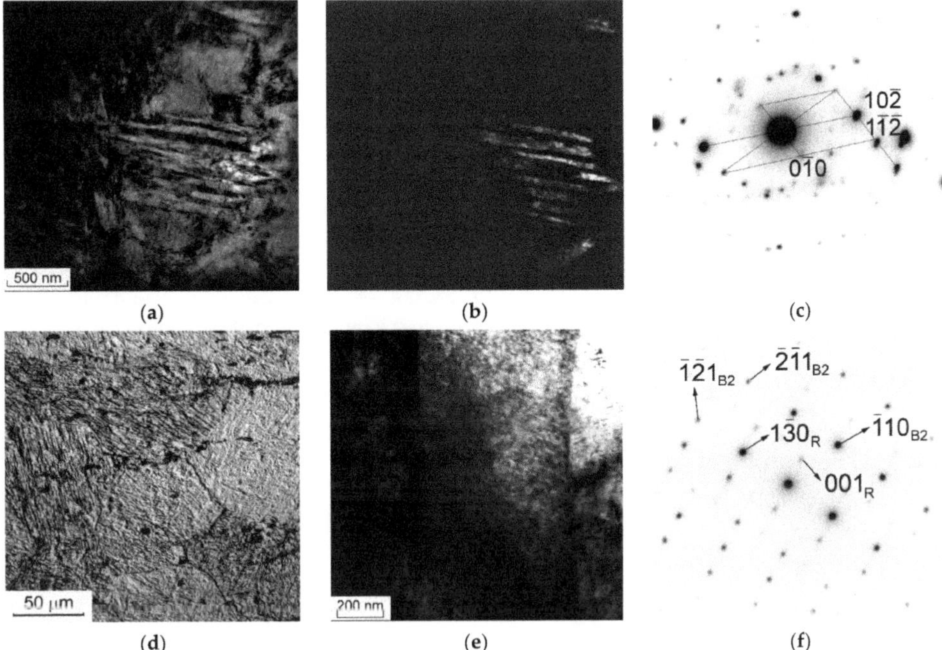

Figure 2. Microstructure of $Ti_{49.8}Ni_{50.2}$ before *abc* pressing: (**a–c**) TEM images and respective diffraction pattern of region with B19′ phase; (**d**) optical image in DIC (Differential Interference Contrast) mode; (**e,f**) TEM images and respective diffraction pattern of region with B2 and R phases.

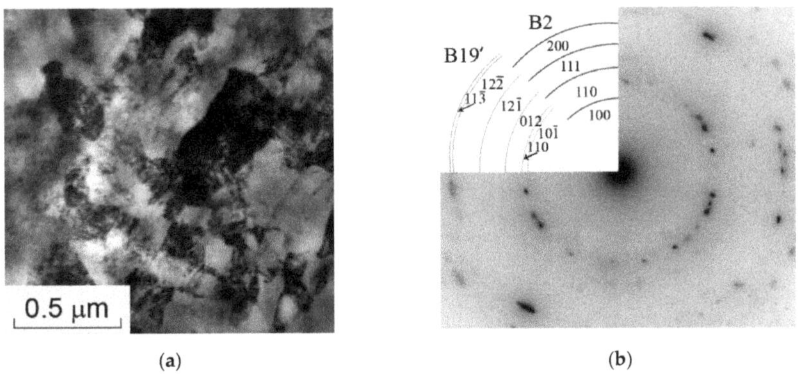

Figure 3. Structure of $Ti_{49.8}Ni_{50.2}$ *abc* pressed to $e = 8.44$: (**a**) TEM image, (**b**) diffraction pattern.

When shaped at 1123 K, the specimens acquire an inhomogeneous grain–subgrain structure (Figure 2). At room temperature, their bulk is represented by grains with an average size of about 40 μm which contain twinned B19′ domains with an average size of about 1.5 μm.

Increasing the true strain in *abc* pressed $Ti_{49.8}Ni_{50.2}$ gradually decreases the average grain size. However, up to $e = 4.15$, the material bulk reveals coarse grains of up to 40 μm with a developed dislocation substructure. At e= 4.15, its structure is represented by coarse and fine grains. The volume fraction of submicrocrystalline grains with $d \leq 1$ μm is no greater than 5 vol.%.

At $e = 8.44$, the material bulk has a submicrocrystalline structure with a grain–subgrain size of 0.1–1.5 µm which shows up as quasi-ring microdiffraction patterns (Figure 3). At the same time, there are individual grains of up to 5 µm. At e = 8.44, the volume fraction of submicrocrystalline grains with $d \leq 1$ µm, including nanograins, increases to ~20 vol.%. The estimated average grain size is ≈ 1.0 µm.

3.3. Mechanical Properties

Figure 4 shows the typical engineering tensile stress–strain curves of Ti$_{49.8}$Ni$_{50.2}$ before and after *abc* pressing to different true strains. There are four stages of the strain–stress curve (as in Reference [32]), the length of which was determined graphically. The initial stage I corresponds to quasi-elastic loading, stage II—pseudo-yield plateau, stage III—quasi-linear stage of strain hardening, ending with a yield stress, stage IV corresponds to parabolic hardening up to the ultimate tensile strength. There is some transition area between the stages.

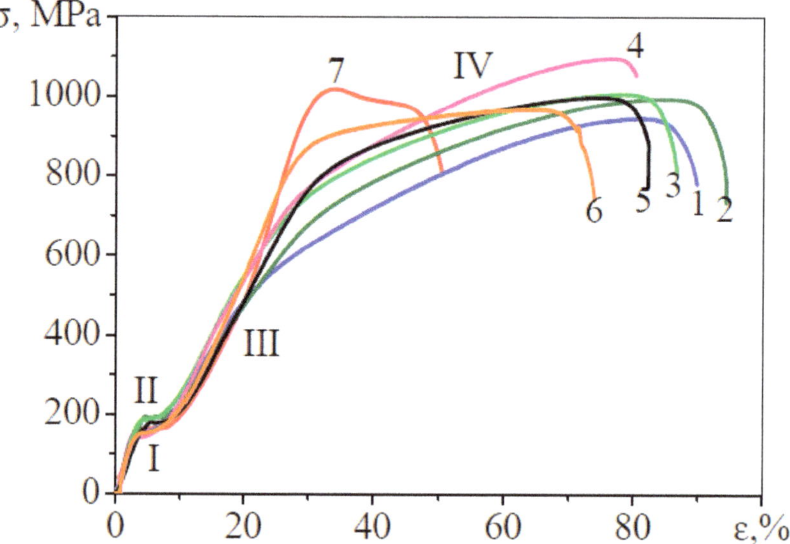

Figure 4. Engineering tensile stress–strain curves of Ti$_{49.8}$Ni$_{50.2}$ before *abc* pressing (*1*) and after *abc* pressing to e = 0.29 (*2*), 0.62 (*3*), 1.82 (*4*), 4.15 (*5*), 6.44 (*6*), and 8.44 (*7*).

It is seen that at early quasi-elastic stage I, the curves are almost coincident. Figure 5 show the average mechanical properties of Ti$_{49.8}$Ni$_{50.2}$ estimated from its respective stress–strain curves. Within the spread of values, the pseudo-yield stress σ_m (martensite shear stress) depends little on the true strain set for *abc* pressing and measures about 180 MPa (Figure 5a, curve 4). The extent of the pseudo-yield plateau (Figure 4, stage II) varies from 3 to 8%, showing no correlation with the true strain.

The average ultimate strength of *abc* pressed Ti$_{49.8}$Ni$_{50.2}$ depends weakly on the true strain and measures σ_{UTS} ~1000 MPa (Figure 5a, curve *1*), which is about 10% higher than its value before *abc* pressing.

The uniform elongation δ_{UTS} and the elongation to fracture δ_{total} vary slightly up to e = 4.15 (Figure 5a curves 2, 3), and as the true strain is increased to e = 8.44, these characteristics decrease steeply (2.5 and 1.5 times, respectively).

The pseudo-yield plateau is followed by a gradual transition to quasi-linear stage III (Figure 4) at which the strain hardening coefficient $\theta = d\sigma/d\varepsilon$ increases with the true strain set for *abc* pressing

(Figure 5b, curve 2). The stress at which linear stage III ends also increases, which corresponds to the yield stress σ_y in Ti$_{49.8}$Ni$_{50.2}$ in its martensite state (Figure 5b, curve *1*).

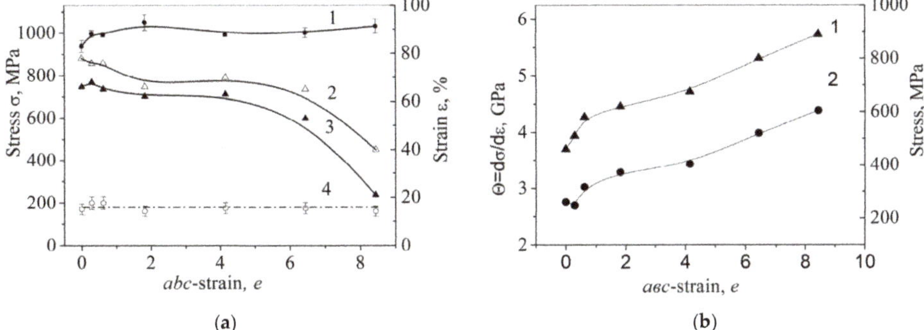

Figure 5. Mechanical characteristics of Ti$_{49.8}$Ni$_{50.2}$ versus true strain after *abc* pressing: (**a**) *1*—ultimate tensile strength, *2*—elongation to fracture, *3*—uniform elongation, *4*—pseudo-yield stress; (**b**) *1*—yield stress σ_y, *2*—strain hardening coefficient θ.

Linear stage III is followed by parabolic hardening stage IV (Figure 4). This deformation stage for all test specimens, except for those *abc* pressed to e = 8.44, is rather long, measuring several tens of percent. In Ti$_{49.8}$Ni$_{50.2}$ *abc* pressed to e = 8.44, the parabolic stage is very short: about 2%.

Once the test specimens reach their maximum stress (ultimate tensile strength σ_{UTS}), they are involved in strain localization with necking such that their tensile stress–strain curves reveal a steep decrease in the stress which culminates in fracture. In Ti$_{49.8}$Ni$_{50.2}$ *abc* pressed to e = 8.44, the final deformation stage differs from what is observed in the other specimens: the stress first decreases noticeably, then weakly and slowly, and then again rapidly up to eventual fracture.

4. Discussion

Our experiments show that increasing the true strain up to e = 8.44 in Ti$_{49.8}$Ni$_{50.2}$ *abc* pressed at 723 K weakly influences the temperatures of its martensite transformations, martensite shear stress σ_m, and ultimate tensile strength σ_{UTS}. At the same time, the true strain in *abc* pressed Ti$_{49.8}$Ni$_{50.2}$ greatly influences its yield stress σ_y, strain hardening coefficient θ at the third stage of its tensile stress–stress curve (Figure 4, stage III), uniform elongation δ_{UTS}, and elongation to fracture δ_{total}.

The same martensite shear stress σ_m irrespective of the true strain is expectable because σ_m depends on the difference between T_{test} and M_S and these temperatures remain the same at all true strains used in *abc* pressing (Figure 1). Analysis of our studies and others available suggests that when deformed by SPD methods to large strains at a temperature of 673–723 K, non-aging binary TiNi alloys with a Ni content of 50.0–50.2 at.% assume mostly a submicrocrystalline grain–subgrain structure [10–16]. At these temperatures, their deformation is accompanied by efficient relaxation which makes them free of any appreciable internal stress [10,15]. For an average grain–subgrain size of 0.5–1.0 μm, the martensite shear stress σ_m lies in the range 180–200 MPa, and for 0.1–0.3 μm, its value increases to about 300 MPa. It is significant that such features are observed not only after *abc* pressing [10,13] but after equal channel angular pressing as well [19–22]. Unfortunately, we have not found any data on the formation of a nanocrystalline structure in Ti$_{49.8}$Ni$_{50.2}$ by one or another SPD method. However, such a structure with an average grain size of 16 nm has been obtained in Ti$_{49.8}$Ni$_{50.2}$ wires after cold drawing and annealing at 623 K [33,34], with the result that the material on its stress–strain dependence at room temperature reveals an increase in the pseudo-yield stress to σ_m = 515 MPa. The increase in σ_m can be explained by substantial hardening during the so appreciable grain–subgrain refinement, which should inevitably decrease the B2→B19′ start temperature and increase σ_m at room temperature.

From the results presented [10–16,19–22,33,34] it follows that in $Ti_{49.8}Ni_{50.2}$ free of appreciable internal stress, the average grain–subgrain size has a critical value $(d_{av})_{cr} \approx 0.5$ μm with respect to the pseudo-yield stress σ_m. When the material structure is refined to an average grain–subgrain size $d_{av} \geq 0.5$ μm, the pseudo-yield stress σ_m remains almost unchanged, and when the average grain–subgrain size goes down to $d_{av} < 0.5$ μm, the pseudo-yield stress σ_m increases such that the lower the size d_{av}, the higher the value of σ_m.

When the pseudo-yield stage ends, all processes associated with martensite transformations to B19 and martensite domain reorientation under external load are mostly completed, showing no effect on the further deformation of the alloy. Each true strain in *abc* pressing provides the alloy with a specific structure which naturally evolves as the true strain is increased. However, if another external load with another temperature and rate is applied to the alloy after *abc* pressing, e.g., uniaxial tension at room temperature (like in our case), the earlier formed structure becomes initial such that those dislocations which participate in its deformation during *abc* pressing are ineffective under further uniaxial tension. After induced phase transformations, the tensile deformation of the alloy at stage III is provided by new dislocations whose nucleation and motion depend on the structure formed by *abc* pressing: on the size of its grains, subgrains, and fragments, on the spatial location and density of dislocations inside its elements and boundaries, and on the state of its interfaces [35]. Our experiments suggest that these characteristics, excepting the average grain–subgrain size, vary little with true strains in *abc* pressing due to recovery processes. Thus, the new dislocation substructure in the *abc* pressed alloy under tension develops against the background of about the same grain–subgrain microstructure at all trues strains used.

For many metals and alloys, including SPD ones, the yield stress σ_y obeys the empirical Hall–Petch relation [36–38]:

$$\sigma_y = \sigma_0 + k\, d_{av}^{-1/2}, \tag{1}$$

where σ_0 is the lattice friction stress of a material (its constant), k is the Hall–Petch coefficient, and d_{av} is the average grain–subgrain size.

If roughly the same grain–subgrain microstructure is formed by *abc* pressing irrespective of the true strain, the dependence of the yield stress on the average grain–subgrain size d_{av} should follow the Hall–Petch relation. The assumption can be verified by estimating d_{av} from our data obtained by different types of microscopy (optical, TEM, SEM, electron back scatter diffraction). The estimates show that for the initial specimens before *abc* pressing, $d_{av} \approx 40$ μm, and for those *abc* pressed to $e = 8.44$, $d_{av} \approx 1.0$ μm. Figure 6a demonstrates how the average grain–subgrain size d_{av} depends on the true strain e in *abc* pressing, which agrees well with data reported elsewhere [19]. Figure 6b shows the $d_{av}^{-1/2}$ dependence of σ_y plotted from the estimates of d_{av}. As can be seen, the dependence is well approximated by a straight line, except for σ_y in the initial specimens and specimens with $e = 0.29$, and this dependence is fitted well by data reported for $Ti_{49.8}Ni_{50.2}$ with $d_{av} = 0.19$ μm and $\sigma_y \approx 1270$ MPa after six cycles of equal channel angular pressing at 723 K [22].

It is considered that behind the main grain size effect on the yield stress of metals and alloys lies dislocation retardation by internal interfaces: the finer the grain size, the larger the number of interfaces and the shorter the free path of mobile dislocations [39]. In our study, the Hall–Petch relation holds for grain refinement in *abc* pressed $Ti_{49.8}Ni_{50.2}$, suggesting that the main deformation mechanism at stage III is the nucleation and motion of new dislocations and that the yield stress σ_y is determined mainly by the average grain–subgrain size.

The $d_{av}^{-1/2}$ dependence of the strain hardening coefficient θ is also well approximated by a straight line, except for θ at the initial true strains in *abc* pressing (Figure 6b). The same behavior of θ in relation to e is observed in submicrocrystalline $Ti_{49.8}Ni_{50.2}$ after equal channel angular pressing [19] and in nanocrystalline $Ti_{49.8}Ni_{50.2}$ with a maximum grain size no greater than 40 nm after cold drawing and annealing [33,34]. In the former case, the strain hardening coefficient is close to our values, and in the latter, it is 3-4 times higher.

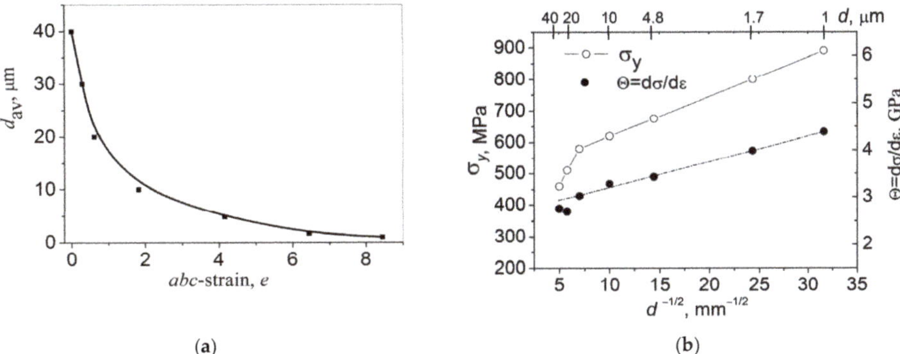

(a)   (b)

Figure 6. (**a**) Average grain–subgrain size d_{av} versus true strain e in *abc* pressing; (**b**) Yield stress σ_y and strain hardening coefficient θ versus $d_{av}^{-\frac{1}{2}}$.

In all test specimens of Ti$_{49.8}$Ni$_{50.2}$, the ultimate strength σ_{UTS} is scarcely affected by the true strain in *abc* pressing. In other words, it remains constant as d_{av} decreases from ≈40 μm to ≈(1.0 ± 0.5) μm. Figure 7 shows the ultimate tensile strength σ_{UTS} versus the average grain–subgrain size d_{av} in Ti$_{49.8}$Ni$_{50.2}$ *abc* pressed in our study (open circles) and in Ti$_{49.8}$Ni$_{50.2}$ exposed to other types of large plastic deformation with attendant recovery processes (solid symbols) [19,22,33,34]. In the cited studies, the average grain sizes were much smaller than those reported here, and in view of their experimental conditions [19,22,33,34], it may be considered that the respective data were obtained for alloy microstructures similar to the one formed by *abc* pressing or for alloy specimens of the very approximate composition but with different average grain–subgrain sizes.

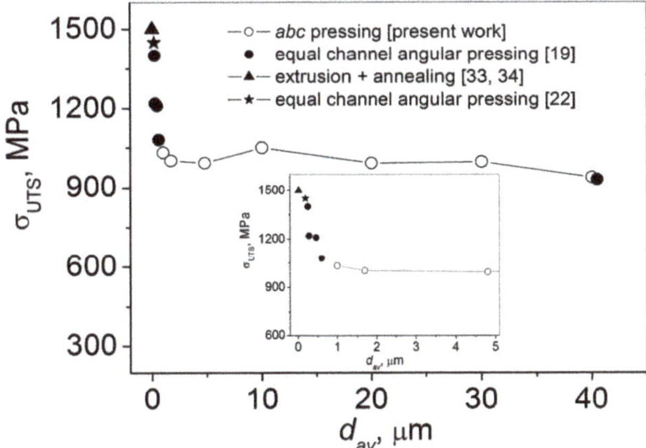

Figure 7. Ultimate strength versus average grain–subgrain size in Ti$_{49.8}$Ni$_{50.2}$ exposed to different types of large plastic deformation.

In view of the above assumptions, the d_{av} dependences of σ_{UTS} suggest the existence of a critical value $(d_{av})_{cr} \approx 0.5$ μm, which is more evident from the inset in Figure 7 at higher magnification. On engineering tensile diagrams, σ_{UTS} means a maximum external stress responsible either for instantaneous failure (brittle fracture) or for macroscale strain localization with necking and eventual failure (ductile fracture). In both cases, the fracture surface of materials may show brittle elements (cleavages, quasi-cleavages) and ductile elements (dimple rupture) but their fracture, irrespective of this, is preceded by the formation of a critical crack which becomes a main one. The nucleation of cracks is due to local internal stresses which result from incompatibilities induced by plastic deformation under external stresses [35]. At a certain level of external stresses, such cracks can merge into a main one. If grains-subgrains of size larger than critical are present in a material, internal stresses sufficient for stable crack nucleation can arise at a certain external stress. Apparently, the critical grain–subgrain size in our case lies between 1 to 5 μm and the external stress measures about 1000 MPa. It is not improbable that the volume fraction of grains-subgrains larger than critical can also be a contributor.

When the size of grains-subgrains become smaller than critical, higher external stresses are needed to provide internal stresses sufficient for stable crack nucleation. The question of why this happens so is a separate research issue.

5. Conclusions

The results of our study into the mechanical properties of $Ti_{49.8}Ni_{50.2}$ at room temperature after *abc* pressing to true strains $e = 0.29$–8.44 at $T = 723$ K and their comparison with other related data suggest the following.

1. As the true strain e is increased, the grain–subgrain structure is gradually refined. At $e = 8.44$, being maximal in our study, the material assumes a microstructure with an average grain size $d_{av} \approx 1$ μm in which the main fraction is submicrocrystalline and the volume fraction of fine-grained and nanocrystalline elements is small (about 20% in total).
2. Refining the grain–subgrain structure of the material to the average size $d_{av} \approx 1$ μm increases its yield stress σ_y and strain hardening coefficient $\theta = d\sigma/d\varepsilon$ at linear stage III of its tensile stress–strain curve, and the dependence of these characteristics on the grain–subgrain size is well described by the Hall–Petch relation.
3. Refining the grain–subgrain structure to the average size $d_{av} \approx 1$ μm does not change the temperatures of martensite transformations, martensite shear stress σ_m, and ultimate strength σ_{UTS}. It is supposed that this is due to the existence of a critical average grain–subgrain size whose value $(d_{av})_{cr} \approx 0.5$ μm has not been attained at the strains used in *abc* pressing.

Author Contributions: Conceptualization, O.K. and A.L.; writing—original draft preparation, O.K.; writing—review and editing, A.L. and V.G.; software, K.K.; investigations, O.K., K.K., and V.G.; project administration, A.L.; funding acquisition, A.L. All authors have read and agreed to the published version of the manuscript.

Funding: This research was funded by Fundamental Research Program of the State Academies of Science for 2013–2020 (priority direct № III.23.2.2).

References

1. Duerig, T.; Pelton, A.; Stoeckel, D. An overview of nitinol medical applications. *Mater. Sci. Eng. A* **1999**, *273–275*, 149–160. [CrossRef]
2. Saadat, S.; Salichs, J.; Noori, M.; Hou, Z.; Davoodi, H.; Bar-on, I.; Suzuki, Y.; Masuda, A. An overview of vibration and seismic applications of NiTi shape memory alloy. *Smart Mater. Struct.* **2002**, *11*, 218–229. [CrossRef]
3. Predki, W.; Knopik, A.; Bauer, B. Engineering applications of NiTi shape memory alloys. *Mater. Sci. Eng. A* **2008**, *481*, 598–601. [CrossRef]
4. Wadood, A. Brief Overview on Nitinol as Biomaterial. *Adv. Mater. Sci. Eng.* **2016**, 4173138. [CrossRef]
5. Valiyev, R.Z.; Aleksandrov, I.A. *Bulk Nanostructure Metal Materials: Obtaining, Structure and Properties;* Akademkniga: Moscow, Russian, 2007; p. 398.

6. Khmelevskaya, I.Y.; Prokoshkin, S.D.; Trubitsyna, I.B.; Belousov, M.N.; Dobatkin, S.V.; Tatyanin, E.V.; Korotitskiy, A.V.; Brailovski, V.; Stolyarov, V.V.; Prokofiev, E.A. Structure and properties of Ti–Ni-based alloys after equal-channel angular pressing and high-pressure torsion. *Mater. Sci. Eng. A* **2008**, *481–482*, 119–122. [CrossRef]

7. Shahmir, H.; Nili-Ahmadabadi, M.; Mansouri-Arani, M.; Langdon, T.G. The processing of NiTi shape memory alloys by equal-channel angular pressing at room temperature. *Mater. Sci. Eng. A* **2013**, *576*, 178–184. [CrossRef]

8. Shri, D.N.A.; Tsuchiya, K.; Yamamoto, A. Surface characterization of TiNi deformed by high-pressure torsion. *Appl. Surf. Sci.* **2014**, *289*, 338–344. [CrossRef]

9. Gunderov, D.; Lukyanov, A.; Prokofiev, E.; Kilmametov, A.; Pushin, V.; Valiev, R. Mechanical properties and martensitic transformations in nanocrystalline $Ti_{49.4}Ni_{50.6}$ alloy produced by high-pressure torsion. *Mater. Sci. Eng. A* **2009**, *503*, 75–77. [CrossRef]

10. Shamsolhodaei, A.; Zarei-Hanzaki, A.; Moghaddam, M. Structural and functional properties of a semi equiatomic NiTi shape memory alloy processed by multi-axial forging. *Mater. Sci. Eng. A* **2017**, *700*, 1–9. [CrossRef]

11. Lotkov, A.I.; Grishkov, V.N.; Dudarev, Y.F.; Girsova, N.V.; Tabachenko, A.N. Formation of ultrafine grain structure, martensitic transformations and unelastic properties of titanium nickelide after abc-pressing. *Vopr. Materialoved.* **2008**, *1*, 161–165.

12. Lotkov, A.I.; Grishkov, V.N.; Dudarev, E.F.; Koval, Y.N.; Girsova, N.V.; Kashin, O.A.; Tabachenko, A.N.; Firstov, G.S.; Timkin, V.N.; Zhapova, D.Y. Ultrafine Structure and Martensitic Transformation in Titanium Nickelide after Warm abc Pressing. *Inorg. Mater. Appl. Res.* **2011**, *2*, 548–555. [CrossRef]

13. Lotkov, A.I.; Grishkov, V.N.; Baturin, A.A.; Dudarev, E.F.; Zhapova, D.Yu.; Timkin, V.N. The effect of warm deformation by abc-pressing method on mechanical properties of titanium nickelide. *Lett. Mater.* **2015**, *5*, 170–174. [CrossRef]

14. Lotkov, A.; Grishkov, V.; Zhapova, D.; Timkin, V.; Baturin, A.; Kashin, O. Superelasticity and shape memory effect after warm abc-pressing of TiNi-based alloy. *Mater. Today Proc.* **2017**, *4*, 4814–4818. [CrossRef]

15. Lotkov, A.I.; Kashin, O.A.; Grishkov, V.N.; Krukovskii, K.V. The influence of degree of deformation under isothermal *abc* pressing on evolution of structure and temperature of phase transformations of alloy based on titanium nickelide. *Inorg. Mater. Appl. Res.* **2015**, *6*, 96–104. [CrossRef]

16. Zhapova, D.Yu.; Lotkov, A.I.; Grishkov, V.N.; Timkin, V.N.; Rodionov, I.S.; Kolevatov, A.S.; Belosludtseva, A.A. Inelastic properties of titanium nickelide after warm abc-pressing. *Izv. Vyssh. Uchebn. Zaved. Fiz.* **2016**, *59*, 60–63.

17. Kreitcberg, A.; Brailovski, V.; Prokoshkin, S.; Gunderov, D.; Khomutov, M.; Inaekyan, K. Effect of the grain/subgrain size on the strain-rate sensitivity and deformability of Ti–50 at %Ni alloy. *Mater. Sci. Eng. A* **2015**, *622*, 21–29. [CrossRef]

18. Prokoshkin, S.D.; Khmelevskaya, I.Y.; Dobatkin, S.V.; Trubitsyna, I.B.; Tat'yanin, E.V.; Stolyarov, V.V.; Prokof'iev, E.A. Structure evolution upon severe plastic deformation of TiNi-based shape-memory alloys. *Phys. Metals Metallogr.* **2004**, *97*, 619–625.

19. Prokofyev, V.A. Structure and properties of ultra-low grain TiNi alloy obtained by intensive plastic deformation. *Vestnik UGATU* **2006**, *8*, 169–174.

20. Yurchenko, L.I.; Dyupin, A.P.; Gunderov, D.V.; Valiev, R.Z.; Kuranova, N.N.; Pushin, V.G.; Uksusnikov, A.N. Mechanical Properties and Structure of High-Strength Nanostructured Nickel-Titanium Alloys Subjected to ECAP and Rolling. *Faz. Perekh. Uporyad. Sost. Novye Mater.* Available online: http://www.ptosnm.ru (accessed on 4 October 2006).

21. Dudarev, E.F.; Bakach, G.P.; Kolobov, Y.R.; Ivanov, K.V.; Lotkov, A.I.; Grishkov, V.N.; Valiyev, R.Z.; Ivanov, M.B. Localization of martensitic deformation on meso- and macroscale levels in large grain and sub-micro grain alloys with shape memory. *Fiz. Mezomekh.* **2004**, *7*, 127–130.

22. Churakova, A.A.; Gunderov, D.V.; Lukyanov, A.V.; Lebedev, Y.A. Effect of thermal cycling in the range of phase transformation B2–B19′ on the microstructure and mechanical properties of UFG alloy $Ti_{49.8}Ni_{50.2}$. *Lett. Mater.* **2013**, *3*, 166–168. [CrossRef]

23. Valiev, R.; Gunderov, D.; Prokofiev, E.; Pushin, V.; Zhu, Y. Nanostructuring of TiNi alloy by SPD processing for advanced properties. *Mater. Trans.* **2008**, *49*, 97–101. [CrossRef]

24. Waitz, T.; Kazykhanov, V.; Karnthaler, H.P. Martensitic phase transformations in nanocrystalline NiTi studied by TEM. *Acta Mater.* **2004**, *52*, 137–147. [CrossRef]
25. Pushin, V.G.; Lotkov, A.I.; Kolobov, Y.R.; Valiev, R.Z.; Dudarev, E.F.; Kuranova, N.N.; Dyupin, A.P.; Gunderov, D.V.; Bakach, G.P. On the nature of anomalously high plasticity of high-strength titanium nickelide alloys with shape-memory effects: I. Initial structure and mechanical properties. *Phys. Metals Metallogr.* **2008**, *106*, 520–530. [CrossRef]
26. Dudarev, E.F.; Valiev, R.Z.; Kolobov, Y.R.; Lotkov, A.I.; Pushin, V.G.; Bakach, G.P.; Gunderov, D.V.; Dyupin, A.P.; Kuranova, N.N. On the nature of anomalously high plasticity in high-strength titanium nickelide alloys with shape-memory effects: II. Mechanisms of plastic deformation upon isothermal loading. *Phys. Metals Metallogr.* **2009**, *107*, 298–311. [CrossRef]
27. Amirkhanova, N.A.; Valiev, R.Z.; Adasheva, S.L.; Prokofiev, E.A. The investigation of corrosion and electrochemical properties of Ni-Ti alloys in large and ultra-fine grained states. *Vestnik UGATU* **2006**, *7*, 143–146.
28. Kuranova, N.N.; Gunderov, D.V.; Ukusnikov, A.N.; Luk'yanov, A.V.; Yurchenko, L.I.; Prokof'ev, E.A.; Pushin, V.G.; Valiev, R.Z. Effect of heat treatment on the structural and phase transformations and mechanical properties of TiNi alloys subjected to severe plastic deformation. *Phys. Metals Metallogr.* **2009**, *108*, 556–568. [CrossRef]
29. Tong, Y.X.; Chen, F.; Guo, B.; Tian, B.; Li, L.; Zheng, Y.F.; Gunderov, D.V.; Valiev, R.Z. Superelasticity and its stability of an ultrafine-grained $Ti_{49.2}Ni_{50.8}$ shape memory alloy processed by equal channel angular pressing. *Mater. Sci. Eng. A* **2013**, *587*, 61–64. [CrossRef]
30. Tong, Y.X.; Hu, K.P.; Chen, F.; Tian, B.; Li, L.; Zheng, Y.F. Multiple-stage transformation behavior of $Ti_{49.2}Ni_{50.8}$ alloy with different initial microstructure processed by equal channel angular pressing. *Intermetallics* **2017**, *85*, 163–169. [CrossRef]
31. Zhang, X.; Song, J.; Huang, C.; Xia, B.; Chen, B.; Sun, X.; Xie, C. Microstructures evolution and phase transformation behaviors of Ni-rich TiNi shape memory alloys after equal channel angular extrusion. *J. Alloys Compd.* **2011**, *509*, 3006–3012. [CrossRef]
32. Pushin, V.G. *Nickel-Titanium Shape Memory Alloys. Part 1. Structure, Phase Transformations and Properties*; UrB RAS: Yekaterinburg, Russia, 2006; p. 438.
33. Shi, X.B.; Hu, Z.C.; Hu, X.W.; Zhang, J.S.; Cui, L.S. Effect of plastic deformation on stress-induced martensitic transformation of nanocrystalline NiTi alloy. *Mater. Charact.* **2017**, *128*, 184–188. [CrossRef]
34. Shi, X.; Yu, M.; Guo, F.; Liu, Z.; Jiang, D.; Han, X.; Cui, L. Effect of deformation on the stability of stress-induced martensite in nanocrystalline NiTi shape memory alloy. *Mater. Lett.* **2014**, *131*, 233–235. [CrossRef]
35. Rybin, V.V. *Large Plastic Deformation and Fracture of Metals*; Metallurgia: Moscow, Russian, 1986; p. 224.
36. Petch, N.J. The Cleavage Strength of Polycrystals. *Iron Steel Inst.* **1953**, *174*, 25–28.
37. Nokhrin, A.V.; Chuvildeev, V.N.; Kopylov, V.I.; Lopatin, Y.G.; Pirozhnikova, O.E.; Sakharov, N.V.; Piskunov, A.V.; Kozlova, N.A. Hall–Petch relation for nano- and microcrystalline metals produced by severe plastic deformation. *Vestn. Lobachevsky State Univ. Nizhni Novgorod* **2010**, *5*, 142–146.
38. Estrin, Y.; Vinogradov, A. Extreme grain refinement by severe plastic deformation: A wealth of challenging science. *Acta Mater.* **2013**, *61*, 782–817. [CrossRef]
39. Malygin, G.A. Plasticity and strength of micro- and nanocrystalline materials. *Phys. Solid State* **2007**, *49*, 1013–1033. [CrossRef]

Article

Structural Defects in TiNi-Based Alloys after Warm ECAP

Aleksandr Lotkov [1,*], **Anatoly Baturin** [1], **Vladimir Kopylov** [2], **Victor Grishkov** [1] and **Roman Laptev** [3]

[1] Institute of Strength Physics and Materials Science of the Siberian Branch of the Russian Academy of Sciences, 634055 Tomsk, Russia; abat@ispms.tsc.ru (A.B.); grish@ispms.tsc.ru (V.G.)
[2] Physical-Technical Institute of the National Academy of Sciences of Belarus State Scientific Institution, 220141 Minsk, Belarus; kopylov.ecap@gmail.com
[3] National Research Tomsk Polytechnic University, 634050 Tomsk, Russia; laptev.roman@gmail.com
* Correspondence: lotkov@ispms.ru; Tel.: +7-913-850-0310

Received: 28 July 2020; Accepted: 22 August 2020; Published: 25 August 2020

Abstract: The microstructure, martensitic transformations and crystal structure defects in the $Ti_{50}Ni_{47.3}Fe_{2.7}$ (at%) alloy after equal-channel angular pressing (ECAP, angle 90°, route B_C, 1–3 passes at T = 723 K) have been investigated. A homogeneous submicrocrystalline (SMC) structure (grains/subgrains about 300 nm) is observed after 3 ECAP passes. Crystal structure defects in the $Ti_{49.4}Ni_{50.6}$ (at%) alloy (8 ECAP passes, angle 120°, B_C route, T = 723 K, grains/subgrains about 300 nm) and $Ti_{50}Ni_{47.3}Fe_{2.7}$ (at%) alloy with SMC B2 structures after ECAP were studied by positron lifetime spectroscopy at the room temperature. The single component with the positron lifetime $\tau_1 = 132$ ps and $\tau_1 = 140$ ps were observed for positron lifetime spectra (PLS) obtained from ternary and binary, correspondingly, annealed alloys with coarse-grained structures. This τ_1 values correspond to the lifetime of delocalized positrons in defect-free B2 phase. The two component PLS were found for all samples exposed by ECAP. The component with $\tau_2 = 160$ ps (annihilation of positrons trapped by dislocations) is observed for all samples after 1–8 ECAP passes. The component with $\tau_3 = 305$ ps (annihilation of positrons trapped by vacancy nanoclusters) was detected only after the first ECAP pass. The component with $\tau_3 = 200$ ps (annihilation of positrons trapped by vacancies in the Ti sublattice of B2 structure) is observed for all samples after 3–8 ECAP passes.

Keywords: TiNi-based alloys; ECAP; microstructure; positron lifetime spectroscopy; nanoclusters; vacancies; dislocations

---

1. Introduction

Severe plastic deformation (SPD) allows one to produce an ultrafine-grained (UFG) structure with a high strength and sufficient ductility in metal materials [1,2]. Among such materials are TiNi-based alloys distinguished for their superelasticity and shape memory effect (SME) and widely used in engineering and medicine [3]. Among the SPD methods is equal channel angular pressing (ECAP), which provides UFG TiNi-based alloys with greatly improved functional properties. It opens the most promising way of producing bulk billets from this type of materials [1,2].

By now, research data are available to judge the microstructure evolution of binary and ternary TiNi-based alloys during ECAP at 623–773 K, and the grain size effect on their mechanical and inelastic properties [4–17], and also the influence of ECAP on the temperatures of martensite transformations (MT) in TiNi-based alloys [9,10,12,17,18]. As has been shown [1,4,12], the grain structure of TiNi-based alloys can be refined to 100–500 nm in six to eight ECAP passes depending on the pressing temperature. Such alloys with an average grain-subgrain size of 300 nm show an unusual behavior with an ultimate strength of up 1200 MPa and plasticity of up to 50–60%. After ECAP at 623–773 K, the temperature of martensite transformations in TiNi-based alloys normally decreases by 20 K and more [17,18]. However, very scant data are available on the evolution of lattice defects in TiNi-based alloys exposed to ECAP. These data are necessary for the understanding both the mechanisms of grain refinement and physical factors affected on MT temperatures in alloys after ECAP. But the sole paper report that the dislocation density in such alloys after SPD increases greatly (to $\sim 10^{15}$–10^{16} m^{-2}) and that its critical value ($\sim 10^{18}$ m^{-2}) exists at which they assume an amorphous structure [19].

However, in addition to dislocations, SPD produces numerous excess vacancies as against thermodynamic equilibrium [20,21]. On the one hand, vacancies greatly accelerate the mass transfer, dissolution or precipitation of secondary phases [22,23], and generation of new grain boundaries in a material [24]. On the other hand, vacancies with a high concentration and low mobility can form clusters [25], and eventually decrease the long-term durability of materials, as it happens, e.g., in Al- and Ti-based alloys after ECAP [26,27]. In this context, studying the formation of vacancy-like free volumes in SPD materials can provide a better understanding of the mechanisms responsible for their properties in UFG states and of the conditions for their more efficient use. Most of the studies of SPD-induced vacancies have been performed on pure metals and their alloys [20,28–32], and almost all of them consider SPD processes at room temperature when the vacancy mobility is low. Among the methods of research in deformation-induced vacancies, their clusters, and dislocation density in metals and alloys are resistometry [21], dilatometry [29], differential scanning calorimetry [21,29], X-ray line profile analysis [33], perturbed γ-γ angular correlation (PAC) [34], and positron annihilation spectroscopy (PAS) [28–32], all of which give comparable vacancy concentrations for pure metals (10^{-2}–10^{-4}). However, little is known about SPD-induced vacancies in intermetallic compounds where their types are more diverse, compared to metals [35]. For example, vacancy-like free volumes classifiable as interface defects are detected by PAS in ball-milled nanocrystalline Fe3Si [36]. In B2 FeAl under high pressure torsion, the vacancy density can reach 10^{-2} as it follows from comparative data of differential scanning calorimetry and calculations [37]. From PAC data [38], the types of defects that dominate after ball milling are Schottky pairs in PdIn, triple defects in NiAl and FeRh, and antisite defects in FeAl. One of the studies shows that after ultrasonic shock treatment at room temperature, the relative concentration of single vacancies in equiatomic TiNi surface layers increases to $\sim 10^{-5}$ [39,40].

This paper presents the results of complex research into the ECAP effect at 723 K on the microstructure, martensite transformations, and lattice defect evolution in Ti$_{50}$Ni$_{47.3}$Fe$_{2.7}$ (at%) alloy combined with experimental results about lattice defects in Ti$_{49.4}$Ni$_{50.6}$ alloy (at%) alloy after eight ECAP passes at the same temperature.

The choice of experimental alloys is due to the following considerations based on the results of studies of the microstructure and martensitic transformations. These alloys have the B2 phase structure at room temperature both in the initial coarse-grained state and in the submicrocrystalline state. In addition, the grain-subgrain structures of these alloys are similar after ECAP. This makes it possible to study only those defects of the crystal structure that are formed during ECAP at 723 K and to avoid the influence of defects that may appear as a result of MT.

2. Materials and Methods

The test Ti$_{50}$Ni$_{47.3}$Fe$_{2.7}$ (at%) alloy were supplied as rotary forged (1220 K) round bars of diameter 25 mm and length 140 mm (MATEK-SMA Ltd., Moscow, Russia). The round bars were forged to a square of 16 × 16 mm^2 at 1073 ± 100 K with further annealing at 773 K for 3 h. Then, the square bars were cut to 14 × 14 mm^2 and were exposed to ECAP (B$_C$ route) with a channel angle 90°. The number of cycles was N$_i$, where i = 1, 2, 3. The test specimens of Ti$_{50}$Ni$_{47.3}$Fe$_{2.7}$ (at%) were pressed at the Physical-Technical Institute NASB (Minsk, Belarus). The initial coarse-grained alloy had a B2 grain size of 20–40 μm. At temperatures above 275 K, its structure was represented by a single B2 phase (the high temperature cubic phase). The sequence and the temperatures of martensite transformations were studied by temperature resistometry (four-point scheme) in cooling–heating cycles.

The test alloy $Ti_{49.4}Ni_{50.6}$ (at%) was supplied by Intrinsic Devices Inc. (San Francisco, CA, USA). In the alloy quenched from 1073 K, the start temperature of direct B2 → B19′ martensite transformation was M_S = 288 K (B19′-the monoclinic martensite phase). At room temperature, its initial coarse-grained state was represented by a B2 structure with an average grain size of 50 μm. Its submicrocrystalline state with an average grain-subgrain size of 300 nm was formed by ECAP at T = 723 K at the Ufa State Aviation Technical University (Ufa, Russia). The channel angle was 110° (B_C route), and the number of passes was $n = 8$ [6].

The samples of dimensions $10 \times 10 \times 1$ mm^3 for the positron annihilation spectroscopy and bars of dimensions $1 \times 1 \times 20$ mm^3 for the electrical resistivity measurements were prepared from bullets of initial alloy and alloy after ECAP. The foils for the transmission electron microscopy (TEM) were prepared by the mechanical polishing with subsequent electrochemical thinning of plates with initial thickness about 0.3 mm.

The microstructure of the test specimens was examined by transmission electron microscopy (JEM-2100, JEOL Ltd., Tokyo, Japan) in Nanotech Shared Use Center, ISPMS SB RAS).

The crystal defects appearing after ECAP were analyzed by positron annihilation spectroscopy (PAS) based on analysis of the positron lifetime spectra. A time resolution of spectrometer is 240 ps [41]. The positron source was ^{44}Ti with an activity of 1 MBq placed between two identical specimens. The positron beam diameter was about 6 mm. The total number of annihilation events recorded per each spectrum was no less than 5×10^6. Each specimen was analyzed from three independent the positron lifetime spectra. Processing of the spectra was carried out using LT software [42]. After subtraction of the component related to annihilation in the positron source and background, the spectra could be reliably decomposed into one lifetime component (initial state) or into two (deformed state),

$$S(t) = (I_1/\tau_1)\exp(-t/\tau_1) + (I_2/\tau_2)\exp(-t/\tau_2), \tag{1}$$

where τ_1, I_1 and τ_2, I_2 are the respective component lifetimes and intensities to judge the type of a defect and its concentration.

The Vickers microhardness was measured on a Duramin-5 microhardness tester (Struers, Ballerup, Denmark) at room temperature under a load of 100 g.

3. Results and Discussion

3.1. Effect of Warm ECAP on Martensite Transformation Temperatures in $Ti_{50}Ni_{47.3}Fe_{2.7}$

Figure 1 shows typical temperature dependences of resistivity on heating and cooling for $Ti_{50}Ni_{47.3}Fe_{2.7}$ (at%) alloy before and after ECAP passes. These $\rho(T)$ dependences are qualitatively similar for both initial alloy state and after ECAP.

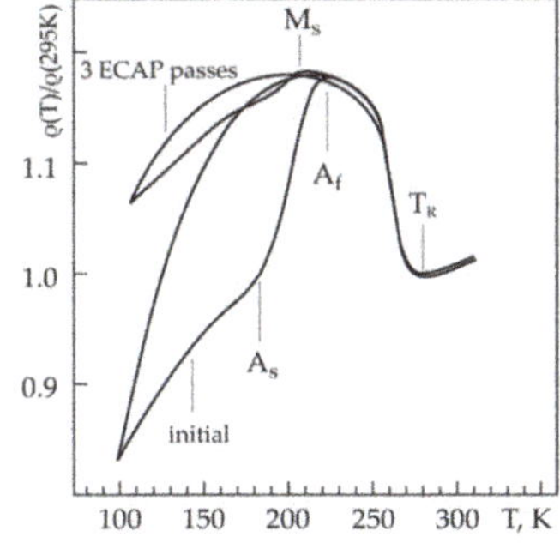

Figure 1. Typical temperature dependences of resistivity for $Ti_{50}Ni_{47.3}Fe_{2.7}$ before (initial), and after, three ECAP passes.

On cooling and heating in the temperature range from 275 K to 77 K (liquid nitrogen), the alloy is involved in the sequence of martensite transformations B2 $\leftrightarrow$ R $\leftrightarrow$ R + B19′ (R and B19′ are rhombohedral and monoclinic martensite phases, respectively). During B2 $\leftrightarrow$ R transformations on cooling and heating, the resistivity of samples varies almost without hysteresis (its value is less than 3 K). The transformation R $\to$ B19′ on cooling to 77 K is incomplete throughout the volumes of all samples. The MT temperatures are denoted as T_R for B2 $\leftrightarrow$ R MT, M_S for the start of R $\to$ B19′ MT (cooling) and A_S and A_f for B19′ $\to$ R MT (heating) The variations of MT temperatures versus ECAP passes are presented in Table 1.

Table 1. Martensite transformation temperatures in $Ti_{50}Ni_{47.3}Fe_{2.7}$.

State	T_R, K	M_s, K	A_s, K	A_f, K
Initial	275	213	184	220
1st ECAP pass	276	215	185	208
3rd ECAP pass	275	195	195	205

The main ECAP effect on MT consists in the following: it decreases M_S by 18 K, compared to the initial state, and narrows the temperature interval of reverse B19′ $\to$ R transformation by 10 K. Another effect, which is more pronounced, is a steep decrease in the value by which the resistivity changes during on cooling to 77 K. Therefore, the B19′ volume fraction appearing at T < Ms decreases greatly (2–3 times) with increasing the number of ECAP passes, i.e., with increasing the true strain. This is likely due to considerable grain-subgrain refinement in the specimens, which causes their hardening and markedly decreases the martensite transformation temperatures of the R $\to$ B19′ MT.

Thus, the B2 structure is observed at room temperature in $Ti_{50}Ni_{47.3}Fe_{2.7}$ (at%) alloy after 1–3 ECAP passes. These data correspond to the results obtained by TEM.

3.2. Microstructure of $Ti_{50}Ni_{47.3}Fe_{2.7}$ after Warm ECAP

Figure 2 shows typical micrographs of $Ti_{50}Ni_{47.3}Fe_{2.7}$ (at%) after one ECAP pass. As can be seen, its grain-subgrain microstructure is rather inhomogeneous. In the bulk of the deformed alloy, a clearly defined multiband structure is observed as evidence of plastic strain localization in the specimens during ECAP. The orientation angles of bands relative to each other are 45° (135°) and 90°. Inside the bands, non-equiaxial fragments of a finer grain-subgrain structure are localized. Their aspect ratio (ratio of minimum to maximum sizes) ranges from 1:5 to 1:10.

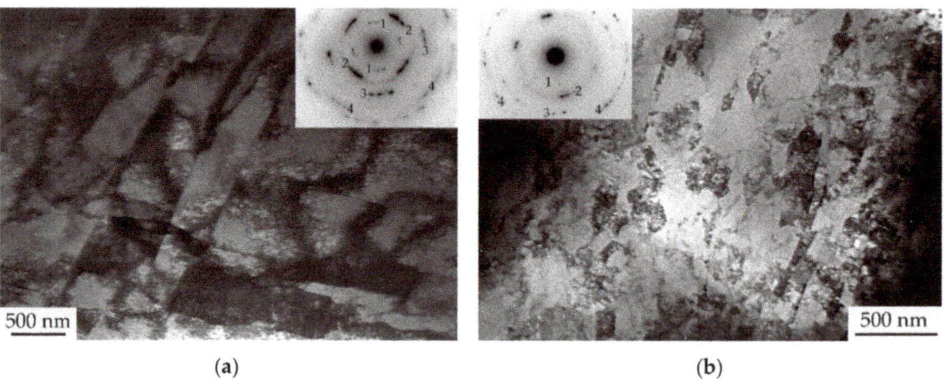

Figure 2. Typical TEM bright-field images of microvolumes with most large-scale subgrains (**a**) and fine-grained structure (**b**) in $Ti_{50}Ni_{47.3}Fe_{2.7}$ after one ECAP pass at 723 K. Details are presented in text. Quasi-rings in microdiffraction patterns include the next types of B2-reflections: 1–(100), 2–(110), 3–(200), 4–(220), 5–(310) in (**a**); 1–(100), 2–(110), 3–(200), 4–(211) in (**b**).

The misorientation of adjacent band fragments reaches 10–12°, and that of adjacent fragments inside each band measures 2–5°. The band width in different microvolumes varies widely, from a microscale size of 400–500 nm (Figure 2a) to a mesoscale one of up to ~5 µm (Figure 2b). In mesoscale bands, one can clearly see a secondary microband structure with a minimum size of grain-subgrain fragments of 100–300 nm (Figure 2b).

The fragments of the grain-subgrain structure are mostly in the state of a B2 phase, as evidenced by electron diffraction patterns in Figure 2. It is observed that, along with quasi-ring reflections, the microdiffraction patterns reveal bright peaks from fragments misoriented to less than 15° with substantial radial broadening for microband volumes (Figure 2a), and with weak radial broadening and rather uniform distribution along the Debye rings for mesoband ones (Figure 2b). From comparison of the electron diffraction patterns in Figure 2a,b, it follows that the microstructure of the alloy after one ECAP pass is spatially inhomogeneous, not only in fragment sizes, but also in internal stresses. Moreover, the stress level in microband volumes (Figure 2a) is much higher than its level in mesoband ones (Figure 2b). Another feature of the electron diffraction patterns is a clear doublet structure of B2-phase reflections typical of deformation twinning. Besides the above microstructure types, individual grains of up to 1 µm occur, but rarely, in the alloy after one ECAP pass.

Therefore, the specimens of $Ti_{50}Ni_{47.3}Fe_{2.7}$ (at%) pressed in one ECAP pass at a channel angle of 90° assumes a grain-subgrain structure with a nonuniform fragment size distribution (100 nm to 1–1.5 µm) in which the main fraction belongs to submicrocrystalline fragments (100–500 nm). Such a grain-subgrain structure is formed through fragmentation on different scales and through B2-phase twinning.

Figures 3 and 4 shows the microstructure of the alloy after three ECAP passes. As can be seen, it is qualitatively similar to, but much finer than, the microstructure after one ECAP pass. The alloy preserves the system of microbands but their width decreases to less than 1 µm. The grain-subgrain structure contains a substantial fraction of nanosized fragments (50–100 nm). The main phase in the alloy is B2, as evidenced by its electron diffraction patterns.

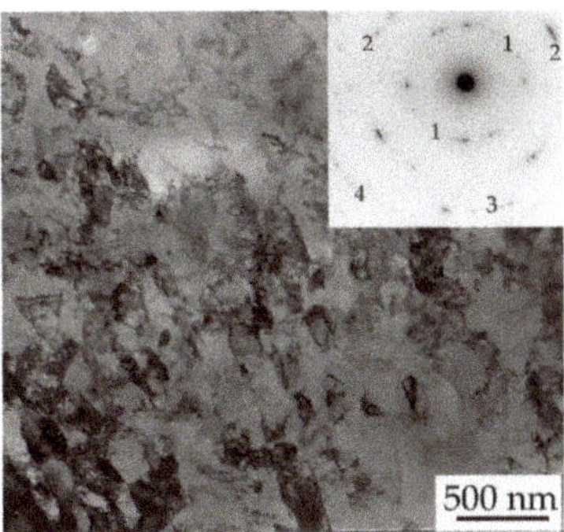

Figure 3. Typical microstructure in $Ti_{50}Ni_{47.3}Fe_{2.7}$ after three ECAP passes. The microdiffraction pattern is presented in the insert. Quasi-rings in microdiffraction pattern include the next types of B2 phase reflections: 1–(110), 2–(211), 3–(310), 4–(312).

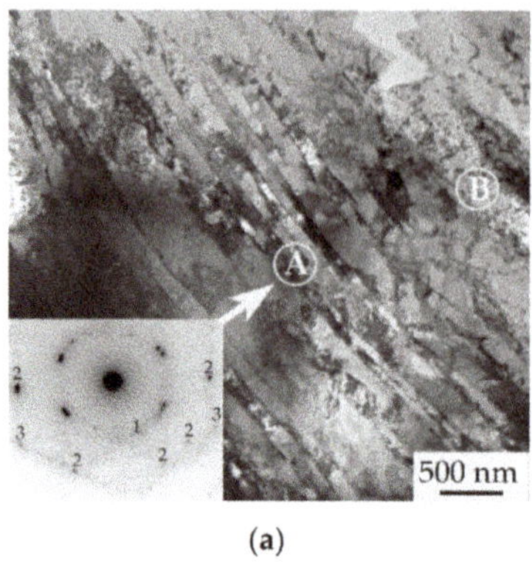
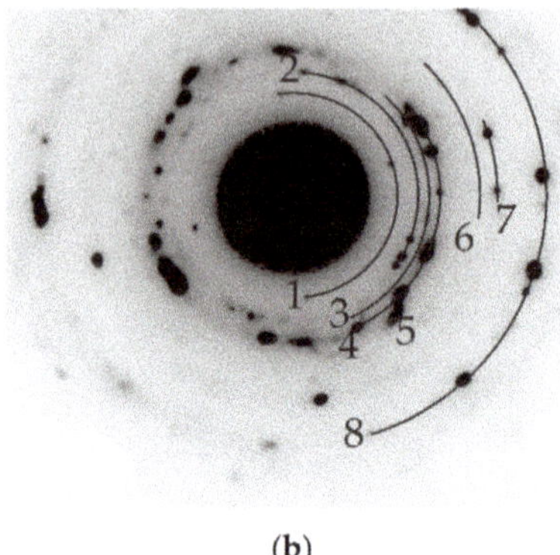

(a)   (b)

Figure 4. Microstructure in $Ti_{50}Ni_{47.3}Fe_{2.7}$ after three ECAP passes with the main B2 phase structure (region A) and small volume fraction of R phase (region B). The microdiffraction patterns are presented in the insert of (**a**,**b**), correspondingly. (**a**) The quasi-rings include the next types of B2 phase reflections: 1–(110), 2–(211), 3–(220); (**b**) The quasi-rings include the next types of B2 phase reflections: 1–(002), 2–(112), 3–(202), 4–(220), 5–(103), 6–(320), 7–(004), 8–(413).

However, several microvolumes with an R phase structure were found in the samples after three ECAP passes (e.g., region B in Figure 4a). It is most likely that the local appearance of the R phase at room temperature is due to the high level of residual stresses in these regions. According to electron microscopy data, the total proportion of regions with the R phase structure is less than 1 vol%. Therefore, the presence of these regions will not be taken into account in the subsequent discussion of the results.

Therefore, a mixed grain-subgrain structure based on a submicrocrystalline (100–500 nm) and a nanostructural fraction (50–100 nm) is formed in $Ti_{50}Ni_{47.3}Fe_{2.7}$ (at%) after ECAP at 723 K (channel angle 90°, strain rate 1 s^{-1}).

According to transmission electron microscopy [6], the microstructure of $Ti_{49.4}Ni_{50.6}$ (at%) after eight ECAP passes is also in the state of a B2 phase. The average grain-subgrain size in the specimen cross section is 300 nm.

3.3. Evolution of Structural Defects in TiNi-based Alloys Exposed to Warm ECAP

The evolution of structural defects was analyzed by positron annihilation spectroscopy. Figure 5 shows the average grain-subgrain size, average positron lifetime, and Vickers microhardness versus the numbers of ECAP passes for the test TiNi-based alloys.

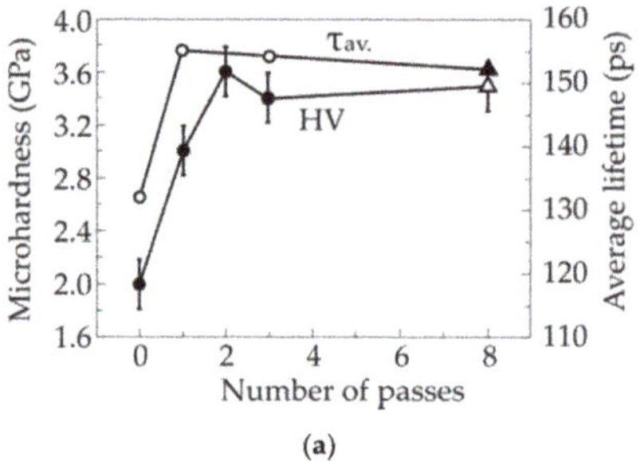
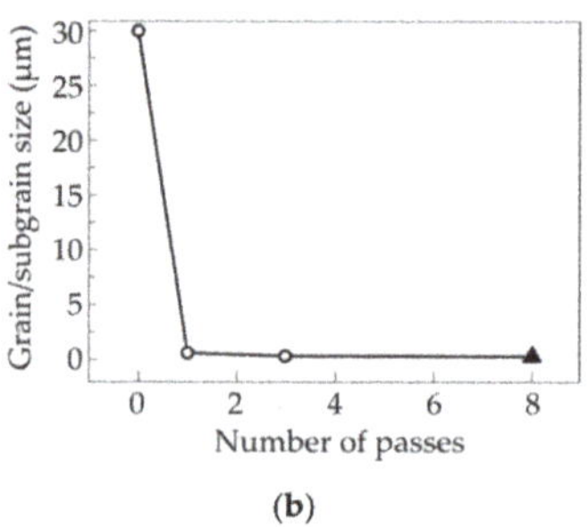

(a)   (b)

Figure 5. Average positron lifetimes, $\tau_{av.}$ (error bars ± 1 ps) and microhardness, HV, (**a**) and average grain/subgrain sizes (**b**) versus number of ECAP passes: $Ti_{49.4}Ni_{50.6}$ (at%) alloy (▲, △) and $Ti_{50}Ni_{47.3}Fe_{2.7}$ (at%) alloy (○, ●).

In the specimens exposed to ECAP, the average positron lifetime τ_{av} increases greatly even after the first ECAP pass and decreases slightly with increasing the number of passes irrespective of the alloy composition (Figure 5a). The Vickers microhardness, HV, behaves in the same way (Figure 5a). The increase in the microhardness can be explained by grain refinement in ECAP (Figure 5b), while the increase in τ_{av} definitely suggests that increasing the strain increases the lattice defect density. The type of defects resulting from ECAP can be identified by decomposing the positron lifetime spectra into components (Figure 6). The initial annealed specimens have a single-component spectrum with a positron lifetime $\tau_1 = 132 \pm 1$ ps for $Ti_{50}Ni_{47.3}Fe_{2.7}$ (at%) and $\tau_1 = 140 \pm 1$ ps for $Ti_{49.4}Ni_{50.6}$ (at%) (Figure 6a). The τ_1 value corresponds to the experimental free positron lifetime 132 ps in TiNi [43] and the theoretical lifetimes of delocalized positrons in defect-free TiNi B2 structure: 120 ps [44] and 129 ps [45]. Such a decomposition shows that, even after the first ECAP pass, the spectra show no evidence of delocalized positrons: All positrons annihilate only from localized states associated with defects (so-called positron saturated trapping).

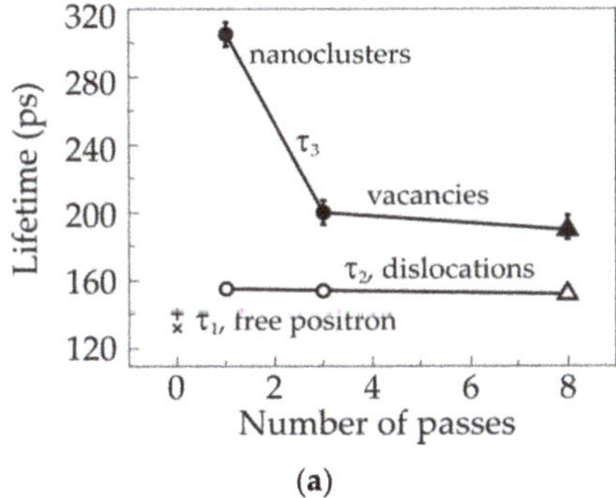

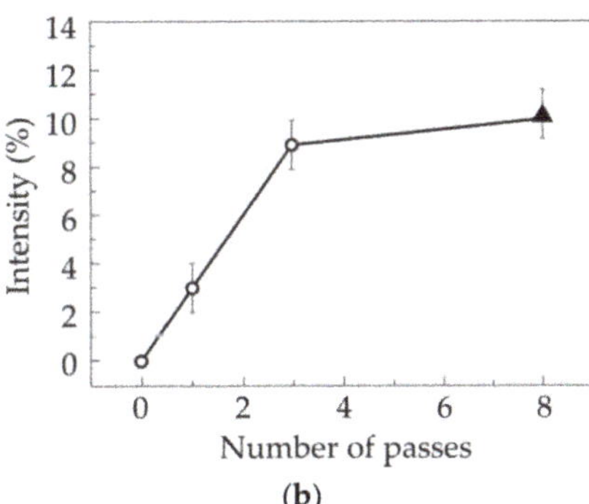

Figure 6. The positron lifetimes, corresponding to the different components of spectra; (**a**) and the intensities of I_3 component; (**b**) versus number of ECAP passes: $Ti_{49.4}Ni_{50.6}$ (at%) alloy (▲, △, +) and $Ti_{50}Ni_{47.3}Fe_{2.7}$ (at %) alloy (○, ●, ×).

Figure 6 demonstrates the evolution of positron lifetime components for $Ti_{50}Ni_{47.3}Fe_{2.7}$ (at%) and $Ti_{49.4}Ni_{50.6}$ (at%) alloys.

Noteworthy is that the positron lifetime component τ_2 is invariant with the number of ECAP passes within experimental error (Figure 6a) and hence, the defect type identified with τ_2 remains intact during ECAP strain accumulation. The same is observed in many experiments [46], and the τ_2 values correspond to the lifetime of positrons trapping by dislocations.

The lifetime of τ_3 component (Figure 6a) is far longer than the lifetime of delocalized positrons and the lifetime of positrons captured by dislocations. According to experimental data [47], the average positron lifetime in TiNi monovacancies is 195 ps. However, as has been shown by experiments [48,49] and theoretical studies [50], two types of vacancies differing in formation energy arise in B2 TiNi: vacancies V_{Ti} and V_{Ni} on its Ti and Ni sublattices, respectively. From first-principles calculations [50], it follows that the positron lifetimes in V_{Ti} and V_{Ni} vacancies differ greatly. The theoretical lifetime of positrons trapped by V_{Ti} and V_{Ni}, vacancies equal 205 ps and 134 ps, correspondingly. This situation is distinct from what is observed in pure metals with only one vacancy type and in B2-structured transition metal aluminides with very close positron lifetimes in two types of sublattice vacancies [47]. The foregoing suggests that the component τ_3 after three and eight ECAP passes corresponds to single V_{Ti} vacancies. After the first ECAP pass, the positron lifetime $\tau_3 = 305$ ps (Figure 6a) being much longer than its value in monovacancies, can be related to nanoclusters of about five to seven vacancies, if judged from its theoretical dependence on the number of cluster vacancies in bcc metals [25]. Therefore, we have a nontrivial result since research data demonstrate that increasing the number of

ECAP passes increases both, the concentration and size of vacancy clusters in pure metals (copper, nickel, titanium) [30,51,52]. However, this conclusion follows from ECAP at room temperature when the diffusion mobility of vacancies is low and they likely fail to escape from the volume of grains to grain boundaries. The increase in vacancy concentration provides the conditions for vacancy clustering. Such an interpretation makes our results clear. After the first pass of warm ECAP, the grain-subgrain sizes remain rather large for vacancy clustering in the grain volume, but after the next passes, its level allows part of the vacancies to reach the boundaries of grains or dislocations, while the rest part falls short of the amount needed for the formation of noticeable concentrations of their complexes.

Figure 6b shows that the vacancy defect concentration associated with the intensity of the component τ_3 increases steeply in three ECAP passes and then varies slightly. The intensities of the positron lifetime components I_2 and I_3 are proportional to the dislocation density and vacancy defect concentration ($I_2 + I_3 = 1$). As the saturated positron trapping, we cannot accurately estimate the dislocation density and vacancy defect concentration from the available positron trapping model [53]. However, we can estimate the vacancy concentration from the ratio $I_2/I_3 = v_d\rho_d/v_vc_v$, considering that the specific positron trapping rates by dislocations, and vacancies are equal, respectively to $v_d \sim 10^{-4}$ m^{2}s^{-1} and $v_v \sim 10^{14}$ s^{-1} in metallic materials [54]. The dislocation density for Ti$_{50}$Ni$_{50}$ (at%) after eight warm ECAP passes is $\rho_d = 1.1 \times 10^{15}$ m^{-2} according to X-ray diffraction estimates [55]. The ratio I_2/I_3 equal to 10.1 and 9.1 after three, and eight ECAP passes, respectively, suggests that dislocations are the main type of defects for positron capture in TiNi-based alloys exposed to ECAP. At $I_2/I_3 \approx 10$, the estimated relative concentration of V_{Ti} vacancies after ECAP is $\sim 10^{-4}$, being many orders of magnitude higher than their thermodynamic equilibrium concentration. Surely, the estimate is rough as the exact values of v_d and v_v are unknown for TiN-based alloys.

It stands to reason that V_{Ni} vacancies are also produced in SPD processes, the more so their formation energy is 0.78 eV [48,49] against 0.97 eV for V_{Ti} [48,49]. However, their concentration is unassessable by positron annihilation spectroscopy because of almost the same theoretical lifetimes of delocalized positrons and positrons in V_{Ni} (134 ps). We expect that defect annealing experiments will help separate the contributions of dislocations and V_{Ni} vacancies as they differ in annealing temperature.

4. Conclusions

Even one warm ECAP pass can provide submicrocrystalline fragmentation in initially coarse-grained TiNi-based alloys, and after three passes, their structure represents a mix of grains-subgrains sized to 100–500 nm and 50–100 nm. The dislocation defects and twins contribute to such refinement. The formation of UFG structures in Ti$_{50}$Ni$_{47.3}$Fe$_{2.7}$ (at%) does not change the martensite transformation sequence B2 $\leftrightarrow$ R $\leftrightarrow$ B19' but it decreases the R $\leftrightarrow$ B19' start temperature by 18 K and narrows the B19' $\rightarrow$ R temperature interval by 10 K. The B2 $\rightarrow$ R start temperature in this alloy after ECAP remains unchanged.

The refinement of the alloys to UFG structures under ECAP involves a substantial increase in the density of dislocations and vacancy-like defects. Their nanoclusters are detected only after the first ECAP pass, and only free volumes close in size to monovacancies are found after the next passes. The relative vacancy concentration estimated for the TiNi-based alloys after ECAP is about 10^{-4} being many orders of magnitude higher than their thermodynamic equilibrium concentration.

Author Contributions: Conceptualization, A.L. and A.B.; writing—original draft preparation, A.B.; writing—review and editing, A.L. and V.G.; software, R.L.; investigations, A.B., V.K., V.G. and R.L.; project administration, A.L.; funding acquisition, A.L. and R.L. All authors have read and agreed to the published version of the manuscript.

Funding: This research was funded by Fundamental Research Program of the State Academies of Science for 2013-2020 (project № III.23.2.2) and the grant of the Competitiveness Enhancement Program of National Research Tomsk Polytechnic University (VIU-OEF-177/2020).

Conflicts of Interest: The authors declare no conflict of interest.

References

1. Valiev, R.Z.; Zhilyaev, A.P.; Langdon, T.G. *Bulk Nanostructured Materials: Fundamentals and Applications*; TMS-Wiley: Hoboken, NJ, USA, 2014.
2. Segal, V.M.; Beyerlein, I.J.; Tome, C.N.; Chuvil'deev, V.N.; Kopylov, V.I. *Fundamentals and Engineering of Severe Plastic Deformation*; Nova: Amityville, NY, USA, 2010.
3. Mohd Jani, J.; Leary, M.; Subic, A.; Gibson, M.A. A review of shape memory alloy research, applications and opportunities. *Mater. Des.* **2014**, *56*, 1078–1113. [CrossRef]
4. Prokoshkin, S.D.; Belousov, M.N.; Abramov, V.Y.; Korotitskii, A.V.; Makushev, S.Y.; Khmelevskaya, I.Y.; Dobatkin, S.V.; Stolyarov, V.V.; Prokof'ev, E.A.; Zharikov, A.I.; et al. Creation of submicrocrystalline structure and improvement of functional properties of shape memory alloys of the Ti–Ni–Fe system with the help of ECAP. *Met. Sci. Heat Treat.* **2007**, *49*, 51–56. [CrossRef]
5. Shahmir, H.; Nili-Ahmadabadi, M.; Langdon, T.G. Shape memory effect of NiTi alloy processed by equal-channel angular pressing followed by post deformation annealing. *IOP Conf. Series: Mater. Sci. Eng.* **2014**, *63*, 012111. [CrossRef]
6. Lukyanov, A.; Gunderov, D.; Prokofiev, E.; Churakova, A.; Pushin, V. Peculiarities of the mechanical behavior of ultrafinegrained and nanocrystalline $Ti_{49.4}Ni_{50.6}$ alloy produced by severe plastic deformation. In Proceedings of the 21st International Conference on Metallurgy and Materials, METAL 2012, Brno, Czech Republic, 23–25 May 2012; pp. 1335–1341.
7. Lotkov, A.I.; Baturin, A.A.; Grishkov, V.N.; Kopylov, V.I.; Timkin, V.N. Influence of equal-channel angular pressing on grain refinement and inelastic properties of TiNi-based alloys. *Izvestiya Visshikh Uchebnikh Zavedenii, Chernaya Metallurgiya* **2014**, *57*, 50–55. [CrossRef]
8. Pushin, V.G.; Valiev, R.Z.; Zhu, U.T.; Gunderov, D.V.; Kourov, N.I.; Kuntsevich, T.E.; Uksusnikov, A.N.; Yurchenko, L.I. Effect of Severe Plastic Deformation on the Behavior of Ti-Ni Shape Memory Alloys. *Mater. Trans.* **2006**, *47*, 694–697. [CrossRef]
9. Fan, Z.; Song, J.; Zhang, X.; Xie, C. Phase Transformations and Super-Elasticity of a Ni-rich TiNi Alloy with Ultrafine-Grained Structure. *Mater. Sci. Forum* **2010**, *667–669*, 1137–1142. [CrossRef]
10. Zhang, D.; Guo, B.; Tong, Y.; Tian, B.; Li, L.; Zheng, Y.; Gunderov, D.V.; Valiev, R.Z. Effect of annealing temperature on martensitic transformation of $Ti_{49.2}Ni_{50.8}$ alloy processed by equal channel angular pressing. *Trans. Nonferrous Met. Soc. Chin.* **2016**, *26*, 448–455. [CrossRef]
11. Karaman, I.; Kulkarni, A.V.; Luo, Z.P. Transformation behaviour and unusual twinning in a NiTi shape memory alloy ausformed using equal channel angular extrusion. *Phil. Mag.* **2005**, *85*, 1729–1745. [CrossRef]
12. Khmelevskaya, I.Y.; Prokoshkin, S.D.; Trubitsyna, I.B.; Belousov, M.N.; Dobatkin, S.V.; Tatyanin, E.V.; Korotitskiy, A.V. Structure and properties of Ti-Ni-based alloys after equal-channel angular pressing and high pressure torsion. *Mater. Sci. Eng. A* **2008**, *481–482*, 119–122. [CrossRef]
13. Prokofyev, E.; Gunderov, D.; Prokoshkin, S.; Valiev, R. Microstructure, mechanical and functional properties of NiTi alloys processed by ECAP technique. In Proceedings of the 8th European Symposium on Martensitic Transformations, ESOMAT 2009, Prague, Czech Republic, 7–11 September 2009. [CrossRef]
14. Zhang, Y.; Jiang, S. The Mechanism of Inhomogeneous Grain Refinement in a NiTiFe Shape Memory Alloy Subjected to Single-Pass Equal-Channel Angular Extrusion. *Metals* **2017**, *7*, 400. [CrossRef]
15. Churakova, A.; Yudahina, A.; Kayumova, E.; Tolstov, N. Mechanical behavior and fractographic analysis of a TiNi alloy with various thermomechanical treatment. In Proceedings of the International Conference on Modern Trends in Manufacturing Technologies and Eguipment: Mechanical Engineering and Materials Science, ICMTMTE 2019, Sevastopol, Russia, 9–13 September 2019. [CrossRef]
16. Lucas, F.L.C.; Guido, V.; Käfer, K.A.; Bernardi, H.H.; Otubo, J. ECAE Processed NiTi Shape Memory Alloy. *Mater. Res.* **2014**, *17* (Suppl. 1), 186–190. [CrossRef]
17. Valiev, R.Z.; Langdon, T.G. Principles of equal-channel angular pressing as a processing tool for grain refinement. *Progr. Mater. Sci.* **2006**, *51*, 881–981. [CrossRef]
18. Li, Z.; Cheng, X.; ShangGuan, Q. Effects of heat treatment and ECAE process on transformation behaviors of TiNi shape memory alloy. *Mater. Lett.* **2005**, *59*, 705–709. [CrossRef]
19. Koike, J.; Parkin, D.M.; Nastasi, M. Crystal-to-amorphous transformation of NiTi induced by cold rolling. *J. Mater. Res.* **1990**, *5*, 1414–1418. [CrossRef]

20. Čížek, J.; Janeček, M.; Vlasák, T.; Smola, B.; Melikhova, O.; Islamgaliev, R.K.; Dobatkin, S.V. The Development of Vacancies during Severe Plastic Deformation. *Mater. Trans.* **2019**, *60*, 1533–1542. [CrossRef]
21. Zehetbauer, M.J.; Steiner, G.; Schafler, G.; Korznikov, A.; Korznikova, E. Deformation Induced Vacancies with Severe Plastic Deformation: Measurements and Modeling. *Mater. Sci. Forum* **2006**, *503–504*, 57–65. [CrossRef]
22. Xue, K.-M.; Wang, B.-X.-T.; Yan, S.-L.; Bo, D.-Q.; Li, P. Strain-induced dissolution and precipitation of secondary phases and synergetic stengthening mechanisms of Al-Zn-Mg-Cu alloy during ECAP. *Adv. Eng. Mater.* **2019**, 1801182. [CrossRef]
23. Straumal, B.B.; Pontikis, V.; Kilmametov, A.R.; Mazilkin, A.A.; Dobatkin, S.V.; Baretzky, B. Competition between precipitation and dissolution in Cu–Ag alloys under high pressure torsion. *Acta Mater.* **2017**, *122*, 60–71. [CrossRef]
24. Farber, V.M. Contribution of diffusion processes to structure formation in intense cold plastic deformation of metals. *Metal Sci. Heat Treat.* **2002**, *44*, 317–323. [CrossRef]
25. Čížek, J.; Melikhova, O.; Barnovská, Z.; Procházka, I.; Islamgaliev, R.K. Vacancy clusters in ultra-fine grained metals prepared by severe plastic deformation. *J. Phys. Conf. Ser.* **2013**, *443*, 012008. [CrossRef]
26. Betekhtin, V.I.; Kadomtsev, A.G.; Kral, P.; Dvorak, J.; Svoboda, M.; Saxl, I.; Sklenička, V. Significance of Microdefects Induced by ECAP in Aluminium, Al 0.2%Sc Alloy and Copper. *Mater. Sci. Forum* **2008**, *567–568*, 93–96. [CrossRef]
27. Betekhtin, V.I.; Kadomtsev, A.G.; Narykova, M.V.; Amosova, O.V.; Sklenicka, V. Defect Structure and Mechanical Stability of Microcrystalline Titanium Produced by Equal Channel Angular Pressing. *Technol. Phys. Lett.* **2017**, *43*, 61–63. [CrossRef]
28. Kuznetsov, P.V.; Mironov, Y.P.; Tolmachev, A.I.; Bordulev, Y.S.; Laptev, R.S.; Lider, A.M.; Korznikov, A.V. Positron spectroscopy of defects in submicrocrystalline nickel after low-temperature annealing. *Phys. Sol. State* **2015**, *57*, 219–228. [CrossRef]
29. Reglitz, G.; Oberdorfer, B.; Fleischmann, N.; Kotzurek, J.A.; Divinski, S.V.; Sprengel, W.; Wilde, G.; Würschum, R. Combined volumetric, energetic and microstructural defect analysis of ECAP-processed nickel. *Acta Mater.* **2016**, *103*, 396–406. [CrossRef]
30. Lukáč, F.; Čížek, J.; Knapp, J.; Procházka, I.; Zháňal, P.; Islamgaliev, R.K. Ultra-fine grained Ti prepared by severe plastic deformation. *J. Phys. Conf. Ser.* **2016**, *674*, 012007. [CrossRef]
31. Bartha, K.; Zháňal, P.; Stráský, J.; Čížek, J.; Dopita, M.; Lukáč, F.; Harcuba, P.; Hájek, M.; Polyakova, V.; Semenova, I.; et al. Lattice defects in severely deformed biomedical Ti-6Al-7Nb alloy and thermal stability of its ultra-fine grained microstructure. *J. Alloys Compd.* **2019**, *788*, 881–890. [CrossRef]
32. Domınguez-Reyes, R.; Savoini, B.; Monge, M.Á.; Muñoz, Á.; Ballesteros, C. Thermal Stability Study of Vacancy Type Defects in Commercial Pure Titanium Using Positron Annihilation Spectroscopy. *Adv. Eng. Mater.* **2017**, *19*, 1500649. [CrossRef]
33. Gubicza, J.; Ungár, T. Characterization of defect structures in nanocrystalline materials by X-ray line profile analysis. *Z. Kristallogr.* **2007**, *222*, 567–579. [CrossRef]
34. Petry, W.; Brussler, M.; Gröger, V.; Müller, H.G.; Vogl, G. The nature of point defects produced by cold working of metals studied with Mössbauer spectroscopy and perturbed γ-γ angular correlation. *Hyperfine Interact.* **1983**, *15–16*, 371–374. [CrossRef]
35. Schaefer, H.-E.; Baier, F.; Müller, M.A.; Reichle, K.J.; Reimann, K.; Rempel, A.A.; Sato, K.; Ye, F.; Zhang, X.; Sprengel, W. Vacancies and atomic processes in intermetallics–From crystals to quasicrystals and bulk metallic glasses. *Phys. Stat. Sol. B* **2011**, *48*, 2290–2299. [CrossRef]
36. Würschum, R.; Greiner, W.; Valiev, R.Z.; Rapp, M.; Sigle, W.; Schneeweiss, O.; Schaefer, H.-E. Interfacial free volumes in ultra-fine grained metals prepared by severe plastic deformation, by spark erosion, or by crystallization of amorphous alloys. *Scr. Metal. Mater.* **1991**, *251*, 2451–2456. [CrossRef]
37. Gammer, C.; Karnthaler, H.P.; Rentenberger, C. Reordering a deformation disordered intermetallic compound by antiphase boundary movement. *J. Alloys Compd.* **2017**, *713*, 148–155. [CrossRef]
38. Collins, G.S.; Sinha, P. Structural, thermal and deformation-induced point defects in PdIn. *Hyperfine Interactions* **2000**, *130*, 151–179. [CrossRef]
39. Lotkov, A.I.; Baturin, A.A.; Grishkov, V.N.; Kopylov, V.I. Possible role of crystal structure defects in grain structure nanofragmentation under severe cold plastic deformation of metals and alloys. *Phys. Mesomech.* **2007**, *10*, 179–189. [CrossRef]

40. Lotkov, A.I.; Baturin, A.A.; Grishkov, V.N.; Kuznetsov, P.V.; Klimenov, V.A.; Panin, V.E. Structural defects and mesorelief of the titanium nickelide surface after severe plastic deformation by an ultrasonic method. *Fizicheskaya Mesomekhanika* **2005**, *8*, 109–112.

41. Laptev, R.S.; Lider, A.M.; Bordulev, Y.S.; Kudiiarov, V.N.; Garanin, G.V.; Wang, W.; Kuznetsov, P.V. Investigation of defects in hydrogen-saturated titanium by means of positron annihilation techniques. *Defect Diffus. Forum* **2015**, *365*, 232–236. [CrossRef]

42. Giebel, D.; Kansy, J. LT10 Program for Solving Basic Problems Connected with Defect Detection. *Phys. Procedia* **2012**, *35*, 122–127. [CrossRef]

43. Würschum, R.; Donner, P.; Hornbogen, E.; Schaefer, H.-E. Vacancy Studies in Melt-Spun Shape Memory Alloys. *Mater. Sci. Forum* **1992**, *105–110*, 1333–1336. [CrossRef]

44. Mizuno, M.; Araki, H.; Shirai, Y. Theoretical calculation of positron lifetimes for defects in solids. *Adv. Quant. Chem.* **2003**, *42*, 109–126. [CrossRef]

45. Kalchikhin, V.V.; Kulkova, S.E. Kinetic properties and characteristics of electron-positron annihilation in NiMn and NiTi. *Russ. Phys. J.* **1992**, *35*, 922–926. [CrossRef]

46. Janeček, M.; Stráský, J.; Cizek, J.; Harcuba, P.; Václavová, K.; Polyakova, V.V.; Semenova, I.P. Mechanical Properties and Dislocation Structure Evolution in Ti6Al7Nb Alloy Processed by High Pressure Torsion. *Met. Mat. Trans. A* **2014**, *45*, 7–15. [CrossRef]

47. Würschum, R.; Badura-Gergen, K.; Kümmerle, E.; Grupp, C.; Schaefer, H.-E. Characterization of radiation-induced lattice vacancies in intermetallic compounds by means of positron-lifetime studies. *Phys. Rev. B* **1996**, *54*, 849–856. [CrossRef]

48. Baturin, A.A.; Lotkov, A.I. Determination of vacancy formation energy in TiNi compound with B2 structure by positron annihilation. *Fiz. Met. Metalloved.* **1993**, *76*, 168–170.

49. Baturin, A.; Lotkov, A.; Grishkov, V.; Lider, A. Formation of vacancy type defects in titanium nickelide. In Proceedings of the 10th European Symposium on Martensitic Transformations, ESOMAT 2015, Antverp, Belgium, 14–18 September 2015. [CrossRef]

50. Lu, J.M.; Hu, Q.M.; Wang, L.; Li, Y.J.; Xu, D.S.; Yang, R. Point defects and their interaction in TiNi from first-principles calculations. *Phys. Rev. B* **2007**, *75*, 094108. [CrossRef]

51. Janeček, M.; Čížek, J.; Dopita, M.; Král, R.; Srba, O. Mechanical properties and microstructure development of ultrafine-grained Cu processed by ECAP. *Mater. Sci. Forum* **2008**, *584–586*, 440–445. [CrossRef]

52. Setman, D.; Schafler, E.; Korznikova, E.; Zehetbauer, M.J. The presence and nature of vacancy type defects in nanometals detained by severe plastic deformation. *Mater. Sci. Eng. A* **2008**, *493*, 116–122. [CrossRef]

53. Hautojärvi, P.; Corbel, C. Positron Spectroscopy in Metals and Superconductors. In *Positron Spectroscopy of Solids, Proceedings of the Internation School of Physics "Enrico Fermi", Course CXXV, Varenna, Italy, 1993*; Dupasquier, A., Mills, A.P., Eds.; IOS: Amsterdam, The Nétherlands, 1995; pp. 491–532. [CrossRef]

54. Čížek, J.; Janeček, M.; Srba, O.; Kužel, R.; Barnovská, Z.; Procházka, I.; Dobatkin, S. Evolution of defects in copper deformed by high-pressure torsion. *Acta Mater.* **2011**, *59*, 2322–2329. [CrossRef]

55. Churakova, A.; Gunderov, D. Increase in the dislocation density and yield stress of the $Ti_{50}Ni_{50}$ alloy caused by thermal cycling. *Mater. Today: Proc.* **2017**, *4*, 4732–4736. [CrossRef]

Article

Effect of Radial Forging on the Microstructure and Mechanical Properties of Ti-Based Alloys

Lev B. Zuev [1], Galina V. Shlyakhova [1] and Svetlana A. Barannikova [1,2,*]

[1]   Institute of Strength Physics and Materials Science, Siberian Branch of Russian Academy of Science, 634055 Tomsk, Russia; lbz@ispms.tsc.ru (L.B.Z.); shgv@ispms.tsc.ru (G.V.S.)

[2]   Physics and Technology Faculty, Tomsk State University, 634050 Tomsk, Russia

*   Correspondence: bsa@ispms.tsc.ru; Tel.: +7-3822-286-923

Received: 22 September 2020; Accepted: 6 November 2020; Published: 8 November 2020

Abstract: Radial forging is a reliable way to produce Ti alloy rods without preliminary mechanical processing of their surface, which is in turn a mandatory procedure during almost each stage of the existing technology. In the present research, hot pressing and radial forging (RF) of the titanium-based Ti-3.3Al-5Mo-5V alloy were carried out to study the specifics of plasticized metal flow and microstructural evolution in different sections of the rods. The structural analysis of these rods was performed using metallography and X-ray diffraction techniques. The X-ray diffraction reveals the two-phase state of the alloy. The phase content in the alloy was shown to vary upon radial forging. Finally, radial forging was found to be a reliable method to achieve the uniform fine-grained structure and high quality of the rod surface.

Keywords: titanium alloys; structure formation; mechanical properties; radial forging

1. Introduction

The need for increasing strength and reliability of constructions, as well as reducing their weight, leads to higher demands for structural materials. In particular, titanium-based alloys that are extensively used in various fields of engineering due to their high specific strength must have high fatigue resistance, corrosion resistance and wear resistance in addition to excellent strength that can be achieved whilst retaining high plasticity [1,2].

According to research conducted over recent decades [3–6], the possibilities of improving the above properties by means of the conventional mechanical and thermal processing methods in order to modify the chemical composition of titanium-based alloys, aimed at introducing alloying elements and varying grain size in the parent material, are almost exhausted. On the other hand, it is evident that strength characteristics along with functional properties of alloys can be noticeably increased via the formation of micro- and submicrocrystalline phases in the bulk of the alloy under severe plastic deformation [7–9].

There is a large number of scientific papers [10–14], in which the processes of evolution of the structure and properties of titanium-based alloys are investigated under deformation by various methods. It is known [15–17] that the formation of structure in a material is largely determined by the degree of deformation, and its scheme and conditions. There is a large amount of experimental data on the effect of deformation on the structural state of titanium-based alloys [18,19]. However, the evolution of the structure of titanium-based alloys during radial forging (RF) has not been studied sufficiently. As is known [20–22], a stressed state is implemented in the dynamic deformation site in the process of compression of a cylindrical rod during radial forging, which is close to the triaxial compression and makes it possible to obtain large degrees of deformation of rods under dynamic influence without the formation of cracks.

According to [3,10,15–17], thermomechanical processing of titanium-based alloys such as Ti-Al-Mo-V and containing Mo and V may lead to the changes in X-ray diffraction patterns owing to texture formation caused by dopant redistribution, forward and backward decomposition of the β-phase, and partial transformation of the α-phase in a metastable α″-phase. These phenomena may arise individually or overlap with each other. Elucidating the complex redistributions of XRD intensities necessitates a careful analysis of XRD ranges with emerging or disappearing uncombined reflexes from the different phases.

The complex problem of the structural strength of Ti alloys consists in obtaining a homogeneous material with a fine-grained structure, hardened by a highly dispersed phase and a high-quality surface of parts. The fine-grained structure and high-quality surface reduce the effect of stress concentrators and, as a result, increase the resistance to brittle fracture when operating under conditions of alternating dynamic loads [1,2,4].

Traditional methods for making alloys include heating the billet to a temperature above the polymorphic transformation temperature in the β-region, rolling at this temperature, cooling to ambient temperature, heating the rolled stock to a temperature 20–50 °C below the polymorphic transformation temperature, and final rolling at this temperature. In the case of titanium products, such a technological scheme does not provide the necessary manufacturability and quality of titanium material due to its high tendency to grain growth, oxidation and gas saturation during heat treatment and, as a consequence, to loss of plasticity and embrittlement of the material. The temperature of polymorphic transformation (TPT) for the Ti-Al-Mo-V alloy is 840–880 °C, at which phase + recrystallization occurs, which is the basic characteristic for the appointment of heat treatment modes, but it is at TPT and above that catastrophic grain growth in titanium alloys and coarsening of the intragranular structure are observed.

It should be noted that, unlike steels, the coarse-grained structure of titanium materials is not corrected by heat treatment. Therefore, during heat treatment, the hardening temperature is set at 80–150 °C below the TPT of the alloy. However, under these conditions, complete recrystallization does not occur, which does not provide complete hardening and does not correct the heredity of the previous processing, as a result of which the material does not have sufficient quality for such critical parts as bolts and springs. Multiple hot forging and air cooling operations will negatively affect the surface quality of the bar. In addition, the method requires an expensive abrasive operation to remove forging defects and surface substandard layers. As a result, the scrap rate rises and the metal yield decreases, which ultimately leads to an increase in the cost of manufacturing rods.

The problem to be solved by this study is to obtain rods from high quality titanium alloys while ensuring high process productivity.

A promising method for the formation of an ultrafine-grained structure is RF, the use of which in the manufacture of products from titanium alloys is still limited. The use of the radial forging process in the production of pipes from Zr alloys is described in monograph [7]. If the processes occurring in hcp alloys during cold rolling and subsequent heat treatment are well studied, then there is still insufficient reliable information regarding the patterns of structure formation and deformation mechanisms acting during radial forging.

This work aims to characterize the Ti-Al-Mo-V rods with optimal structural and mechanical properties achieved through severe plastic deformation of a Ti-Al-Mo-V alloy using radial forging machines (RFMs). Radial forging is a reliable way to produce Ti alloy rods without preliminary mechanical processing of their surface, which is in turn a mandatory procedure during almost every stage of the existing technology. In order to meet the technical requirements of semiproducts, the development of a new technology needs to solve some important problems such as:

-   The choice of thermal treatment modes for forged semiproducts, which meet the technical requirements to the structure and mechanical characteristics;

-   The elaboration of algorithms for the deformation of rods and alignment of the RFM tool using data acquired during the analysis of structural and mechanical properties, which obeys requirements to the geometrical sizes and surface quality.

2. Materials and Methods

High and ultrahigh degrees of deformation and the subsequent or concomitant thermal effect lead to the implementation of boundary-substructure and dispersion mechanisms of strengthening. Among the industrial methods that allow us to form a more dispersed structural condition than after hot forming and heat treatment, including quenching and tempering in a wide temperature range, is the technology of RF. The billet material is in overall uniform compression in the deformation center when using four strikers under RF, which leads to the formation of an extremely dispersed and homogeneous structure [7]. It should be noted that the treatment of the billet is implemented at the expense of multiple ultrafast simultaneous compressions. This scheme of deformation allows us to form a high degree of localization of the center of the deformation. Basically, the technology of the radial forging is used to improve the strength, reliability, and durability of the special pipe billets, the main task of which is the long-term operation under normal and extreme conditions.

The investigation of structure and properties of the alloys were performed at various stages of manufacture using mechanical, metallographic, and X-ray diffraction techniques. Samples were rods with diameters Φ of 25 mm, produced from a Ti-Al-Mo-V alloy via hot pressing, as well as rods with diameters $\Phi20$, $\Phi16$, $\Phi12$, $\Phi10.5$, and $\Phi8.5$ mm, using the SXP-16 machine of GFM firm (Steyr, Austria). The chemical composition of a Ti-Al-Mo-V workpiece used in testing the technology is given in Table 1.

Table 1. Chemical composition of Ti-Al-Mo-V alloy (wt %).

Ti	Al	Mo	V	C	N	Fe	Si	H_2	O_2
base	3.3	4.6	4.5	0.05	0.01	<0.01	≤0.10	0.007	0.014

Radial forging of Ti-Al-Mo-V alloy rods, implemented on a SXP-16 machine of GFM firm (Steyr, Austria), changes their manufacturing technology to a large extent. Radial forging allows one to achieve the better plastic processing of the material, but also to ensure high quality of the rod surface, which enables to avoid facing at intermediate stages of the technological process and to facilitate the process itself. The forging modes applied in this work are given in Table 2. According to the results, the surface roughness of the rod with a diameter Φ of 10 mm was within a range of 3–9 μm, which met the technical requirements of less than 10 μm; the ovality of the rod at a length of 2.5 m was below 30 μm.

Table 2. Forging modes of Ti-Al-Mo-V alloy rods.

Transition	Number of Beats Per Minute	Feed Rate, mm/s
$\Phi20 \rightarrow \Phi16$ mm	580	5...7
$\Phi16 \rightarrow \Phi10$ mm	580	5...7

The alloy structure in rods exposed to radial forging was inspected in two transition zones—i.e., $\Phi20 \rightarrow \Phi16$ mm and $\Phi16 \rightarrow \Phi10$ mm (Figure 1).

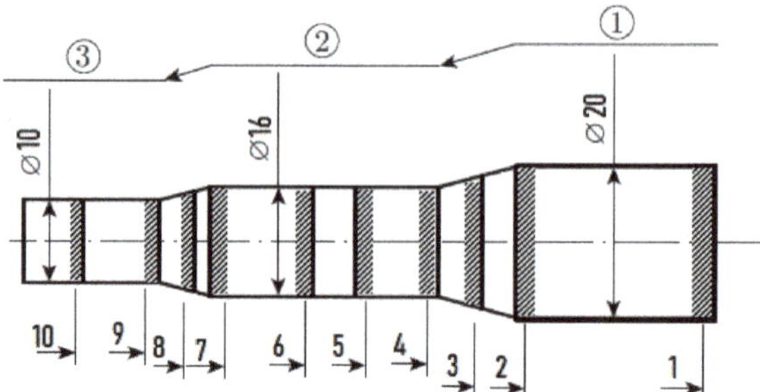

Figure 1. Schematics of cutting-out of samples for metallographic analysis in case of $\Phi20\rightarrow\Phi16$ mm and $\Phi16\rightarrow\Phi10$ mm transitions.

The microstructure of Ti-Al-Mo-V alloy in these zones was probed using a Neophot-21 metallographic microscope equipped with a digital camera (Genius VideoCam Smart 300) and a special software intended for digital image processing. The microstructure of the alloy was determined via chemical etching in a solution of 3% hydrofluoric acid and 3% nitric acid in water [23].

To implement the metallographic analysis of metallic state, as well as to assess the efficiency of the used deformation method and to optimize the alignment of strikers of the SXP-16 RFM, the samples were selected from the deformation center. The schematics of cutting-out of samples for metallographic tests are shown in Figure 1.

Table 3 contains information about specimens that were cut from deformation zones in a cross section according to the scheme in Figure 1, so that each sample corresponds to a certain degree of reduction in diameter, which increases with increasing number.

Table 3. Tested samples according Figure 1.

Sample No.	Workpiece Diameter, µm	Location along the Workpiece	Reduction K, %
1	$\Phi20$	Input (initial state)	0
3	$\Phi20$	Deformation center (RF)	17
5	$\Phi16$	output	20
7	$\Phi16$	Deformation center (RF)	23
8	$\Phi16$	Deformation center (RF)	38
10	$\Phi10$	Output (final state)	50

X-ray diffraction measurements of Ti-Al-Mo-V alloy were conducted on a sample with $\Phi8.5$ mm, obtained from the rod exposed to hot pressing, as well as on sample 6 made by radial forging at a $\Phi16\rightarrow\Phi10$ mm transition (see Figure 1). The data were recorded in the Bragg—Brentano geometry (Θ-2Θ mode) within a range of angles of $10 \leq \Theta \leq 125$ deg [24] using a DRON-3 (Burevestnik, Russia) diffractometer and Cu K_α radiation with the filtered K_β line. The X-ray diffractograms were recorded in a stepwise mode with a step of 0.1 deg; the exposure time was 5 s. The XRD data were processed using a special computational software. The phase analysis was performed in ARFA-7000 software (Russia), including the database for 40,000 compounds.

The mechanical characteristics of Ti-Al-Mo-V alloy were studied on the flat samples, whose working element dimensions were 2 mm × 6 mm × 40 mm. Samples were stretched at room temperature on an Instron-1185 (US) universal testing machine at a rate of 8.3×10^{-5} s^{-1}.

3. Results

3.1. Mechanical Properties of Ti-Al-Mo-V Alloy after Hot Pressing

The mechanical characteristics of Ti-Al-Mo-V alloy were studied on the flat samples for initial state (sample 1) with heads cut off from the center of the workpiece (1), as well as from the surface (2) and transition (3) layers. The appropriate stress–strain diagrams of the tested samples are given in Figure 2. The strength characteristics of all three samples are quite close to each other, as can be seen from the values of strength limit $(\sigma_B)_1 = 1060$ MPa, $(\sigma_B)_3 = 1040$ MPa, and $(\sigma_B)_2 = 980$ MPa. At the same time, the plasticity drastically varies from one sample to the other. So, the uniform elongation (before the formation of the neck) was found to be $(\varepsilon_B)_1 = 0.06$, $(\varepsilon_B)_2 = 0.08$, and $(\varepsilon_B)_3 = 0.105$. The elongation at failure was even more different for each sample, being $\delta_1 = 0.09$, $\delta_2 = 0.11$, and $\delta_3 = 0.16$.

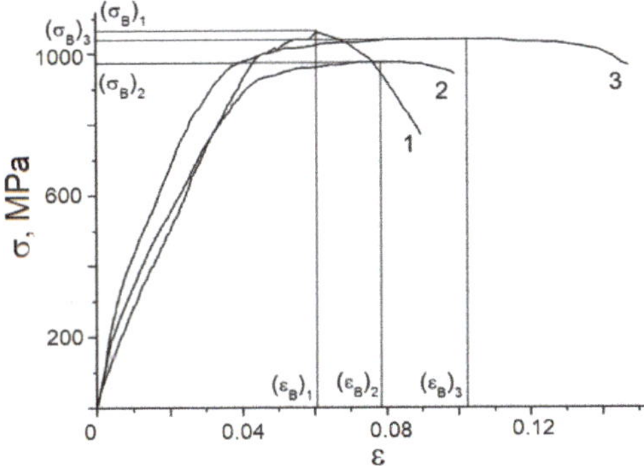

Figure 2. Stress–strain diagrams of Ti-Al-Mo-V alloy: center of the workpiece (1), surface layer (2), transition layer (3).

This is owing to the large cross-sectional heterogeneity of the alloy structure. According to the metallographic data acquired on the cross-section of the workpiece, the phase composition within the scanned area from the center to the surface is presented by a mixture of α and β phases of titanium, where the volume fraction of the α-phase exceeds that of the β-phase (see Figure 3).

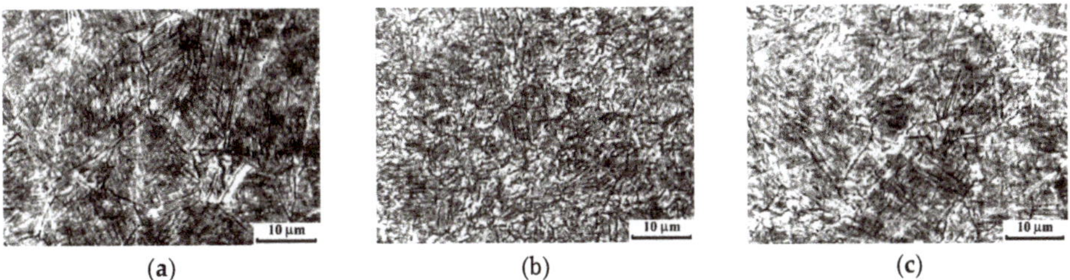

(a)   (b)   (c)

Figure 3. Microstructure of Ti-Al-Mo-V alloy in the initial state before radial forging: center of the workpiece (**a**), surface layer (**b**), transition layer (**c**).

Meanwhile, the average grain size of the α-phase near the surface is ~2.7 µm, which is almost five times lower than in the center, as seen in Figure 3 for initial state before radial forging. The finest grains attributed to the α-phase were detected in the 3-mm-thick surface layer. However, their size gradually increased while moving toward the center, attaining a value of ~16.5 µm, which is consistent with that for pristine workpieces. Hence, a pronounced difference in plasticity of samples cut off from the different parts of the alloy seems to be due to structural heterogeneity of Ti-Al-Mo-V, which persists under hot pressing of rods.

The microhardness of the alloy was measured in the indicated zones using a PMT-3M (Russia) hardness tester at a load of 1 N. In comparison with a central part of the sample, the microhardness near the surface is slightly reduced and continues to decrease from 1750 to 1690 MPa while going away from the center. The reason of such a change in structure and properties can be surface layer disordering caused by degassing of surface layers during processing.

3.2. Structure of of Ti-Al-Mo-V Alloy after Radial Forging

Radial forging allows one to suppress porosity that is usually present in axial sections of rods obtained via hot pressing. According to the experimental data, the highest pore concentration was found in sample 1 (Φ20 mm) (see Figure 4). The pores at the central part of the sample form the axial porosity. The surface pores have the round shape. The pores at the workpiece outlet (Φ20→Φ16 mm) (sample 5) and at the inlet (Φ16→Φ10 mm) (sample 6) may form the small linear clusters. The smallest density and pore sizes were observed in sample 10 (with the outlet of Φ16→Φ10 mm). Figure 4 shows the presence of pores for the initial state (sample 1) and after radial forging (final state, sample 10).

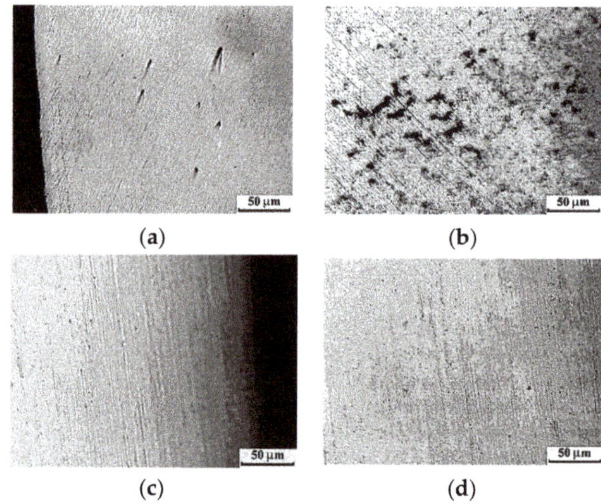

Figure 4. Microstructure of Ti-Al-Mo-V alloy in the initial state before radial forging—sample 1 (**a,b**) and after radial forging in sample 10 (**c,d**). From left to right: surface layer (**a,b**), center of the workpiece (**b,d**).

Table 4 presents data on the change in the parameters of the structure depending on the reduction.

Table 4. Porosity and grain size (in points) depending on the degree of reduction.

Ti-Al-Mo-V Sample No.	Pore Density, mm⁻²		Maximum Pore Size, μm		Grain Score (in Points)	
	Center of the Workpiece	Surface Layer	Center of the Workpiece	Surface Layer	Center of the Workpiece	Surface Layer
1	771	325	18	3,5	6	4
3	542	274	9	3	5-4	3
5	362	244	2.5	3	4	3-2
7	297	202	2	2	3-2	2-1
8	183	149	1.5	1	2	1
10	102	98	<1	1	2	1

The metallographic data are illustrated in Figures 5 and 6 for subsequent $\Phi20\to\Phi16$ mm and $\Phi16\to\Phi10.5$ mm transitions. A comparative analysis of images reveals that noticeable structural changes at the deformation center of the workpiece during the $\Phi20\to\Phi16$ mm transition are manifested by the grain refinement near the surface and close to the axis of the workpiece.

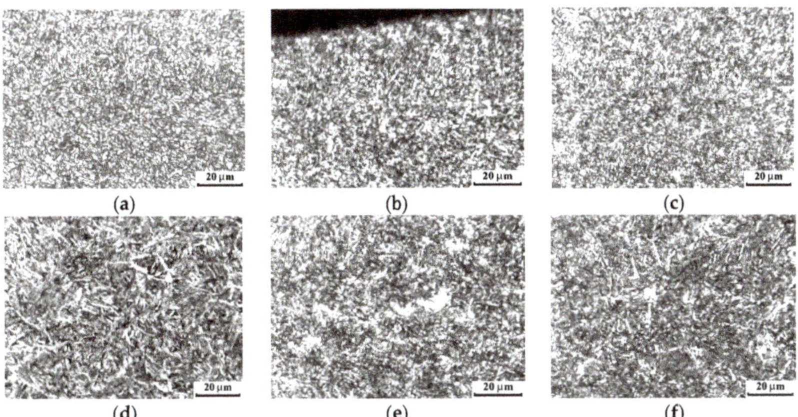

Figure 5. Microstructure of Ti-Al-Mo-V alloy at the $\Phi20\to\Phi16$ mm transition. From left to right: samples 1, 3, and 5 (see Figure 1); (**a–c**) surface structure; (**d–f**) structure at the central part of the rod.

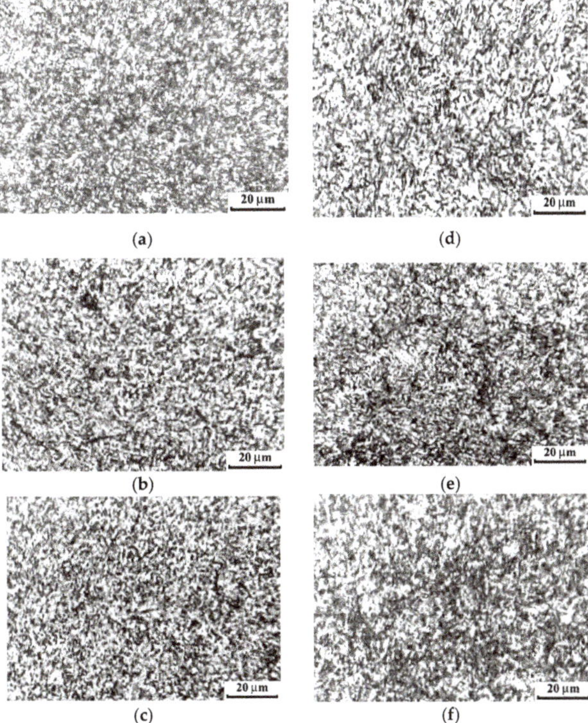

Figure 6. Microstructure of Ti-Al-Mo-V alloy at the $\Phi16\to\Phi10$ mm transition. From top to bottom: samples 6, 8, and 10 (see Figure 1); (**a–c**) surface structure; (**d–f**) structure at the central part of the rod.

Meanwhile, the structural changes at the central part of the workpiece are less pronounced. At a small rod diameter during the Φ20→Φ16 mm transition it observes the drastic refinement of the structure and the variation in grain morphology. The structure near the rod surface is 80% globular. The average grain size at the axis of the workpiece is much greater and the amount of globular grains achieves a value of 50–60%. As follows from the metallographic study of samples cut off from the Φ16→Φ10 mm transition, the alloy structure at this stage of deformation gradually transforms from a duplex (globular-lamellar) state with a high degree of inhomogeneity (sample 6) to a more uniform globular (sample 10) state. However, there is still partial heterogeneity: while the surface of sample 10 exhibits the uniform and globular structure, the central part of the rod contains up to 25% lamellar grains, which results in a stronger structural heterogeneity. Nevertheless, the central porosity in samples obtained by radial forging is less pronounced in comparison with the porosity near the axis of hot-pressed Ti-Al-Mo-V rods. Figure 7 shows the porosity and grain score (in points) depending on the degree of reduction.

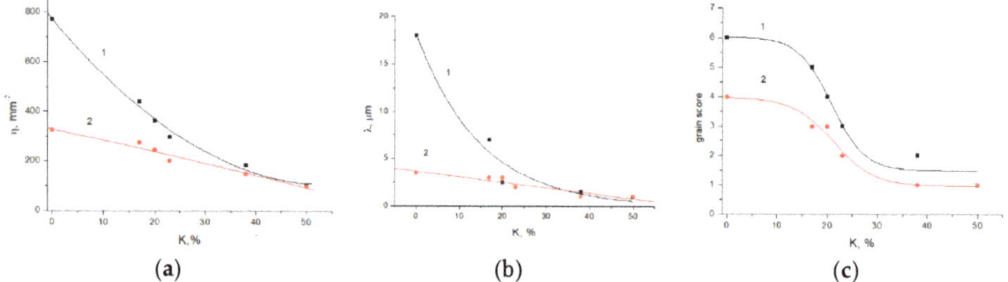

(a)   (b)   (c)

Figure 7. Changes in pore density (**a**), maximum pore size (**b**), and grain score (**c**) depending on the degree of reduction in radial forging: 1—center of the workpiece; 2—surface layer.

The results of transmission electron microscopy showed that in the initial state (sample 1) in the initially inhomogeneous microstructure, there is a deformation-induced bending of the alpha-plates and beta-interlayers, as well as the process of destruction of colonies of the alpha and beta phases with dispersion of the structure. Solid solutions disintegrate inside the structural elements. At a reduction of 20%, the mixture of alpha and beta phases occupies most of the near-surface layer. The average transverse dimension of the structural elements was ~0.3 μm. The microstructure formed after radial forging as a result of a strong deformation effect is noticeably dispersed, homogeneous, and is a mixture of alpha and beta phases with a size of 0.05–0.3 μm.

3.3. X-ray Diffraction Analysis of Ti-Al-Mo-V Alloy

X-ray diffraction measurements of Ti-Al-Mo-V alloy were conducted on a sample with Φ8.5 mm, obtained from the rod exposed to hot pressing, as well as on sample 6 made by radial forging at a Φ16→Φ10 mm transition (see Figure 1). Figure 6 displays the X-ray diffractogram that is characteristic of a biphase alloy, as seen from the presence of reflexes from the coexisting hcp α-phase and bcc β-phase.

The X-ray diffractogram of sample 6 after radial forging at a Φ16→Φ10.5 mm transition (after the Φ20→Φ16 mm transition) is given in Figure 8. The observed pattern is strongly different from that for a hot-pressed sample. First of all, there are the differences between the intensities of the main peaks attributed to the α-phase and those associated with a quantity of the β-phase. Moreover, the (002) and (004) reflexes of the α-phase vanish, indicating the emergence of anisotropy in the properties along various crystallographic directions, which is due to the alloy processing.

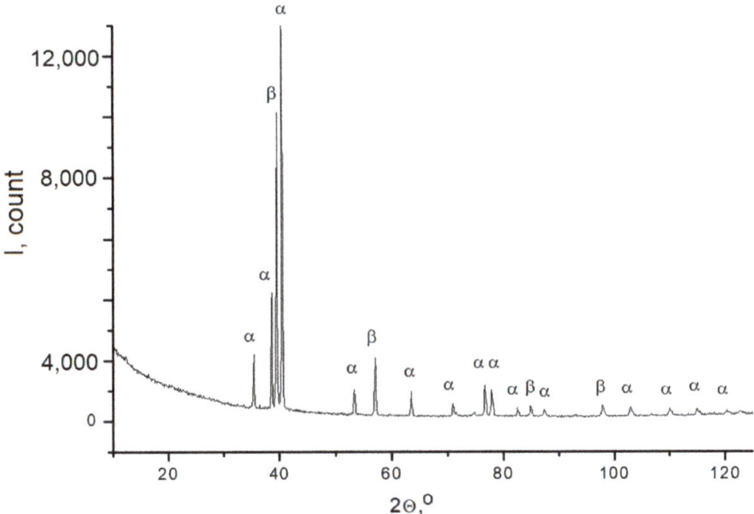

Figure 8. X-ray diffraction of a hot-pressed biphase Ti-Al-Mo-V alloy.

To confirm these changes, the X-ray diffractograms from Figures 8 and 9 were merged in Figure 10. One can see that $(002)_\alpha$ line is almost absent in the case of sample 6 after forging.

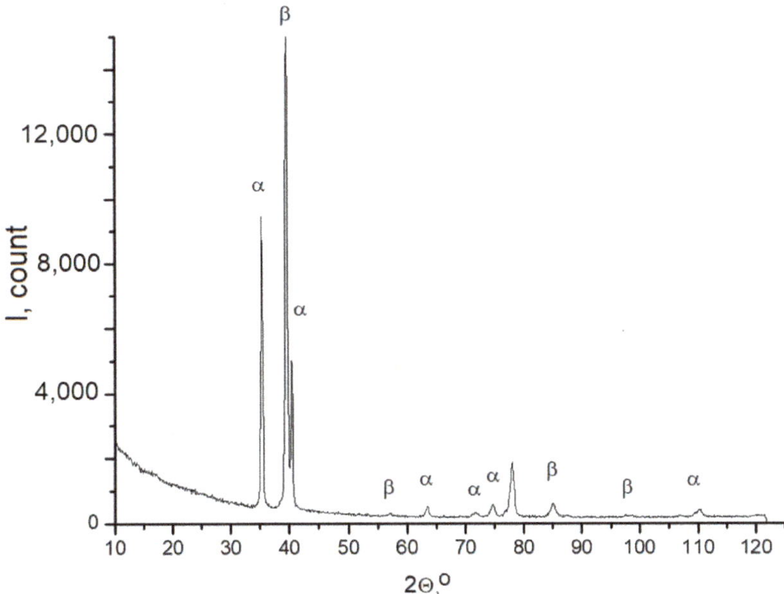

Figure 9. X-ray diffraction of sample 6 of Ti-Al-Mo-V alloy (see Figure 1).

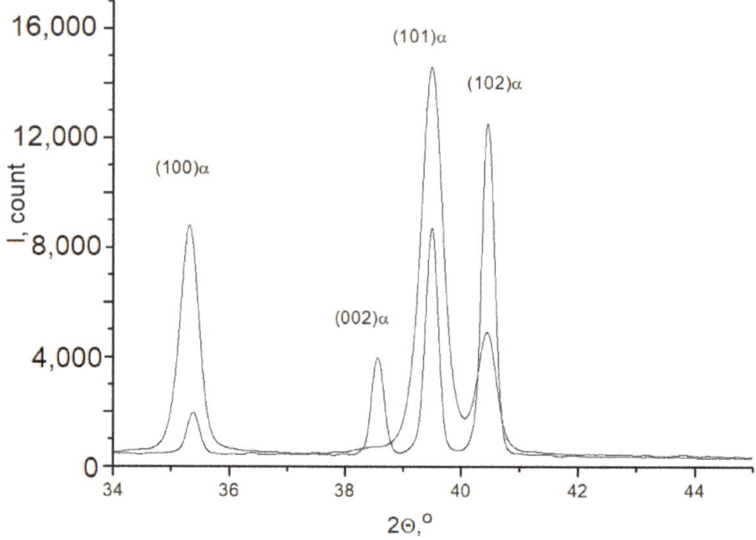

Figure 10. Merged X-ray diffractions from Figures 8 and 9.

4. Discussion

As established via X-ray diffraction with regard to the behavior of the fcc β-phase during radial forging, there was no decrease in its reflex intensity, but an increase for all samples. This evidences the absence of the β→α transformation. The analysis of other XRD peaks reveals that some lines from the β-phase become larger during the processing and lose their intensity as compared with the initial state. This testifies to a decrease in crystallite size of the β-phase in the VT16 alloy undergoing plastic deformation during radial forging.

An important phenomenon that arises under severe plastic deformation in radial forging can be the β→α″ martensitic transformation, because, as found, there is a small amount of the martensitic α″-phase in all samples. However, in the case of the initial state, there is also a high content of a stable α-phase. The amount of the α-phase is lower in forged samples, where the parent phase is a metastable α″-phase. At the same time, the α-phase has a nonstoichiometric composition, as follows from its lattice parameter values. Moreover, it appears from the XRD data that the lattice parameters of the hcp α-phase of Ti-Al-Mo-V alloy are $a = b = 0.29487$ nm and $c = 0.46857$ nm, whereas the lattice parameter of a metastable β-phase of titanium (bcc) is $a = 0.3231$ nm [1,2].

The most significant change in the phase composition of the alloy caused by radial forging is the change in the volume fraction of the beta-phase shown in the Figure 11 during this process. It is calculated from the ratio of the experimentally determined intensities of the lines (102) of the alpha phase and (200) of the beta phase for various degrees of compression. As can be seen, at the beginning of radial forging with a reduction of up to 20% by volume, the bulk content of the beta phase remains at a level of ~1/3, and then the proportion of the beta phase decreases to ~$\frac{1}{4}$ at a final reduction of 50%. In the process of radial forging, there is a gradual change in the texture formed during hot pressing. With an increase in the degree of reduction, the ratio of the intensities of the diffraction maxima approaches the ratio for a fine-grained material after annealing.

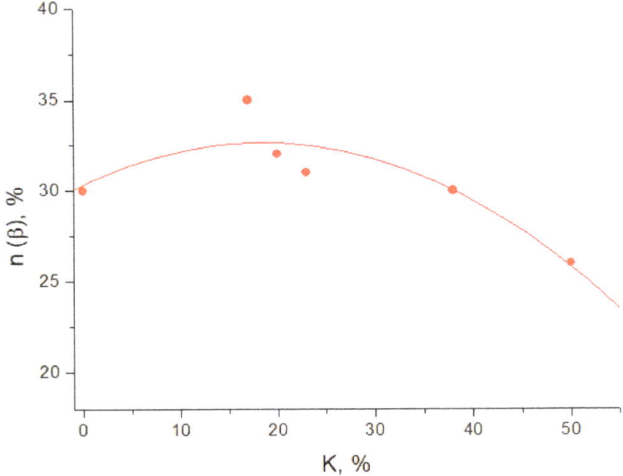

Figure 11. Change in the volume fraction of the beta phase depending on the degree of reduction in radial forging.

5. Conclusions

- Radial forging of Ti-3.3Al-5Mo-5V alloy workpieces leads to the structural refinement and the changes in grain morphology from predominately lamelar to uniform globular, as well as the variations in phase composition of the alloy and the supression of powder formation.
- The methods of rod deformation and alignment of the RFM working tool, applied in this work, ensure fulfillment of technical requirements to geometrical sizes and surface quality.
- Radial forging is found to allow one to produce semiproducts from titanium-based Ti-Al-Mo-V alloys without using multiple mechanical processing of workpieces and rods.

Author Contributions: Project administration, supervision, and conceptualization, L.B.Z.; methodology, investigation, and visualization, G.V.S.; formal analysis, writing—original draft preparation, and writing—review and editing, S.A.B. All authors have read and agreed to the published version of the manuscript.

Funding: The work was performed according to the government research assignment for ISPMS SB RAS (project No. III.23.1.2) and in the framework of the Program of Competitiveness by the National Research Tomsk State University.

Acknowledgments: The authors are gratefully thanks to Danilov Vladimir for the results discussion.

Conflicts of Interest: The authors declare no conflict of interest.

References

1. Leyens, C.; Peters, M. *Titanium and Titanium Alloys: Fundamentals and Applications*; John Wiley & Sons: Weinheim, Germany, 2006; pp. 250–302.
2. Lütjering, G.; Williams, J.C. *Titanium*, 2nd ed.; Springer: Berlin/Heidelberg, Germany, 2007; pp. 203–250.
3. Kabra, S.; Ya, K.; Mayer, S.; Schmoelzer, T.; Reid, M.; Dippenaar, R.; Clemens, H.; Liss, K. Phase transition and ordering behavior of ternary Ti–Al–Mo alloys using in-situ neutron diffraction. *Int. J. Mater. Res.* **2011**, *102*, 697–702. [CrossRef]
4. Xiao, D.-H.; Yuan, T.-C.; Ou, X.-Q.; He, Y.-H. Microstructure and mechanical properties of powder metallurgy Ti-Al-Mo-V-Ag alloy. *Trans. Nonferrous Met. Soc. China* **2011**, *21*, 1269–1276. [CrossRef]
5. Prakash, D.G.L.; Honniball, P.; Rugg, D.; Withers, P.J.; da Fonseca, J.Q.; Preuss, M. The effect of beta phase on microstructure and texture evolution during thermo-mechanical processing of alpha plus beta Ti alloy. *Acta Mater.* **2013**, *61*, 3200–3213. [CrossRef]

6. Singh, A.K.; Schwarzer, R.A. Evolution of texture during thermo-mechanical processing of titanium and its alloys. *Trans. Ind. Inst. Met.* **2008**, *61*, 371–387. [CrossRef]

7. Zavodchikov, S.Y.; Zuev, L.B.; Kotrekhov, V.A. On the Use of the Radial Forging Process in Pipe Production. In *Physical Metallurgy Aspects of the Production of Zirconium Alloy Components*; Glezer, A.M., Ed.; Nauka: Novosibirsk, Russia, 2012; pp. 165–191.

8. Barannikova, S.; Schlyakhova, G.; Maslova, O.; Li, Y.; Lev, Z. Fine structural characterization of the elements of a Nb-Ti superconducting cable. *J. Mater. Res. Technol.* **2019**, *8*, 323–332. [CrossRef]

9. Barannikova, S.; Shlyakhova, G.; Zuev, L.; Malinovskiy, A. Microstructure of superconducting cable components. *Int. J. Geom.* **2016**, *10*, 1906–1911. [CrossRef]

10. Gloaguen, D.; Girault, B.; Courant, B.; Dubos, P.A.; Moya, M.J.; Edy, F.; Kornmeier, J.R. Study of residual stresses in additively manufactured Ti-6Al-4V by neutron diffraction measurements. *Met. Mat. Trans. A* **2020**, *51*, 951–961. [CrossRef]

11. Wagner, F.; Ouarem, A.; Richeton, T.; Toth, L.S. Improving mechanical properties of cp titanium by heat treatment optimization. *Adv. Eng. Mater.* **2018**, *20*, 1700237. [CrossRef]

12. Lei, X.F.; Dong, L.M.; Zhang, Z.Q.; Liu, Y.J.; Hao, Y.L.; Yang, R.; Zhang, L.C. Microstructure, Texture Evolution and Mechanical Properties of VT3-1 Titanium Alloy Processed by Multi-Pass Drawing and Subsequent Isothermal Annealing. *Metals* **2017**, *7*, 131. [CrossRef]

13. Hu, Y.; Randle, R. An electron backscatter diffraction analysis of misorientation distributions in titanium alloys. *Scr. Mater.* **2007**, *56*, 1051–1054. [CrossRef]

14. Meng, L.; Kitashima, T.; Tsuchiyama, T.; Watanabe, M. Effect of α precipitation on β texture evolution during β-processed forging in a near-β titanium alloy. *Mater. Sci. Eng. A* **2020**, *771*, 138640. [CrossRef]

15. Qin, D.; Lu, Y.; Guo, D.; Zheng, L.; Liu, Q.; Zhou, L. On preparation of bimodal Ti–5Al–5V–5Mo–3Cr–0.4Si (Ti-5553s) alloy: α+β forging and heat treatment. *Mater. Sci. Eng. A* **2014**, *609*, 42–52. [CrossRef]

16. Li, T.R.; Liu, G.H.; Xu, M.; Wang, B.X.; Misra, R.D.K.; Wang, Z.D. High temperature deformation and control of homogeneous microstructure during hot pack rolling of Ti-44Al-5Nb-(Mo, V, B) alloys: The impact on mechanical properties. *Mater. Sci. Eng. A* **2019**, *751*, 1–9. [CrossRef]

17. Li, T.R.; Liu, G.H.; Xu, M.; Wang, B.X.; Fu, T.L.; Wang, Z.D.; Misra, R.D.K. Flow stress prediction and hot deformation mechanisms in Ti-44Al-5Nb-(Mo, V, B) Alloy. *Materials* **2018**, *11*, 2044. [CrossRef] [PubMed]

18. Liu, G.; Wang, Z.; Fu, T.; Li, Y.; Liu, H.; Li, T.; Gong, M.; Wang, G. Study on the microstructure, phase transition and hardness for the TiAl–Nb alloy design during directional solidification. *J. Alloy. Compd.* **2015**, *650*, 45–52. [CrossRef]

19. Zhang, S.Z.; Zhao, Y.B.; Zhang, C.J.; Han, J.C.; Sun, M.J.; Xu, M. The microstructure, mechanical properties, and oxidation behavior of beta-gamma TiAl alloy with excellent hot workability. *Mater. Sci. Eng. A* **2017**, *700*, 366–373. [CrossRef]

20. Nematzadeh, F.; Akbarpour, M.R.; Kokabi, A.H.; Sadrnezhaad, S.K. Structural changes of radial forging die surface during service under thermo-mechanical fatigue. *Mater. Sci. Eng. A* **2009**, *527*, 98–102. [CrossRef]

21. Wu, Y.; Dong, X.; Yu, Q. An upper bound solution of axialmetal flow in cold radial forging process of rods. *Int. J. Mech. Sci.* **2014**, *85*, 120–129. [CrossRef]

22. Sheremetyev, V.; Kudryashova, A.; Cheverikin, V.; Korotitskiy, A.; Galkin, S.; Prokoshkin, S.; Brailovski, V. Hot radial shear rolling and rotary forging of metastable beta Ti-18Zr-14Nb (at. %) alloy for bone implants: Microstructure, texture and functional properties. *J. Alloy. Compd.* **2019**, *800*, 320–326. [CrossRef]

23. Greaves, R.H.; Wrighton, H. *Practical Microscopical Metallography*; Chapman and Hall: London, UK, 1957; pp. 112–160.

24. Taylor, A. *An Introduction to X-Ray Metallography*; Chapman & Hall Ltd.: London, UK, 1945; pp. 170–185.

Publisher's Note: MDPI stays neutral with regard to jurisdictional claims in published maps and institutional affiliations.

Article

Correlation between Microstructure and Mechanical Properties of Heat-Treated Ti–6Al–4V with Fe Alloying

Yongwei Liu [1], Fuwen Chen [1,2,*], Guanglong Xu [1], Yuwen Cui [1] and Hui Chang [1]

[1]   Tech Institute for Advanced Materials and College of Materials Science and Engineering,
    Nanjing Tech University, Nanjing 210009, China; liuyongwei@njtech.edu.cn (Y.L.);
    guanglongxu@njtech.edu.cn (G.X.); ycui@njtech.edu.cn (Y.C.); ch2006@njtech.edu.cn (H.C.)
[2]   State Key Laboratory of Powder Metallurgy, Central South University, Changsha 410083, China
*   Correspondence: fuwenchen@njtech.edu.cn; Tel.: +86-25-83587270

Received: 29 May 2020; Accepted: 25 June 2020; Published: 28 June 2020

Abstract: The microstructure and mechanical properties of a newly developed Fe-microalloyed Ti–6Al–4V titanium alloy were investigated after different heat treatments. The volume fraction and the morphological features of the lamellar α phase had significant effects on the alloy's mechanical performance. A dataset showing the relationship between microstructural features and tensile strength, elongation, and fracture toughness was developed. A high aging temperature resulted in high plasticity and fracture toughness, but relatively low strength. The high strength favored the fine α and the slender β. The high aspect ratio of lamellar α led to high strength but low fracture toughness. The alloy with ~84 vol % α exhibited the highest strength and lowest fracture toughness because the area of its α/β-phase interface was the highest. Optimal comprehensive mechanical performance and heat-treatment procedures were thus obtained from the dataset. Optimal tensile strength, yield strength, elongation, and fracture toughness were 999 and 919 MPa, 10.4%, and 94.4 MPa·m$^{1/2}$, respectively.

Keywords: Fe-microalloyed Ti–6Al–4V titanium alloy; microstructure; mechanical properties

---

1. Introduction

Ti–6Al–4V, the most widely used titanium alloy in the aerospace industry, has been used to manufacture compressor disks and blades for gas turbines in advanced jet engines [1–4]. To further improve its properties and reduce costs, a series of techniques, such as alloying [5–7], processing optimization for melting and hot working [8,9], and additive manufacturing [10–14], have been employed to modify the Ti–6Al–4V alloy. From these techniques, alloying is considered a low-cost but efficient method for improving its properties. Fe is the strongest β-phase stabilizer and the cheapest alloying element that can strengthen titanium alloys via solid solution and aging treatment [6,15]. Fe alloying refines β grains and facilitates subsequent forging [6,15]. The optimal amount of Fe should be controlled within the range of 0.1–1.0 wt% [15]. The authors' preliminary studies [7] found that the trace addition of Fe into Ti–6Al–4V alloy could remarkably improve fracture toughness and increase hardness while maintaining the alloy's strength. However, the optimal comprehensive properties and the corresponding heat-treatment procedures for Fe-microalloyed Ti–6Al–4V have not been found.

It is well known that an alloy's mechanical properties, such as strength, ductility, creep resistance, and fracture toughness are affected by its microstructure (e.g., phase fraction and morphology). The high mole fraction of the fine α phase is conducive to high strength [4,16–18]. The formation of coarse α phase generally leads to good plasticity [4,16–18]. The introduction of α phase with a lamellar shape is better for fracture toughness than the equiaxed shaped α [4,16–18]. The microstructure varies with heat-treatment parameters. Annealing at high temperatures in the β-phase region for a long time enlarges the β grain [4,16–18]. A high cooling rate before aging promotes the precipitation of fine α during subsequent aging [4,16–18]. Aging at high temperatures in the α + β-phase region for a long time leads to coarse α precipitation [4,16–18]. Guo et al. [19] found that the fracture toughness of the TC4–DT alloy was improved through appropriate heat treatment to adjust its microstructural parameters. Therefore, for a given alloy, optimal properties can be assessed using a developed dataset of correlative heat-treatment parameters, microstructures, and properties. The objectives of this work were to develop a dataset showing the correlation between microstructural features and mechanical properties, and to optimize the mechanical properties of Fe-microalloyed Ti–6Al–4V alloy by adjusting heat-treatment procedures.

2. Materials and Methods

2.1. Materials

The as-received Fe-microalloyed Ti–6Al–4V (TC4F) sheet had a thickness of 14 mm and was produced by vacuum-arc remelting, followed by hot rolling in the β- and α + β-phase regions. Chemical composition was measured by inductively coupled plasma mass spectrometry (ICP-MS; iCAP Qc, Thermo Fisher, Waltham, MA, USA), and results are listed in Table 1. The microstructure of the hot-rolled TC4F alloy is shown in Figure 1. It consisted of an elongated α phase (white), an equiaxed α phase (white), and a small amount of transformed $β_T$ (gray).

Table 1. Chemical composition of as-received Fe-microalloyed Ti–6Al–4V (TC4F).

Ti	Al	V	Fe	C	N	O	H	Si
Balance	6.20	4.14	0.537	0.020	0.020	0.13	0.001	0.016

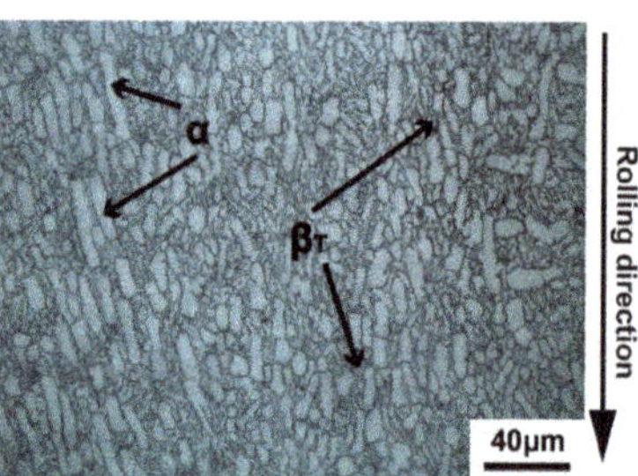

Figure 1. Optical-microscope (OM) image of as-received TC4F.

2.2. Heat Treatment

Before heat treatment, the β-transus temperature of the as-received TC4F was determined by the metallographic method specified in GB/T 23605-2009 [20]. The TC4F was isothermally treated at a 970–1010 °C temperature range with an interval of 5 °C for 40 min, and then quenched in water. The α phase was observed with an optical microscope (OM; Axio Observer A1m, ZEISS, Jena, Germany). When the temperature reached and exceeded 980 °C, the α phase disappeared. At 975 °C, the volume fraction of α was less than 3%, so the β-transus temperature was determined to be 975 ± 5 °C.

The BASCA heat treatment (β-annealing, cooling, and aging) was selected [21]. Five heat-treatment procedures (marked as Samples 1–5) could be divided into two groups: high-temperature β-annealing + low-temperature aging (Samples 1 and 2), and low-temperature β-annealing + high-temperature aging (Samples 3–5), as shown in Table 2. For the first group (high-temperature β-annealing + low-temperature aging; Samples 1 and 2), alloys were subjected to β-annealing at 1005 °C for 70 min, followed by air cooling, aging for 2 h, and air cooling. The only difference between Samples 1 and 2 was the aging temperature, namely, 722 °C for Sample 1, and 732 °C for Sample 2. For the second group (low-temperature β-annealing + high-temperature aging; Samples 3–5), alloys were subjected to β-annealing at 1100 °C for 40 min, followed by cooling, aging at 712 °C for 2 h, and air cooling. The difference between Samples 3, 4, and 5 was cooling rate. A water-cooling was used for Sample 3, air cooling for Sample 4, and furnace cooling for Sample 5.

Table 2. Heat-treatment procedures.

Sample	β-Annealing	Cooling after β-Annealing	Aging	Cooling after Aging
1	1005 °C for 70 min	Air cooling to room temperature	722 °C for 2 h	Air cooling to room temperature
2	1005 °C for 70 min	Air cooling to room temperature	732 °C for 2 h	Air cooling to room temperature
3	1100 °C for 40 min	Water cooling to room temperature	712 °C for 2 h	Air cooling to room temperature
4	1100 °C for 40 min	Air cooling to room temperature	712 °C for 2 h	Air cooling to room temperature
5	1100 °C for 40 min	Furnace cooling to aging temperature	712 °C for 2 h	Air cooling to room temperature

2.3. Mechanical Properties

Room-temperature tensile and mode-I fracture-toughness tests were carried out on the heat-treated TC4F. The tensile test was performed on an INSTRON 5581 (INSTRON, USA) at a rate of 0.02 mm/s. Sample processing and testing adhered to ASTM E8M-04 [22], and the treatment was the same as in our previous research [7]. The average values of the tensile strength, yield tensile strength, and elongation for TC4F were obtained from three valid data points.

Mode-I fracture toughness was measured on an INSTRON 651 (INSTRON, Norwood, MA, USA). The processing of the chevron-notched half compact-tension specimens, and the Mode-I fracture toughness test were performed in accordance with ASTM E399-06 [23]. The dimensions of the compact-tension specimen were thickness B = 12 mm, width W = 48 mm, and notch length a = 21 mm. The 24 mm precracks were preset at a constant ΔK of 21 MPa·m$^{1/2}$, with a sinusoidal waveform at a frequency of 10 Hz. The fracture-toughness value of TC4F after individual heat treatment was obtained by averaging four measurements.

2.4. Microstructure Observation

The microstructure was observed with an OM and a scanning electron microscope (SEM; Scios, FEI, Hillsboro, OR, USA). Samples examined with the OM were polished using standard metallographic techniques to obtain a mirrorlike surface, and then etched with $HF:HNO_3:H_2O = 1:2:7$. The prior-β size was obtained through OM observation. In addition to the above polishing procedures, the samples examined with the SEM were electrochemically polished and cleaned with alcohol. For the SEM and OM images, the size of prior-β grains, the phase fractions, and the size of the α and retained β phases were obtained using Image-Pro Plus software. The phase fraction and size of the phases were measured repeatedly in different areas and/or samples; thus, the values of the average phase fraction and the average size of the α phases were statistically significant.

3. Results and Discussion

3.1. Microstructure

Figure 2 shows the optical images of the microstructure of TC4F after heat treatments, in which the equiaxed prior-β grain could be identified. Figure 3 shows the phase-indexing image of Sample 4 after heat treatment taken by electron backscatter diffraction (EBSD). According to Kikuchi patterns, the main cubic β phase, with a body-centered cubic (BCC) structure, and a small amount of the α phase, with a hexagonal close-packed (HCP) structure, were indexed. The distribution of the two phases is presented in blue and red. Figure 4 shows the backscattered SEM images of the TC4F samples' microstructures. Corresponding to the phase morphology in Figure 3, the dark region in Figure 4 is the α phase, and the light region is the β phase. These regions appeared that way because β stabilizers are heavy elements. The region rich in β stabilizers, which appeared light in color in the backscattered SEM image, was the β phase. Conversely, the light α stabilizers dwelled in the α phase, appearing dark in color [19]. The five samples all contained the lamellar α and retained β phases, and the retained β delineated the lamellar α. Figures 5–9 show frequency vs. microstructural features (namely, length and width of the α lamellae, the α aspect ratio, and the width of the retained β) of each sample based on SEM images (Figure 4). These comparisons of frequency vs. microstructural features could be fitted to normal distribution curves.

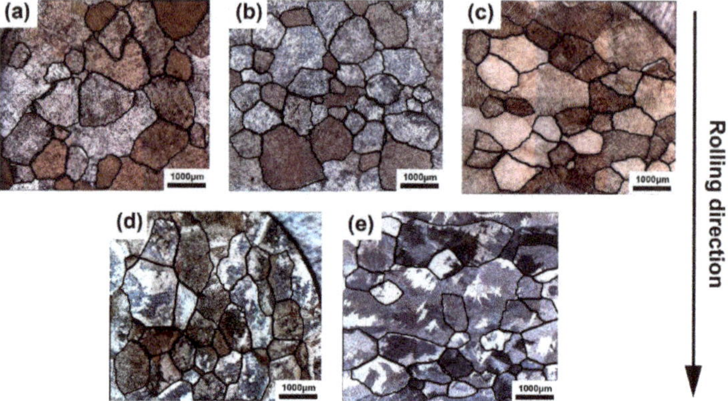

Figure 2. OM images of TC4F after treatment. Samples (**a**) 1, (**b**) 2, (**c**) 3, (**d**) 4 and (**e**) 5.

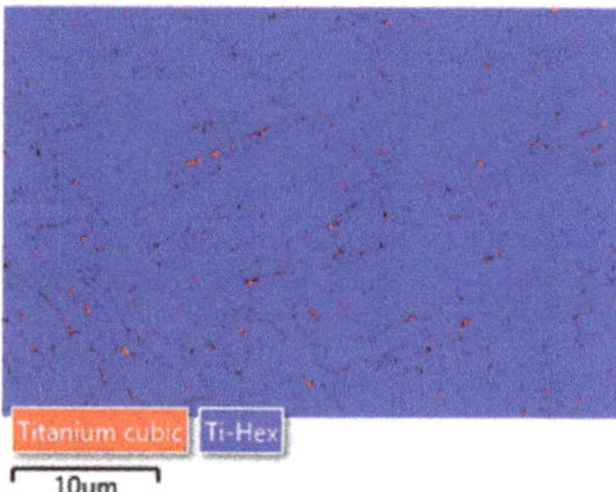

Figure 3. Phase-indexing image of Sample 4 after heat treatment taken by electron backscatter diffraction (EBSD). Regions with hexagonal close-packed (HCP) Kikuchi diffraction pattern identified as α phase and represented in blue; those with body-centered cubic (BCC) pattern identified as β phase and represented in red.

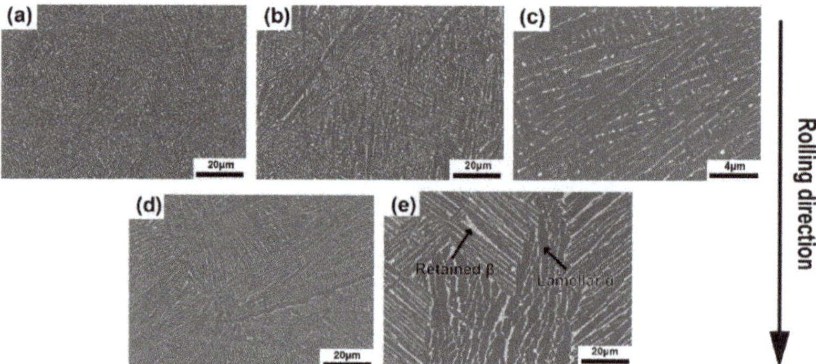

Figure 4. Backscattered scanning-electron-microscope (SEM) images of TC4F after treatment. Samples (**a**) 1, (**b**) 2, (**c**) 3, (**d**) 4, and (**e**) 5. Dark lamellar areas, α phase; light ones, β phase.

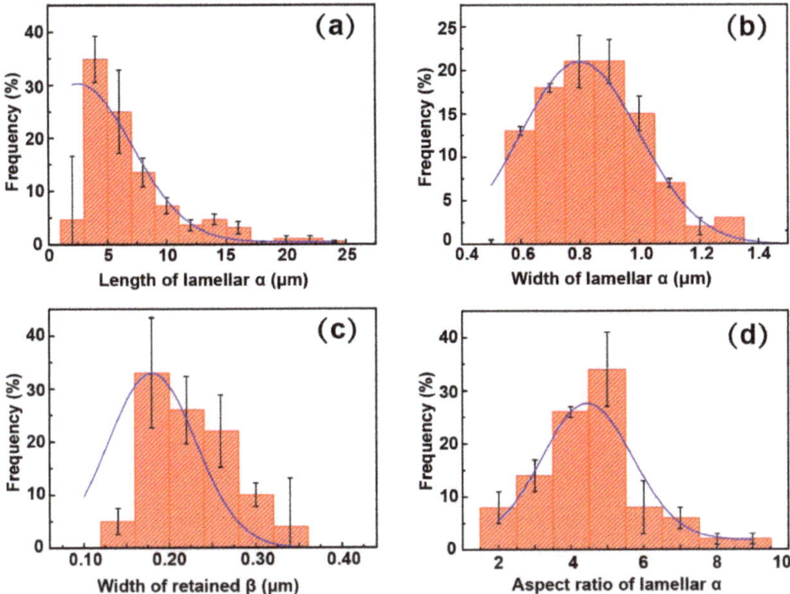

Figure 5. Distribution of microstructural features of TC4F Sample 1 after treatment. (**a**) Length of lamellar α, (**b**) width of lamellar α, (**c**) width of retained β, and (**d**) aspect-ratio distribution of lamellar α. Microstructural features followed Gaussian distribution. Most probable values of α length, width, aspect ratio, and retained β were ~4, ~0.85, ~5, and ~0.175 μm, respectively.

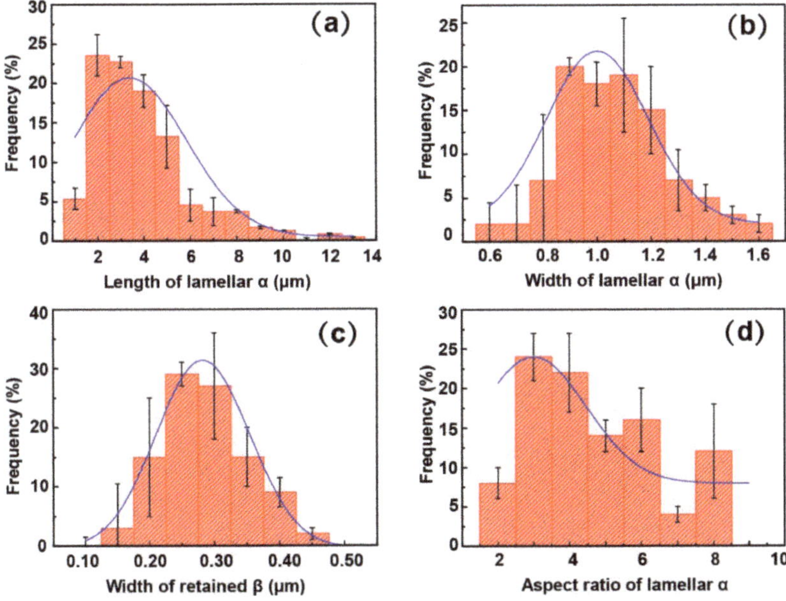

Figure 6. Distribution of microstructural features of TC4F Sample 2 after treatment. (**a**) Length of lamellar α, (**b**) width of lamellar α, (**c**) width of retained β, and (**d**) aspect-ratio distribution of lamellar α. Microstructural features followed Gaussian distribution. Most probable values of α length, width, aspect ratio, and retained β were ~3, ~0.9, ~3, and ~0.25 μm, respectively.

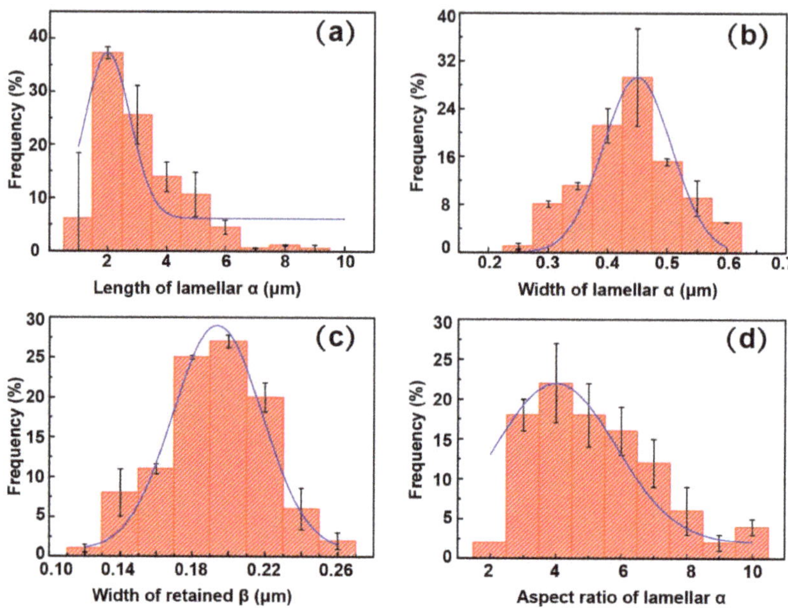

Figure 7. Distribution of microstructural features of TC4F Sample 3 after treatment. (**a**) Length of lamellar α, (**b**) width of lamellar α, (**c**) width of retained β, and (**d**) aspect-ratio distribution of lamellar α. Microstructural features followed Gaussian distribution. Most probable values of α length, width, aspect ratio, and retained β were ~2, ~0.45, ~4, and ~0.20 μm, respectively.

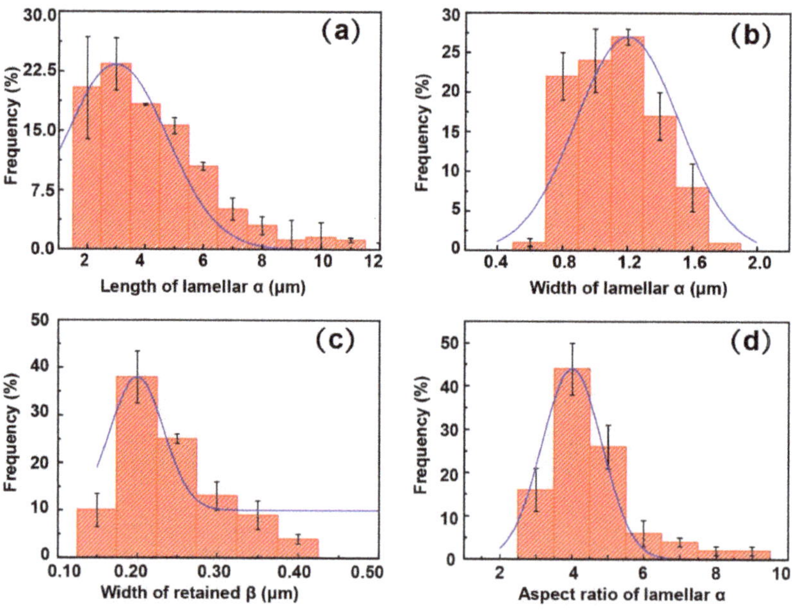

Figure 8. Distribution of microstructural features of TC4F Sample 4 after treatment. (**a**) Length of lamellar α, (**b**) width of lamellar α, (**c**) width of retained β, and (**d**) aspect-ratio distribution of lamellar α. Microstructural features followed Gaussian distribution. Most probable values of α length, width, aspect ratio, and retained β were ~3, ~1.2, ~4, and ~0.20 μm, respectively.

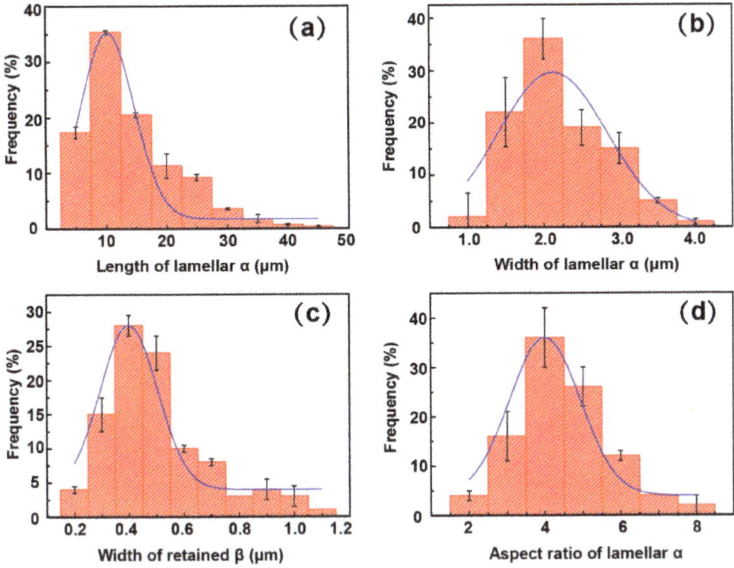

Figure 9. Distribution of microstructural features of TC4F Sample 5 after treatment. (**a**) Length of lamellar α, (**b**) width of lamellar α, (**c**) width of retained β, and (**d**) aspect-ratio distribution of lamellar α. Microstructural features followed Gaussian distribution. Most probable values of α length, width, aspect ratio, and retained β were ~10, ~2.0, ~4, and ~0.4 μm, respectively.

Figure 10 shows the statistical average values of TC4F's microstructural features after heat treatments. Data are summarized in Table 3. The aging temperature of Sample 2 was higher than that of Sample 1, which led to a lower α fraction in Sample 2 than in Sample 1 (as shown in Figure 10c and Table 3). A higher aging temperature facilitated the growth of the α phase, as evidenced by the higher average width of lamellar α in Sample 2 than that in Sample 1 (see Figure 10b and Table 3), although the average length of lamellar α was almost identical. Cooling after the solution treatment mainly affected the size of the α phase precipitated during the subsequent aging process. The higher the cooling rate was, the smaller the α phase size was. The average length and width of lamellar α in Sample 4 were larger than those in Sample 3 but smaller than those in Sample 5 (see Figure 10a,b and Table 3). The prior-β grain size was controlled by the β-annealing treatment. A high temperature and a long time accelerated the growth of the prior-β grain. After β-annealing at 1005 °C for 70 min or 1100 °C for 40 min, the β-grain size was around 900 μm (see Figure 10d and Table 3). The difference in β-grain size of Samples 3–5 was mainly due to the result of the cooling rate after solution treatment; 1100 °C was more than 100 °C higher than the β-transus temperature. The slow-cooling rate caused β-phase grains to continue growing at temperatures higher than 975 °C during the cooling process. Similar conclusions were also found for other α + β titanium alloys [10–19]. β-Annealing at a low temperature for a short time in the β-phase region facilitated small prior-β grains. Fast water cooling and aging at a low temperature for a short time facilitated the formation of the finest lamellar α precipitates.

Table 3. Statistical microstructural features of TC4F after heat treatments.

Sample	Average Width of Lamellar α /μm	Average Length of Lamellar α /μm	Average Width of Retained β /μm	Average Aspect Ratio of Lamellar α	Volume Fraction of Lamellar α/%	Average Prior-β Size /μm
1	0.81 ± 0.02	3.37 ± 0.21	0.21 ± 0.04	4.22 ± 0.13	82.71 ± 1.63	890.00 ± 2.48
2	0.93 ± 0.10	3.48 ± 0.03	0.23 ± 0.03	4.20 ± 0.33	80.47 ± 1.87	891.39 ± 4.32
3	0.45 ± 0.05	2.35 ± 0.50	0.18 ± 0.01	4.39 ± 0.32	84.05 ± 0.96	863.97 ± 7.23
4	1.00 ± 0.04	4.32 ± 0.60	0.24 ± 0.02	4.13 ± 0.16	84.57 ± 0.29	900.26 ± 6.34
5	1.83 ± 0.34	11.81 ± 0.20	0.44 ± 0.02	3.87 ± 0.37	85.07 ± 1.16	938.53 ± 12.22

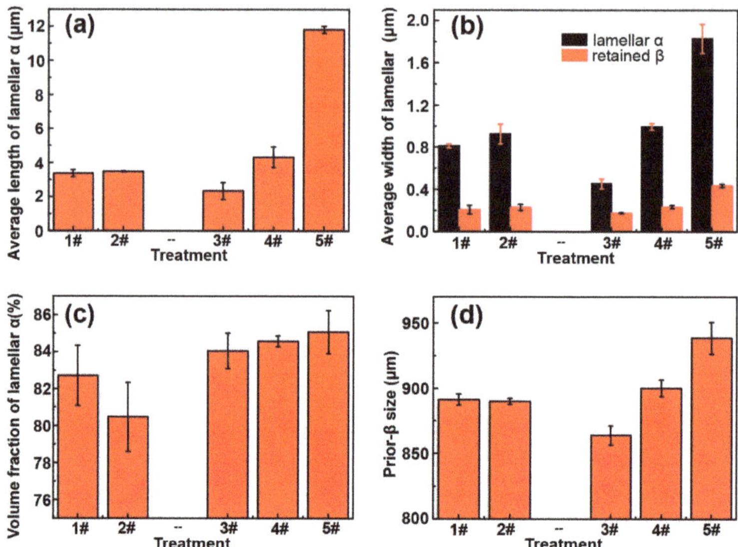

Figure 10. Statistical microstructural features of TC4F after heat treatments. (**a**) Average length of lamellar α, (**b**) average width of lamellar α and retained β, (**c**) volume fraction of lamellar α, and (**d**) average prior-β size.

3.2. Mechanical Properties

Figure 11 shows the tensile strength, elongation, and fracture toughness of the heat-treated TC4F samples; data are summarized in Table 4. Results showed that TC4F had a wide performance window to optimize microstructure and heat treatments. Tensile strength, yield tensile strength, elongation, and fracture toughness were in the range of 878–1038 MPa, 864–1016 MPa, 4.0–10.7%, and 89.8–109.5 MPa·m$^{1/2}$, respectively. When the mechanical properties of Samples 1 and 2 were compared, the alloy treated at a higher aging temperature was found to have lower strength, and higher plasticity and fracture toughness. When the mechanical properties of Samples 3–5 were compared, the high cooling temperature was found to result in high strength, but low plasticity and toughness.

Table 4. Tensile properties of TC4F alloy after different heat treatments.

Sample	Tensile Strength /MPa	Yield Strength /MPa	Elongation /%	Fracture Toughness /MPa·m$^{1/2}$
1	999^{+15}_{-35}	919^{+17}_{-31}	$10.4^{+.3}_{-2.4}$	$94.4^{+7.4}_{-2.0}$
2	993^{+7}_{-13}	879^{+9}_{-8}	$10.7^{+3.1}_{-1.5}$	$96.4^{+2.1}_{-5.0}$
3	1038^{+9}_{-8}	1016^{+2}_{-4}	$4.0^{+.8}_{-1.0}$	$89.8^{+2.9}_{-6.8}$
4	904^{+10}_{-7}	876^{+9}_{-18}	$5.8^{+2.4}_{-0.7}$	$99.4^{+1.7}_{-7.9}$
5	878^{+5}_{-5}	864^{+11}_{-6}	$6.8^{+.7}_{-0.6}$	$109.9^{+.5}_{-0.9}$

Figure 12 shows the fracture surface of the heat-treated TC4F alloys after the fracture-toughness test. The fracture surfaces of TC4F in Samples 1 and 2 showed plastic fractures (see Figure 12a,b). The cracked surfaces contained large dimples and the formation of secondary cracks. The fracture surfaces of Samples 3 and 4 exhibited brittle fractures (see Figure 12c,d). There were micropores and cleavage steps with strip patterns. More cracking along the lamella could be observed on the fracture surfaces. The fracture surface of Sample 5 showed a typical mixed-mode fracture (see Figure 12e). Dimples, ravines, and micropores could be observed on the fracture surface. When the fracture surfaces were compared after three treatments, Sample 5 had the largest dimples. It was evident that the high numbers of relatively large dimples and higher undulation depth were beneficial to fracture toughness since Sample 5 had the highest value in the experiment.

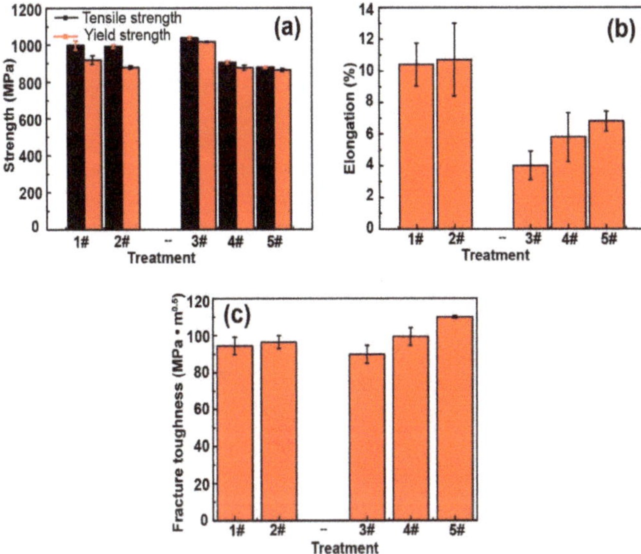

Figure 11. Mechanical properties of TC4F after heat treatments. (**a**) Strength, (**b**) elongation, and (**c**) fracture toughness.

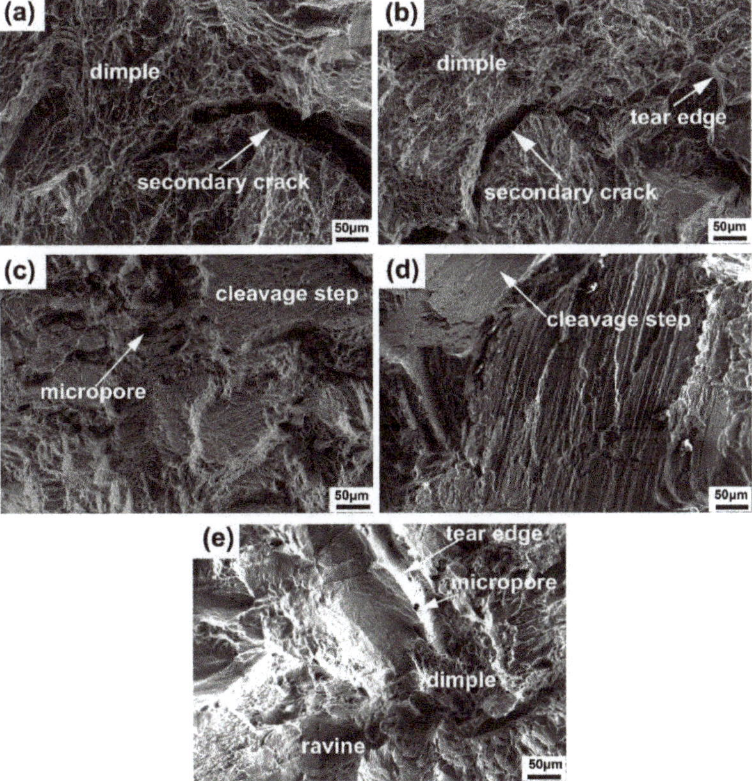

Figure 12. SEM images of fracture surface of heat-treated TC4F. Samples (**a**) 1, (**b**) 2, (**c**) 3, (**d**) 4, and (**e**) 5.

3.3. Discussion on Correlation between Microstructure and Mechanical Properties

Figure 13 depicts the correlation between microstructural features and strength. Tensile and yield strength decreased as the length and width of α, the width of the retained β, and the prior-β size increased. However, tensile and yield strength increased with the aspect ratio of α. The effect of the α fraction on strength was complicated. The TC4F with ~84 vol % α showed the highest strength values. The α phase strengthening is actually the α/β phase interface strengthening. The large area of the α/β-phase interface contributed to high strength values. In this work, due to differences in α-phase width, a high α-phase fraction did not mean a large area of the α/β-phase interface. The α/β-phase interface could be suggested by the α-fraction-to-width ratio. Figure 14 shows the plots for strength and α-fraction-to-width ratio vs. the α fraction. Variations in strength and α-fraction-to-width ratio were consistent. Therefore, the TC4F with the highest areas of α/β-phase interface had the highest strength value, although the α fraction was not maximal.

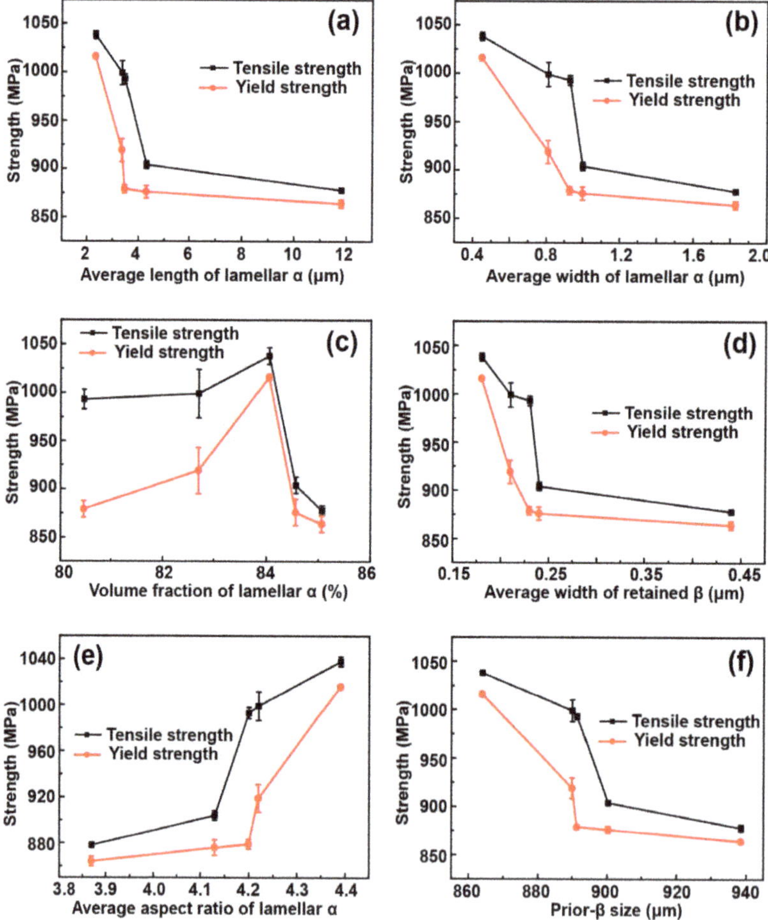

Figure 13. Variations in strength and microstructural features. (**a**) Average length, (**b**) average width, and (**c**) volume fraction of lamellar α; (**d**) average width of retained β; (**e**) average aspect ratio of lamellar α; (**f**) prior-β size.

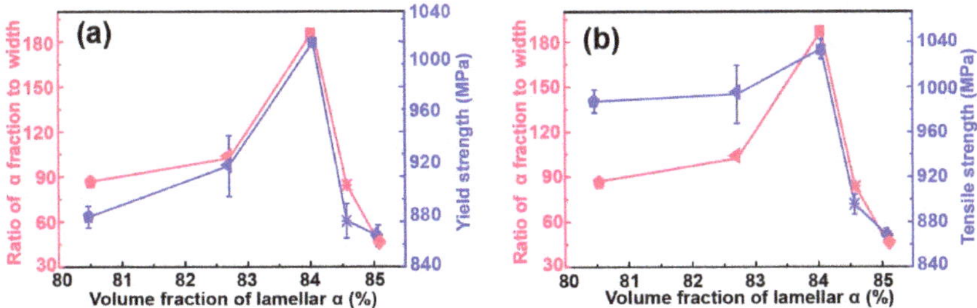

Figure 14. (**a**) Yield strength and α-fraction-to-width ratio varied with volume fraction of lamellar α. (**b**) Tensile strength and α-fraction-to-width ratio varied with volume fraction of lamellar α. Yield and tensile strengths were more closely related to α-fraction-to-width ratio.

Figure 15 depicts the relationship between microstructural features and elongation. The dependence of elongation on the microstructure was not obvious compared to that of strength. Figure 16 shows the relationship between microstructural features and fracture toughness. A coarse lamellar α, wide retained β, and large prior-β grain were beneficial to fracture toughness (see Figure 16a,b,d). The elongated fine-α lamellae might have caused stress concentration at the lamellar tips. The greater the aspect ratio of α was, the lower the fracture toughness that was exhibited (Figure 16e). Fracture toughness changed nonmonotonically with the α fraction. The TC4F with ~84 vol % α exhibited the lowest fracture toughness. The reason for this low fracture toughness could be because it had the highest area of α/β-phase interface.

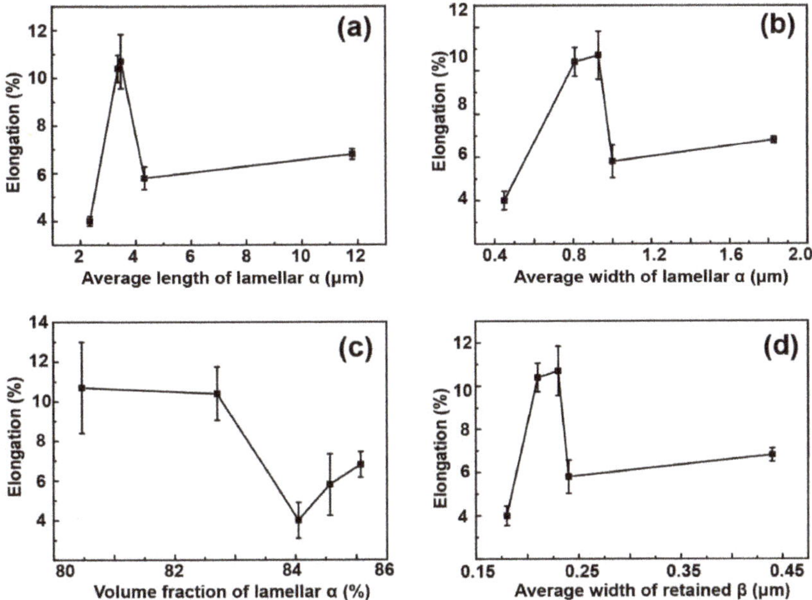

Figure 15. *Cont.*

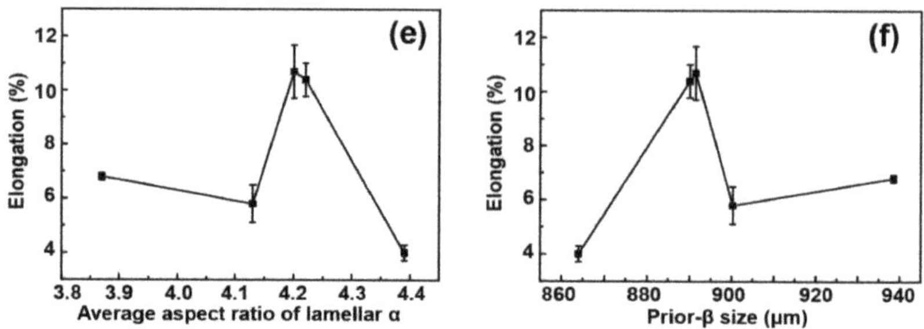

Figure 15. Variations in elongation and microstructural features. (**a**) Average length, (**b**) average width, and (**c**) volume fraction of lamellar α; (**d**) average width of retained β; (**e**) average aspect ratio of lamellar α; and (**f**) prior-β size.

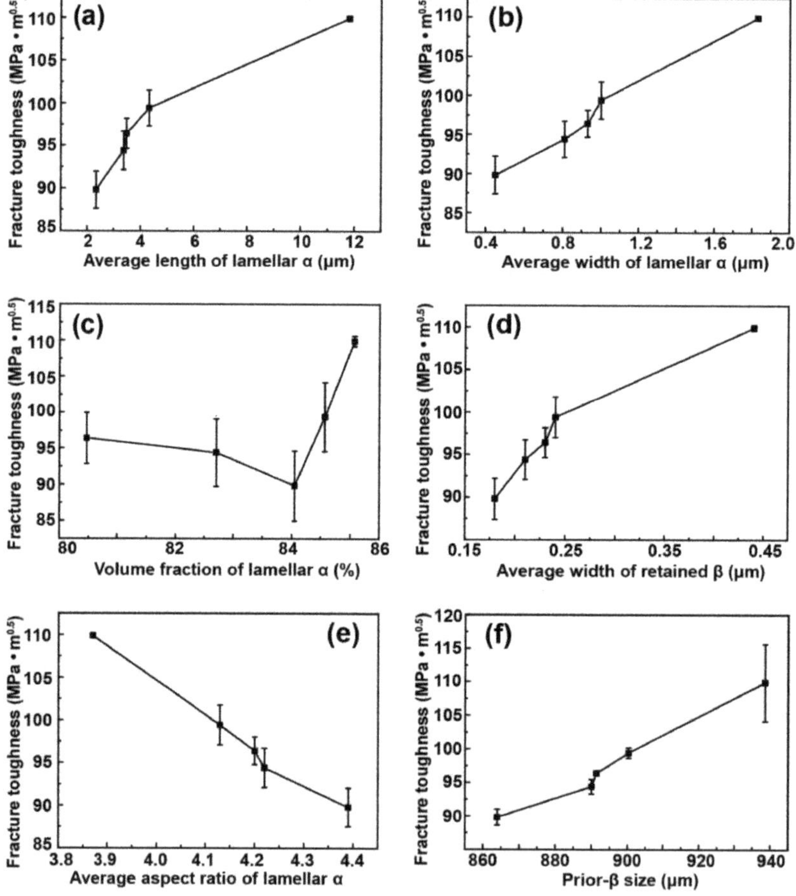

Figure 16. Variations in fracture toughness and microstructural features. (**a**) Average length, (**b**) average width, and (**c**) volume fraction of lamellar α; (**d**) average width of retained β; (**e**) average aspect ratio of lamellar α; and (**f**) prior-β size.

Therefore, optimal comprehensive mechanical performance could be assessed on the basis of the above analysis. Optimal performance was obtained from Sample 1 with respect to tensile strength, yield strength, elongation, and fracture toughness, which were measured at 999 and 919 MPa, 10.4%, and 94.4 MPa·m$^{1/2}$, respectively. Performance was generated by β-annealing at 1005 °C for 70 min, followed by air cooling to room temperature, aging at 722 °C for 2 h, and air cooling to room temperature again (treatment for Sample 1).

Figure 17 compares the combinations of yield strength/fracture toughness of the heat-treated samples in this work with BASCA heat-treated TC4 [24]. The fracture toughness of TC4F was superior to that of TC4. The fracture toughness of BASCA heat-treated TC4 was below 80 MPa·m$^{1/2}$, while the fracture toughness of TC4F could reach 110 MPa·m$^{1/2}$. This further demonstrates that fracture toughness was improved in TC4F through the addition of Fe. Thus, the TC4F shows promise in marine and aeronautical applications.

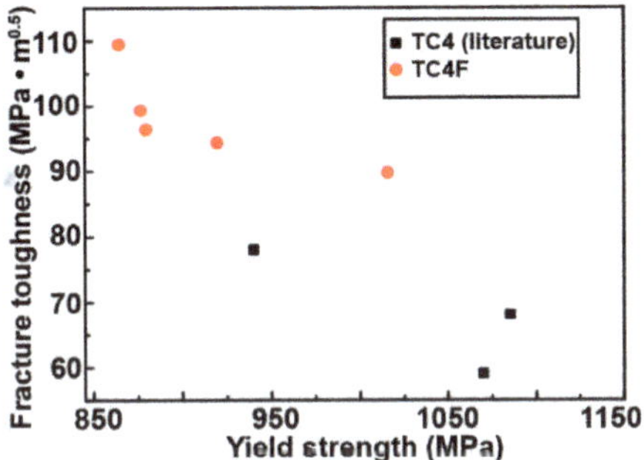

Figure 17. Comparison of mechanical properties BASCA heat-treated TC4F and TC4 [24].

4. Conclusions

1. The heat-treatment procedure had a significant effect on the microstructure. A higher aging temperature facilitated the growth of the α phase. The cooling period after solution treatment mainly affected the size of the precipitated α phase during the subsequent aging process. The higher the cooling rate was, the smaller the α-phase size was.

2. Aging temperature and cooling rate had an impact on mechanical properties. High aging temperatures and low cooling rates led to low strength, and high plasticity and fracture toughness.

3. Tensile and yield strength decreased as the length and width of α, the width of retained β, and prior-β size increased. However, strengths increased with the aspect ratio of lamellar α. Fracture toughness showed the opposite tendency, that is, it increased as the length and width of α, the width of the retained β, and prior-β size increased, and as the aspect ratio of lamellar α decreased. The effect of the α fraction on strength and fracture toughness was complex. The TC4F alloy with ~84 vol % α exhibited the highest strength and lowest fracture toughness because it had the highest area of α/β-phase interface. The dependence of plasticity on the microstructure was not obvious compared to that of the strength and fracture toughness.

4. Optimal comprehensive mechanical performance was achieved via β-annealing at 1005 °C for 70 min, followed by air cooling to room temperature, aging at 722 °C for 2 h, and air cooling to room temperature again. Tensile strength, yield strength, elongation, and fracture toughness reached values of 999 and 919 MPa, 10.4%, and 94.4 MPa·m$^{1/2}$, respectively.

Author Contributions: Conceptualization, H.C. and Y.C.; methodology, F.C. and G.X.; validation, F.C. and G.X.; formal analysis, Y.L.; investigation, Y.L.; data curation, Y.L.; writing—original-draft preparation, Y.L.; writing—review and editing, F.C.; visualization, F.C. and G.X.; project administration, Y.C. and H.C.; funding acquisition, F.C. and H.C. All authors have read and agreed to the published version of the manuscript.

Funding: This research was funded by the National Natural Science Foundation of China, grant number 51801101; and by the Primary Research and Development Plan of Jiangsu Province, grant number BE2019119.

Conflicts of Interest: The authors declare no conflict of interest.

References

1. Leyens, C. Advanced materials and coatings for future gas turbine applications. In Proceedings of the 24th International Congress of the Aeronautical Sciences, Yokohama, Japan, 29 August–3 September 2004.
2. Nalla, R.K.; Ritchie, R.O.; Boyce, B.L.; Campbell, J.P.; Peters, J.O. Influence of microstructure on high-cycle fatigue of Ti-6Al-4V: Bimodal vs. lamellar structures. *Metall. Mater. Tran. A* **2002**, *33*, 899–918. [CrossRef]
3. Kermanpur, A.; Amin, H.S.; Ziaei-Rad, S.; Mosaddeghfar, M. Failure analysis of Ti-6Al-4V gas turbine compressor blades. *Eng. Fail. Anal.* **2008**, *15*, 1052–1064. [CrossRef]
4. Leyens, C.; Peters, M. *Titanium and Titanium Alloys: Fundamentals and Applications*; Wiley-VCH: Weinheim, Germany, 2003.
5. Sen, I.; Tamirisakandala, S.; Miracle, D.B.; Ramamurty, U. Microstructural effects on the mechanical behavior of B-modified Ti–6Al–4V alloys. *Acta Mater.* **2007**, *55*, 4983–4993. [CrossRef]
6. Meier, M.L.; Lesuer, D.R.; Mukherjee, A.K. The effects of the α/β phase proportion on the superplasticity of Ti-6Al-4V and iron-modified Ti-6Al-4V. *Mater. Sci. Eng. A* **1992**, *154*, 165–173. [CrossRef]
7. Fuwen, C.; Yulei, G.; Guanglong, X.; Yuwen, C.; Hui, C.; Lian, Z. Improved fracture toughness by microalloying of Fe in Ti-6Al-4V. *Mater. Design.* **2020**, *185*, 108251.
8. Shen, G.; Furrer, D. Manufacturing of aerospace forgings. *Mater. Process. Tech.* **2000**, *98*, 189–195. [CrossRef]
9. Liu, C.; Liu, D.; Zhang, X.; Liu, D.; Ma, A.; Ao, N.; Xu, X. Improving fatigue performance of Ti-6Al-4V alloy via ultrasonic surface rolling process. *Mater. Sci. Tech.* **2019**, *35*, 1555–1562. [CrossRef]
10. Carroll, B.E.; Palmer, T.A.; Beese, A.M. Anisotropic tensile behavior of Ti–6Al–4V components fabricated with directed energy deposition additive manufacturing. *Acta Mater.* **2015**, *87*, 309–320. [CrossRef]
11. Baufeld, B.; Van der Biest, O.; Gault, R. Additive manufacturing of Ti–6Al–4V components by shaped metal deposition: Microstructure and mechanical properties. *Mater Design.* **2010**, *31*, S106–S111. [CrossRef]
12. Tan, X.; Kok, Y.; Tan, Y.J.; Descoins, M.; Mangelinck, D.; Tor, S.B.; Leong, K.F.; Chua, C.K. Graded microstructure and mechanical properties of additive manufactured Ti–6Al–4V via electron beam melting. *Acta Mater.* **2015**, *97*, 1–16. [CrossRef]
13. Zhao, X.; Li, S.; Zhang, M.; Liu, Y.; Timothy, B.; Wang, S.; Hao, Y.; Yang, R.; Murr, L.E. Comparison of the microstructures and mechanical properties of Ti–6Al–4V fabricated by selective laser melting and electron beam melting. *Mater. Design.* **2016**, *95*, 21–31. [CrossRef]
14. Harun, W.S.W.; Manam, N.S.; Kamariah, M.S.I.N.; Sharif, S.; Zulkifly, A.H.; Ahmad, I.; Miura, H. A review of powdered additive manufacturing techniques for Ti-6Al-4V biomedical applications. *Powder Technol.* **2018**, *331*, 74–97. [CrossRef]
15. Bermingham, M.J.; McDonald, S.D.; StJohn, D.H.; Dargusch, M.S. Segregation and grain refinement in cast titanium alloys. *Int. J. Mater. Res.* **2009**, *24*, 1529–1535. [CrossRef]
16. Lütjering, G.; Williams, J.C. *Titanium*; Springer Science & Business Media: Berlin/Heidelberg, Germany, 2007.
17. Donachie, M.J. *Titanium: A Technical Guide*; ASM International: Cleveland, OH, USA, 2000.
18. Banerjee, S.; Mukhopadhyay, P. *Phase Transformations: Examples from Titanium and Zirconium Alloys*; Elsevier: Amsterdam, The Netherlands, 2007.
19. Guo, P.; Zhao, Y.; Zeng, W.; Hong, Q. The effect of microstructure on the mechanical properties of TC4-DT titanium alloys. *Mat. Sci. Eng. A* **2013**, *563*, 106–111. [CrossRef]
20. Securities Association of China (SAC). *Method of β Transus Temperature Determination of Titanium Alloys*; Securities Association of China: Beijing, China, 2010.
21. Boyer, R.R. *Titanium and Its Alloys: Metallurgy, Heat Treatment and Alloy Characteristics. Encyclopedia of Aerospace Engineering*; John Wiley & Sons, Ltd.: Hoboken, NJ, USA, 2010.
22. American Society for Testing and Materials (ASTM). *Standard Test Methods for Tension Testing of Metallic Materials*; American Society for Testing and Materials International: West Conshohocken, PA, USA, 2009.
23. American Society for Testing and Materials (ASTM). *Standard Test Method for Linear-Elastic Plane-Strain Toughness KIC of Metallic Materials*; American Society for Testing and Materials International: West Conshohocken, PA, USA, 2006.
24. Welsch, G.; Boyer, R.; Collings, E.W. *Materials Properties Handbook: Titanium Alloys*; ASM International: West Conshohocken, PA, USA, 1993.

 *metals*

Article

Effect of *abc* Pressing at 573 K on the Microstructure and Martensite Transformation Temperatures in Ti$_{49.8}$Ni$_{50.2}$ (at%)

Oleg Kashin *, Aleksandr Ivanovich Lotkov, Victor Grishkov, Konstantin Krukovskii, Dorzhima Zhapova, Yuri Mironov, Natalia Girsova, Olga Kashina and Elena Barmina

Institute of Strength Physics and Materials Science of the Siberian Branch of the Russian Academy of Sciences, 634055 Tomsk, Russia; lotkov@ispms.ru (A.I.L.); grish@ispms.tsc.ru (V.G.); kvk@ispms.tsc.ru (K.K.); dorzh@ispms.tsc.ru (D.Z.); myp@ispms.ru (Y.M.); girsova@ispms.tsc.ru (N.G.); ocash@ispms.tsc.ru (O.K.); barmina@ispms.tsc.ru (E.B.)
* Correspondence: okashin@ispms.ru; Tel.: +7-382-228-69-20

Abstract: This paper presents experimental data on the microstructure and martensite transformation temperatures of Ti$_{49.8}$Ni$_{50.2}$ (at%) after *abc* pressing (multi-axial forging) to different true strains *e* from 1.84 to 9.55 at 573 K. The data show that increasing the true strain results in grain–subgrain refinement on different scales at a time. With *e* = 9.55 at 573 K, the average grain–subgrain size measured approximately 130 nm. Decreasing the *abc* pressing temperature from 723 to 573 K caused a decrease in all martensite transformation temperatures, a change in the lattice parameters, R phase formation, and angular shifts of diffraction peaks and their broadening. The largest change in the microstructure of Ti$_{49.8}$Ni$_{50.2}$ was provided by *abc* pressing to *e* = 1.84. Increasing the true strain to *e* = 9.55 resulted in a much smaller effect, suggesting that the alloy obtained a high density of structural defects even at *e* = 1.84. Two possible mechanisms of grain–subgrain refinement are discussed.

Keywords: TiNi; isothermal *abc* pressing; microstructure; grain–subgrain refinement; martensite transformation temperatures

Citation: Kashin, O.; Lotkov, A.I.; Grishkov, V.; Krukovskii, K.; Zhapova, D.; Mironov, Y.; Girsova, N.; Kashina, O.; Barmina, E. Effect of *abc* Pressing at 573 K on the Microstructure and Martensite Transformation Temperatures in Ti$_{49.8}$Ni$_{50.2}$ (at%). *Metals* **2021**, *11*, 1145. https://doi.org/10.3390/met 11071145

Academic Editor: Martin Bache

Received: 24 June 2021
Accepted: 17 July 2021
Published: 20 July 2021

Publisher's Note: MDPI stays neutral with regard to jurisdictional claims in published maps and institutional affiliations.

1. Introduction

The mechanical and functional properties of polycrystalline metals and alloys are determined in many respects by the degree of their grain–subgrain refinement [1,2] and, hence, the average size of grains–subgrains can be considered one of the key microstructural factors responsible for almost all characteristics of metals and alloys. Among the promising methods of grain refinement is severe plastic deformation (SPD). By now, abundant experimental data are available showing that such refinement increases the hardness of materials [1–8]. The main outcome desired from SPD is an optimum average grain size that can provide a material with a high strength and acceptable plasticity. Another no less important outcome is a minimum possible grain and average grain–subgrain size, which is of interest for fundamental research in grain refinement mechanisms and for practical use in SPD materials. Obviously, achieving such outcomes needs knowledge of grain refinement mechanisms in metals and alloys exposed to severe plastic deformation. As has been shown [8], the refinement of grains under severe plastic deformation is controlled by two main sets of parameters: (1) SPD parameters, including the strain, its rate, temperature, and path; (2) material parameters, including the initial average grain size and staking fault energy.

Certainly, because of numerous factors of influence, it is challenging to propose a unified grain refinement mechanism for all metals and alloys and for all SPD methods. In available reviews [1–8], one can find brief descriptions of grain refinement models and mechanisms in metals and alloys exposed to different SPD methods and references to respective original works.

79

Some of these models describe the division of a grain in terms of partial-disclination nucleation and motion and associated misorientation of adjacent grain fragments due to the plastic strain accommodation in the grain. It is assumed that on approaching a certain critical grain size, diffusion accommodation of disclinations starts to dominate and no further grain refinement is possible. According to estimates [9], the relative temperature of transition from one mechanism of partial-disclination elastic energy accommodation to another during equal channel angular pressing (ECAP) is $\approx 0.65 T_m$ for fcc metals and $0.33 T_m$ for bcc ones (T_m is the melting temperature). Increasing the SPD temperature increases the minimum possible grain size.

The most widespread grain refinement models are those for metals and alloys that acquire a cellular dislocation substructure early in their deformation [1,8]. This type of model suggests that dislocations and other crystalline defects (vacancies, interstitial atoms) appear, move, and interact in SPD materials. Increasing the dislocation density in a grain results in a cellular dislocation substructure with low-angle cell boundaries and increasing the dislocation density in cell walls causes misorientation accumulation between adjacent cells with gradual transformation of low-angle boundaries or at least its large part into high-angle grain boundaries. As the strain is further increased, this process occurs in newly formed fine grain (finer than initial), ensuring additional grain refinement. It is assumed that the average dislocation cell size, d, is inversely proportional to the square root of the total dislocation density, ρ [1]:

$$d \sim \rho^{-1/2}, \tag{1}$$

From this assumption, the average size of grains during their refinement cannot reach values smaller than 100 nm [1]. Besides, at a certain critical grain size d_c, diffusion accommodation of dislocations in cell walls dominates [8], and their density increases slightly. The critical grain size, d_c, decreases with an increasing strain rate and increases with increasing temperature.

Actually, the minimum grain size attainable in an SPD process is determined by a certain dynamic balance between the generation of dislocations into the volume of a grain and their recovery from the volume via their escape to grain–subgrain boundaries or annihilation of dislocations with Burgers vectors of opposite sign. One of the mechanisms of dynamic dislocation recovery from cell walls can be non-conservative transverse dislocation glide with vacancy diffusion.

In materials highly prone to twinning at high strain rates and/or low strain temperatures, the mechanism of grain refinement includes the formation of packets of nanotwins that transform into nanograins via twin or matrix lamella fragmentation due to the twin boundary interaction with dislocations or shear bands [8]. In such refinement, the minimum grain size is much smaller than that ensured by gradual transformation of dislocation cell structures into an ultrafine grain structure due to the dislocation glide and diffusion motion of point defects.

The above examples of SPD-induced grain refinement are not exhaustive and other models are surely available. Nevertheless, these examples allow quite definite conclusions about the influence of individual factors on the mechanisms of grain structure transformation under SPD. In particular, there is no doubt that decreasing the SPD temperature leads to more efficient grain refinement. However, at relatively low temperatures, the load on equipment increases steeply and failure of test specimens becomes a risk. Therefore, in most related studies, the strain temperature is somewhat lower or higher than $T = 0.4 T_m$, where T_m is the melting temperature of a metal or alloy, i.e., the deformation is warm.

Note that the developed models are based mainly on experimental studies of copper, aluminum, nickel, titanium, iron, and their alloys, and the applicability of the models to TiNi is left almost untouched [10–39]. In our opinion, this is due to the insufficient systematic experimental data on TiNi alloys, and, in particular, on their structure and properties at different SPD temperatures. Bridging this gap needs experiments on TiNi alloys of specified compositions for analyzing the formation of their microstructure and physico-mechanical properties under the same SPD conditions but at different strain temperatures.

By now, a number of studies of TiNi alloys, including near-equiatomic binary ones, are available to grasp the idea of their grain refinement at different SPD temperatures. For example, research data on TiNi alloys exposed to high strains by ECAP at 673–823 K [10–13] show that the alloys at temperatures above 773 K experience an intense dynamic recrystallization characteristic for increased temperatures, which does not allow for the formation of their ultrafine-grained (UFG) structure, and that the optimum ECAP temperature for their efficient grain refinement with no risk of failure of technological equipment is 723 K. Increasing the number of ECAP passes to eight provides the formation of an UFG structure with an average grain size of $\leq$0.25 μm, but the formation of a substantial nanocrystalline fraction with a grain size of $\leq$100 nm at this temperature fails [13]. It is supposed that for the formation of a nanocrystalline structure in bulk TiNi alloys, the ECAP temperature should be decreased to below 623 K [12]. Another paper reports that even in stainless-steel shells, Ti–50Ni and Ti–50.1Ni (at%) fail when deformed by ECAP at room temperature [14]. Unfortunately, no data on grain sizes are reported in the cited paper.

Research data are also available on the grain–subgrain structure of $Ti_{49.8}Ni_{50.2}$ specimens exposed to *abc* pressing (multi-axial forging) with their true strain set from 2.2 to 1.5 or equal to $\approx$3.6 at each step of the strain temperature successively decreased from 873 K to 573 K [15–18]. Actually, such pressing conditions mean that each temperature step of *abc* pressing is applied to specimens with a different initial grain–subgrain structure. The research data show that after *abc* pressing at 673 K, the alloy is dominated by a submicrocrystalline structure with a grain–subgrain size of 300–700 nm, and after *abc* pressing at 573 K (total true strain $e = 7.7$), the submicrocrystalline structure is refined to a grain–subgrain size of 100–500 nm, and at strain band intersections, to 2–100 nm with an estimated nanocrystalline fraction of no more than 30%.

After *abc* pressing to $e = 8.44$ at 723 K [19–22], this type of alloy acquires an inhomogeneous grain–subgrain structure with an average grain size of approximately 1 μm that comprises fine-grained, submicrocrystalline, and nanocrystalline fractions. It is noted that intense recovery processes develop in the alloy *abc* pressed at 723 K.

Thus, the previous studies show that equal channel angular pressing and *abc* pressing at a temperature of 773–673 K provide binary TiNi-based alloys mostly with a submicrocrystalline structure having an average grain–subgrain size of <1 μm.

Reasoning from these studies, it might be expected that isothermal severe plastic deformation at temperatures lower than the temperature range of warm deformation can provide more substantial grain refinement in TiNi-based alloys and a larger effect on their physico-mechanical properties, including inelastic ones. Here, we analyzed the grain–subgrain structure, phase state, and martensite transformation temperatures in $Ti_{49.8}Ni_{20.2}$ (at%) *abc* pressed to different true strains at 573 K.

2. Materials and Methods

The test material was $Ti_{49.8}Ni_{50.2}$ (at%) supplied by MATEK-SMA Ltd. (Moscow, Russia). Its specimens were made as described in detail elsewhere [22]. The initial specimens, which were shaped as a cube with a side of 20 mm via a single *abc* pressing cycle at 1073 K, were placed in dies, aged at 573 K for 10 min, and compressed on a hydraulic press with a rate of 0.16–0.18 s^{-1}. The decrease in the specimen temperature at the end of each compression event was no greater than 10 K. Each *abc* pressing cycle included compressive hits in three mutually perpendicular directions. After each compression event, the specimens were removed from their dies and again placed in them without cooling to deform the specimens in a direction perpendicular to the previous one. The cycle of three compression events was repeated several times to bring the specimens to their specified strain. After single pressing in one direction, the true strain was from $e \approx 0.15$ to $e \approx 0.30$. Thus, the specimens were deformed to a true strain of 1.84, 3.60, 5.40, 7.43, and 9.55.

The temperatures of martensite transformations were determined at an experimental set (ISPMS SB RAS, Tomsk, Russia) by a four-point method of measuring the temperature dependence of the electrical resistivity. The samples were made in the form of rods with a square cross-section of 1×1 mm^2 and a length of 20 mm. Two current and two potential electrodes were fixed to the sample by spot welding. The sample temperature was measured with a chromel–alumel thermocouple. The sample temperature was changed at a cooling/heating rate of 3 K/min. The temperatures of martensitic transformations were determined by tangents to the graphs of the temperature dependence of the electrical resistivity. The accuracy of estimating the martensitic transformations temperatures using this method was ± 2 K.

The microstructure of the alloy was analyzed on an Axiovert-200M inverted light microscope (Carl Zeiss AG, Oberkochen, Germany), a LEO EVO 50 XVP scanning electron microscope (Carl Zeiss AG, Oberkochen, Germany), and a JEM-2100 transmission electron microscope (JEOL Ltd., Tokyo, Japan). All equipment was provided by the NANOTECH Shared Use Center of ISPMS SB RAS (Tomsk, Russia). For the analysis, specimen slices were cut on electro-erosion machine. The slices were mechanically polished on a Saphir 350 grinding and polishing machine (Audit Diagnostics, Business & Technology Park, Carrigtwohill, Co., Cork, Ireland) using a SiC abrasive with a gradual decrease in its grit to 1200, finished with a diamond paste grained to 3 µm and etched in a mixture of nitric acid, hydrofluoric acid, and water with a ratio of 1:4:5.

For transmission electron microscopy (TEM), thin foils were prepared for which thin plates of a thickness of 0.5 mm were cut from the *abc* pressed specimens by electro-erosion cutting and were then either polished in an electrolyte containing sulfuric, nitric, and hydrofluoric acids in a ratio 6:1:3 or etched on an EM-09 100 15 ion slicer (JEOL Ltd., Tokyo, Japan). The crystalline structure of the alloy and its phase state were also analyzed by X-ray diffraction (XRD) on a DRON-7 diffractometer with the PDWin software (JSC Burevestnik, Russia) in Co-Kα radiation at room temperatures. The phase compositions of the samples were analyzed using crystallographic data on the structures of the phases B19$'$ (space group P2$_1$/m [40]) and R (space group P3 [41]). In the analysis, the ratio of the intensity of the doublet lines K$_{\alpha 2}$ and K$_{\alpha 1}$ was assumed at 0.52. The asymmetry of the intensity of the reflex profile was assumed at 1.1. The separation of the doublet $(112)_R$ and $(300)_R$ R phases was determined by the minimum value of the standard deviations of the sum of the profiles of these components from the experimental profile. The parameters of the crystal lattices of the phases were calculated using the positions of the maxima of the reflex profiles after separation. The B19$'$ lattice parameters were estimated from 10 reflections in the deformed alloy and from 14 reflections in the initial one. The R lattice parameters were estimated from reflections (11-2) and (300). Errors in the lattice parameters are shown in corresponding dependences on true strain.

3. Results

3.1. Temperatures of Martensite Transformation

Figure 1 shows the temperatures of martensite transformations in Ti$_{49.8}$Ni$_{50.2}$ versus the true *abc* strain. Here, M$_S$ and M$_f$ are the start and finish temperatures of direct B2$\rightarrow$R$\rightarrow$B19$'$ transformation on cooling, A$_S$ and A$_f$ are the start and finish temperatures of reverse martensite transformation on heating, T$_R$ is the start temperature of B2$\rightarrow$R transformation, and T$_{room}$ is the room temperature (dashed line).

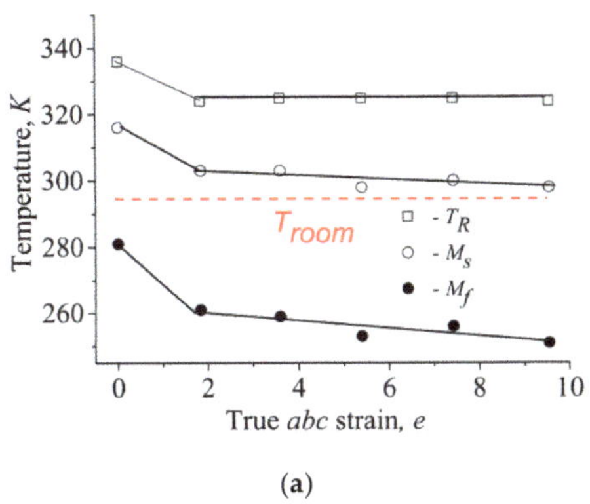
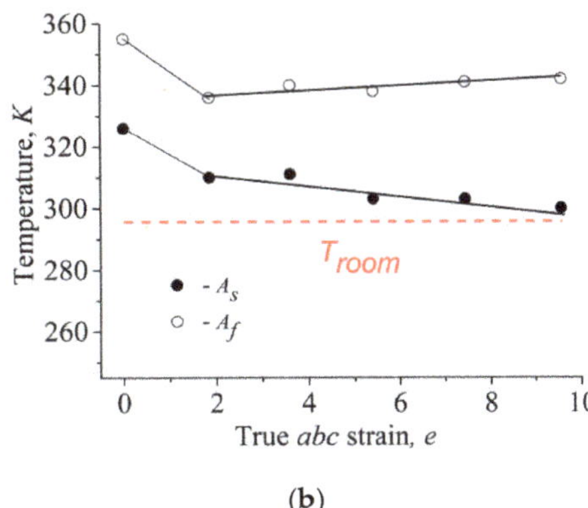

(a)

(b)

Figure 1. Martensite transformation temperatures versus true *abc* strain in Ti$_{49.8}$Ni$_{50.2}$: (**a**) on cooling; (**b**) on heating.

As can be seen from Figure 1a, the sequence of direct transformations in all specimens was B2→R→B19′. It can also be seen that the martensite transformation temperatures decreased steeply, even after *abc* pressing to $e = 1.84$. As the true strain further increased, the temperature T$_R$ remained almost unchanged, and the temperatures M$_S$ and M$_f$ decreased linearly with *e*. The temperature interval between M$_S$ and M$_f$ remained nearly constant for all values of *e*.

From Figure 1b, it can be seen that the temperatures A$_S$ and A$_f$ first decreased steeply, and as the strain increased, the temperature A$_S$ decreased slightly from 310 K in the specimens with $e = 1.84$ to 300 K in those with $e = 9.55$, whereas the temperature A$_f$ increased somewhat. Thus, increasing the true strain *e* increased the temperature interval between A$_S$ and A$_f$.

3.2. Phase States

Figure 2 shows fragments of the XRD spectra recorded in Ti$_{49.8}$Ni$_{50.2}$ at room temperature before and after *abc* pressing to different true strains *e*.

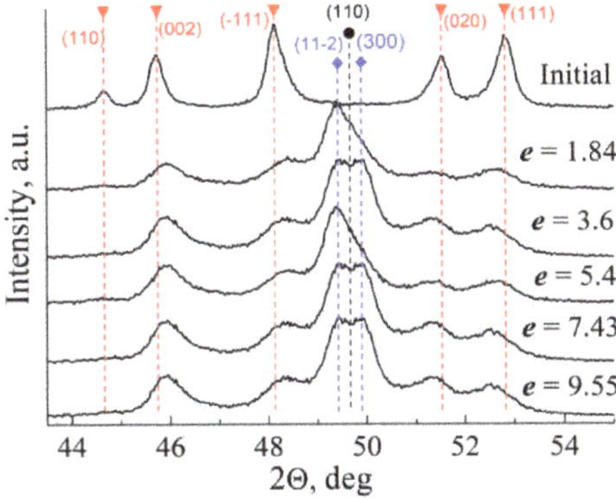

Figure 2. Fragments of XRD spectra of Ti$_{49.8}$Ni$_{50.2}$ at room temperature (Co-Kα radiation) after *abc* pressing to different true strains: ▼—B19′, ♦—R, and ●—B2.

At room temperature, the initial alloy revealed B19′ reflections and traces of R reflections, while any B2 reflections were absent. Another situation was observed after *abc* pressing to $e = 1.84$. First, at room temperature, the main volume of all specimens contained

R and B19′ phases. Second, the positions of B19′ peaks were markedly shifted compared to those in the initial alloy. As the strain *e* increased, their angular position changed little.

Figure 3 shows the R and B19′ lattice parameters at different true strains. It can be seen that at *e* = 1.84, the B19′ lattice parameter *a* decreased somewhat, and further, it weakly depended on *e* (Figure 3a). At the same time, the parameters *b*, *c*, and the angle β at *e* = 1.84 changed steeply, and further, their *e* dependence weakened (Figure 3b–d). The R lattice parameters *a* and *b* after *abc* pressing remained almost the same as before *abc* pressing (Figure 3e), whereas the parameter *c* decreased markedly at *e* = 1.84, and further, it weakly depended on *e* (Figure 3f).

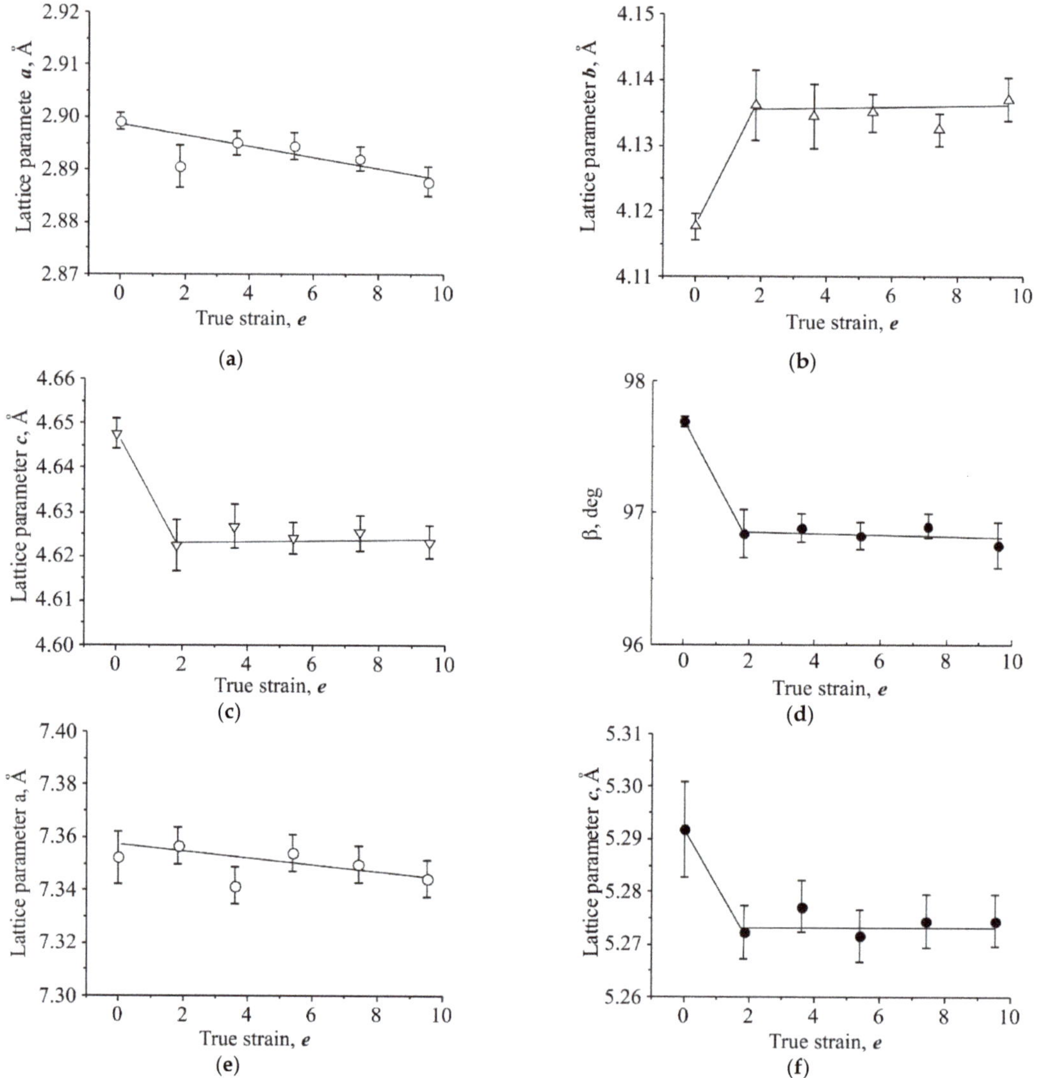

Figure 3. Effect of true *abc* strain in Ti$_{49.8}$Ni$_{50.2}$ on lattice parameters *a* (**a**), *b* (**b**), *c* (**c**), and monoclinic angle β (**d**) of the B19′ phase and the lattice parameters *a* (**e**) and *b* (**f**) of the R phase.

Figure 4 shows the full-width at half maximum (FWHM) of the most intense B19′ peaks at different true strains. It can be seen that at *e* = 1.84, the FWHM of these peaks increased steeply, and further, their weak dependence on *e* can be observed.

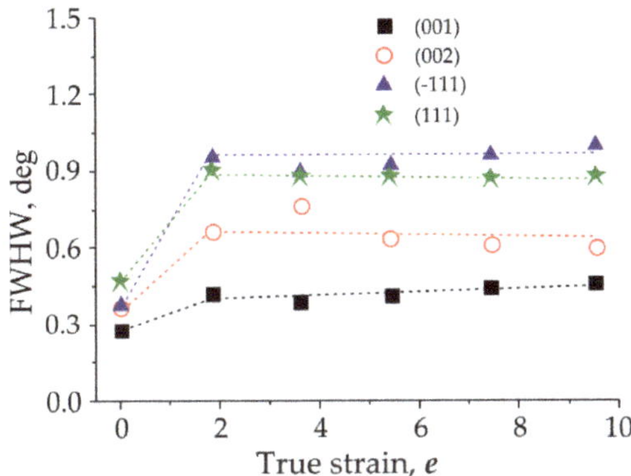

Figure 4. The full-width at half maximum (FWHM) of the most intense B19′ peaks versus true *abc* strain in Ti$_{49.8}$Ni$_{50.2}$.

3.3. Microstructure

The microstructure of Ti$_{49.8}$Ni$_{50.2}$ before and after *abc* pressing with different true strains can be evaluated from its images in Figure 5. Such images taken by differential interference contrast (DIC) microscopy were used to estimate the average grain size. First, the number of grains within an image was determined and the average grain area was calculated; then, the grain shape was approximated by squares with a side that was taken equal to the average grain size. The initial alloy revealed rather clear-cut grain boundaries (Figure 5a) such that the number of grains was well identifiable. In some grains, the deformation relief was represented by rather thin similarly directed bands, and in the others, by two systems of intersecting bands (red lines). In their adjacent grains, the band direction was different. It is not improbable that martensite plates are in the images. As the *abc* strain increased, the identification of grain boundaries by DIC microscopy became difficult (Figure 5b), and at *e* = 9.55, they were almost invisible (Figure 5c). Therefore, highly magnified SEM images were used for their identification at such strains.

Figure 5. DIC images of Ti$_{49.8}$Ni$_{50.2}$ before (**a**) and after *abc* pressing to *e* = 3.6 (**b**) and *e* = 9.55 (**c**).

The grain size in Ti$_{49.8}$Ni$_{50.2}$ at $e = 5.4$ can be evaluated from its SEM image in Figure 6a and the dependence of the average grain size on the true strain from Figure 6b. The dependence suggests that the most noticeable change in the average grain size falls on $e = 1.84$. As e increased, the average grain size decreased linearly with e, and this decrease was slight compared to that at $e = 1.84$. Such changes occurred in the coarse grain structure of the alloy at $e = 9.55$, and the average grain size was $D_{av} \approx 12$ µm.

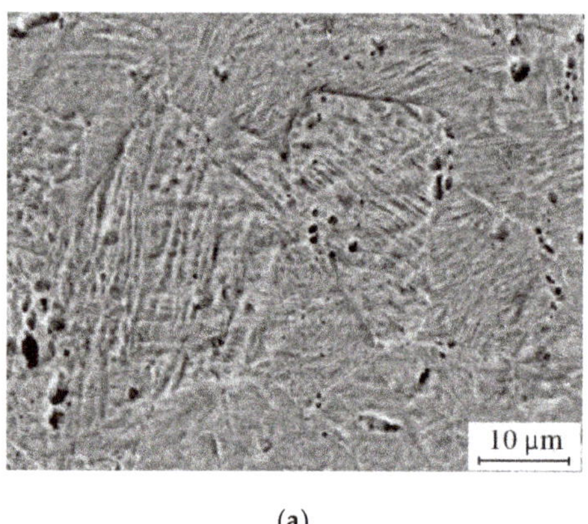

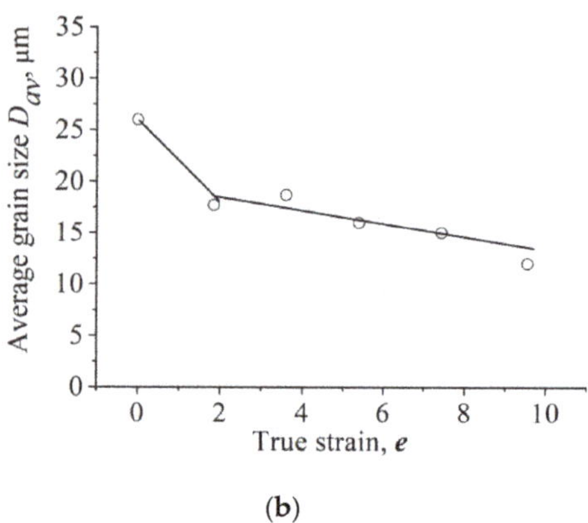

(a)
(b)

Figure 6. SEM image of Ti$_{49.8}$Ni$_{50.2}$ with a true *abc* strain $e = 5.4$ (**a**) and e dependence of the average grain size D_{av} (**b**).

The influence of *abc* pressing on the microstructure and phase state of the alloy was analyzed by transmission electron microscopy and selected area diffraction, which confirmed the results of our XRD analysis. At room temperature, the initial and *abc* pressed specimens, irrespective of the true strain, contained R and B19′ phases. Their reflections in the initial alloy and alloy *abc* pressed at $e = 9.55$ are marked in Figure 7c and Figure 7g,j, respectively. In all specimens, no B2 phase was found.

As can be seen from bright-field TEM images (Figure 7a,d–f,i), the microstructure of all specimens, whether initial or *abc* pressed, featured differently directed bands of length 1–4 µm and width 0.5 µm, which were particularly evident at rather low magnification (Figure 7e). Such bands were also distinguished on many dark-field TEM images (Figure 7b). From comparison of the characteristic band sizes on DIC and SEM images (Figures 5 and 6a) it follows that such a band microstructure is the internal structure of bands of the deformation relief. In this sense, the bands of the deformation relief can be considered as mesoscale structural elements.

Our analysis shows that the initial specimens are normally characterized by R and B19′ point reflections on their microdiffraction patterns. Such a pattern from the R region, circled in Figure 7a, is shown in Figure 7c. As the true strain increased, the microdiffraction pattern from the region selected with a diaphragm with a diameter ≈ 1 µm (circled in Figure 7a) gradually took the form of a ring (Figure 7e,g), which is indicative of microscale grain–subgrain refinement. With a diaphragm with a diameter of 115 nm, the corresponding region (circled in Figure 7i) showed only point reflections (Figure 7j), suggesting that the size of the microscale grains–subgrains formed at $e = 9.55$ was at least 100 nm.

The average grain–subgrain size, d_{av}, was determined from dark-field images such as those shown in Figure 7b,h,k. First, the absolute area of each luminous region was determined and then all such regions were approximated by a square with a side length that was taken equal to d_{av}. The results are presented in Figure 8. As can be seen from the figure, the average grain–subgrain size d_{av}, according to TEM, decreased almost linearly with the increase in the true strain, and at $e = 9.55$, it measured approximately 130 nm.

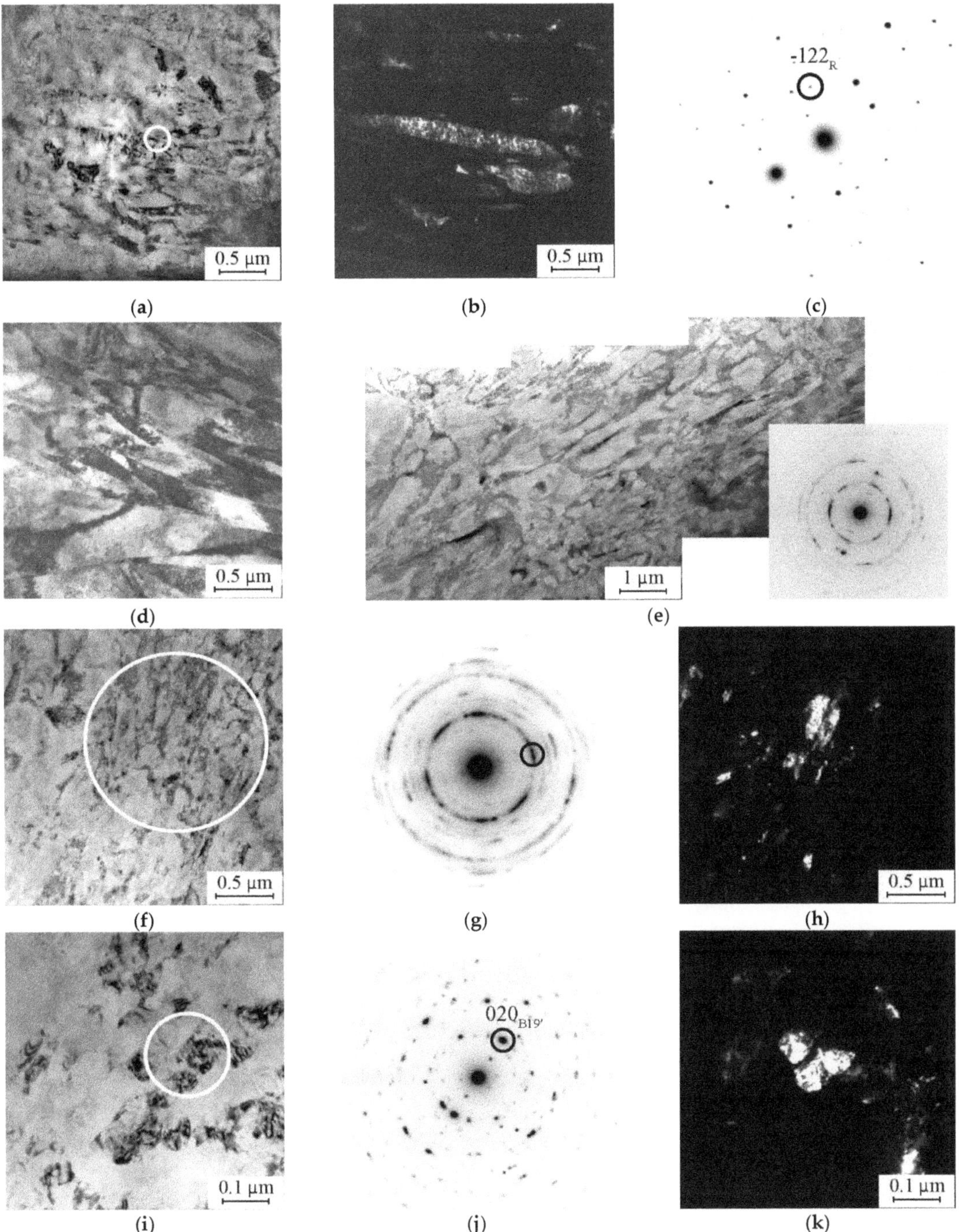

Figure 7. TEM images and diffraction patterns of Ti$_{49.8}$Ni$_{50.2}$ before (**a**–**c**) and after *abc* pressing to e = 1.84 (**d**), e = 5.4 (**e**), and e = 9.55 (**f**–**k**).

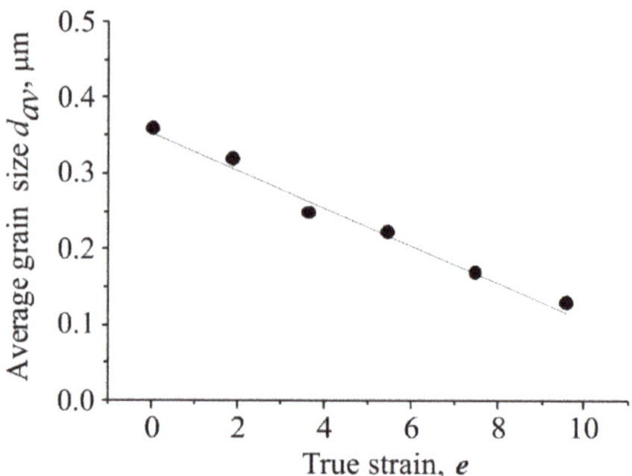

Figure 8. Average grain–subgrain size d_{av} versus true *abc* strain in Ti$_{49.8}$Ni$_{50.2}$ according to TEM.

4. Discussion

Our study shows that after *abc* pressing at 573 K, the temperatures of direct and reverse martensite transformations in Ti$_{49.8}$Ni$_{50.2}$ was 15–20 K lower than those after *abc* pressing at 723 K [19]. In other words, the microstructure of the alloy *abc* pressed at 573 K stabilized its B2 phase and impeded the transition to martensite on cooling. After *abc* pressing at 573 K, no sign of aging and Ti$_3$Ni$_4$ precipitation were found in the alloy to explain the observed decrease in the temperatures of martensite transformations. Thus, this decrease can be provided only by crystalline defects and associated internal stresses induced by *abc* pressing. Likely, the number of defects and the value of internal stresses after *abc* pressing at 573 K was higher than those after *abc* pressing at 723 K due to the less intense recovery processes [19–21,29].

The steep decrease in the temperatures M$_S$ and M$_f$ after *abc* pressing to e = 1.84 suggests that even this true strain provided the formation of a microstructure with a high defect density in the alloy. Their weaker dependence on the true strain at e > 1.84 indicates that although such a strain does increase the crystalline defect density in the volume of grains–subgrains, this increase is insignificant compared to its level at e = 1.84.

The formation of crystalline defects during *abc* pressing was confirmed by our XRD data (Figures 2–4), which revealed an increase in the peak halfwidth as evidence that the density of the crystalline defects increased and the size of coherent scattering regions decreased. Even after *abc* pressing to e = 1.84, the material showed a steep increase in the amount of its R phase induced by internal stresses. Moreover, it was the effect of internal stresses from crystalline defects and associated lattice distortion that are responsible for the observed substantial change in the angular position of B19$'$ peaks and in the lattice parameters.

According to our TEM data, the microstructure of Ti$_{49.8}$Ni$_{50.2}$ after *abc* pressing is, in many ways, similar to that of fcc alloys after severe plastic deformation. Figure 9 shows an enlarged fragment of the TEM image in Figure 7e with arrows pointing to five characteristic elements of the alloy microstructure after *abc* pressing to e = 5.4: a grain boundary (GB), a subgrain boundary (SGB), dislocation walls (DWs), tangled dislocation structure (TDS), and microshear bands (MSBs). Such a representation can also be found elsewhere [42]. The data suggest that the mechanisms of grain–subgrain refinement in Ti$_{49.8}$Ni$_{50.2}$ are essentially no different from those in other materials deformed by SPD methods [1,4,8]. Reasoning from our study, we think that two parallel mechanisms of grain–subgrain refinement operate in Ti$_{49.8}$Ni$_{50.2}$ during *abc* pressing. One of the mechanisms is the formation of a cellular dislocation substructure that transforms into a grain–subgrain structure due to the generation of dislocations and to their escape into the cell walls. The other mechanism of grain–subgrain refinement is the formation of intersecting microshear bands in the volume of grains (Figure 10). Although the dislocation substructure in our study was not addressed at length, some features of its behavior during *abc* pressing were evident in our experiments. In particular, after the first cycle of *abc* pressing, the scalar density of dislocations increased greatly, and after the next ones, its increase was small. Now, experiments with an XRD analysis of dislocation densities are being conducted. Their results will soon be published.

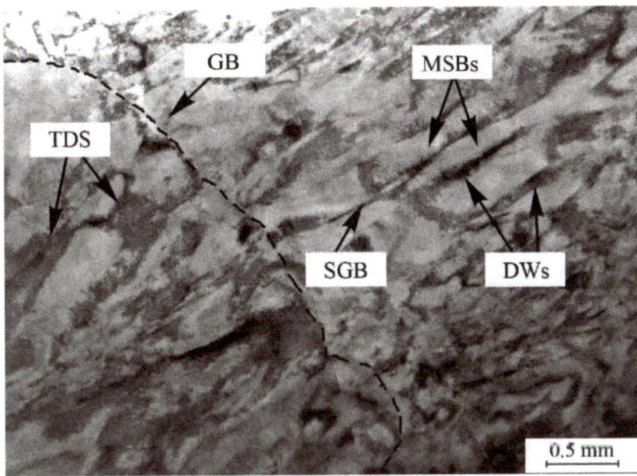

Figure 9. Characteristic microstructural elements of Ti$_{49.8}$Ni$_{50.2}$ after *abc* pressing to $e = 5.4$: GB—grain boundary, SGB—subgrain boundary, DWs—dislocation walls, TDS—tangled dislocation structure, MSBs—microshear bands; (enlarged fragment of TEM image in Figure 7e).

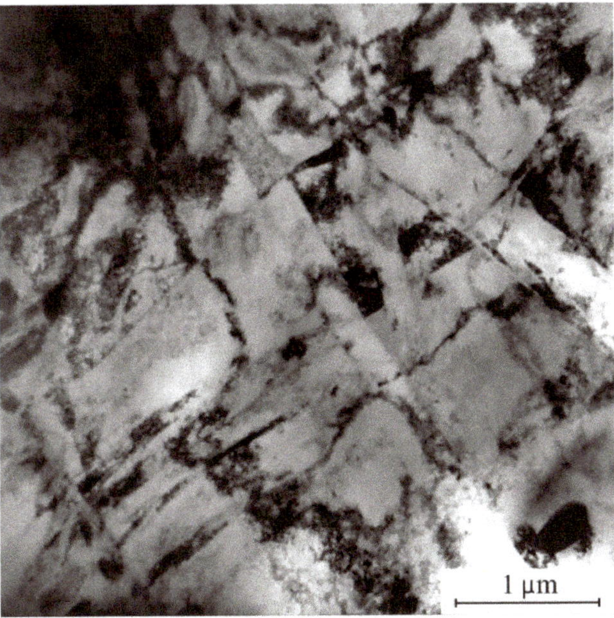

Figure 10. Intersecting microshear bands in grain volume after *abc* pressing to $e = 1.84$.

There is no doubt that the microstructural elements formed in the alloy during *abc* pressing influenced its mechanical and functional properties to one or another extent. However, the available research data are insufficient for any unambiguous conclusions on the efficiency of these elements, which necessitates further experimental studies on the mechanical and functional properties of the alloy and precision analyses of its microstructure.

The effect of severe plastic deformation on the properties of metals and alloys is associated with grain–subgrain refinement. The criterion is the size of grains–subgrains. In most studies, the initial materials are annealed (recrystallized) ones with clearly visible high angle grain boundaries on their metallographic sections. The average grain size at small true strains is determined by light microscopy, and the size of grains–subgrains at high true strains is determined from dark-field TEM images almost without regard for differences between grains with high-angle boundaries and subgrains with low-angle boundaries. However, as demonstrated by our experimental data in Section 3.3, even small true strains can provide the formation of micron and submicron elements (microshear bands, subgrains and possibly grains), and on dark-field TEM images, such elements can be broken into fine fragments. As a result, any quantitative estimates of the average grain–subgrain size are uncertain. In our opinion, the refinement of grain–subgrain structures under severe plastic deformation can conveniently be described in the context of structural scales. On the macroscale, we have a decrease in the average size of parent grains which can be measured by light microscopy, and on the microscale, the volume of parent grains acquires a grain–subgrain structure that can be measured by transmission electron microscopy. On the microscale, the deformation of metals and alloys results in crystalline defects (vacancies, interstitial atoms, dislocations, microshear bands), and their motion and interaction can lead to the formation of various hierarchical structures [43].

In our study, the most significant decrease in the grain size on the macroscale was provided by the first cycle of *abc* pressing (Figure 6), and the refinement of parent grains in the next cycles was less intense. At the same time, the average grain–subgrain size at the microscale decreased linearly with the increase in the true strain (Figure 8).

Thus, our experimental results show that even the first cycle of *abc* pressing at 573 K produced substantial changes in the microstructure of $Ti_{49.8}Ni_{50.2}$. The results can be used in models descriptive of the mechanisms by which the microstructure of TiNi alloys influences their mechanical and functional properties.

5. Conclusions

The results of our study and their comparison with other related data allow for the following conclusions on the microstructure of $Ti_{49.8}Ni_{50.2}$ *abc* pressed at T = 573 K.

1. As the true strain *e* increased, the grain–subgrain structure of $Ti_{49.8}Ni_{50.2}$ was refined on different scales at a time: through decreasing the average size of parent grains on the macroscale and the size of grains–subgrains in the volume of parent grains on the microscale;

2. The average grain–subgrain size attained after *abc* pressing to *e* = 9.55 at 573 K was approximately 130 nm;

3. The alloy *abc* pressed at 573 K revealed a decrease in all martensite transformation temperatures, a change in the lattice parameters, R phase formation, and angular shifts of diffraction peaks and their broadening because the number of structural defects became larger;

4. The largest change in the microstructure of $Ti_{49.8}Ni_{50.2}$ on all scales was provided by *abc* pressing to *e* = 1.84. Increasing the true strain to *e* = 9.55 gave a much smaller effect, suggesting that the alloy obtained a high density of structural defects even at *e* = 1.84;

5. Likely, the grain–subgrain structure of $Ti_{49.8}Ni_{50.2}$ during *abc* pressing was refined by two parallel mechanisms: through the formation of a cellular dislocation substructure with its further transformation into a grain–subgrain structure and through the formation of intersecting microshear bands in the volume of grains.

Author Contributions: Conceptualization, O.K. (Oleg Kashin), A.I.L., and V.G.; writing—original draft preparation, O.K. (Oleg Kashin); writing—review and editing, A.I.L. and V.G.; software, K.K.; investigation, O.K. (Oleg Kashin), K.K., V.G., D.Z., Y.M., N.G., O.K. (Olga Kashina), and E.B.; project administration, A.I.L.; funding acquisition, A.I.L. All authors have read and agreed to the published version of the manuscript.

Funding: The work was performed according to the government research assignment for ISPMS SB RAS, project FWRW-2021-0004.

Data Availability Statement: Not applicable.

Conflicts of Interest: The authors declare no conflict of interest. The funders had no role in the design of the study; in the collection, analyses, or interpretation of data; in the writing of the manuscript, or in the decision to publish the results.

References

1. Estrin, Y.; Vinogradov, A. Extreme grain refinement by severe plastic deformation: A wealth of challenging science. *Acta Mater.* **2013**, *61*, 782–817. [CrossRef]
2. Meyers, M.A.; Mishra, A.; Benson, D.J. Mechanical properties of nanocrystalline materials. *Prog. Mater. Sci.* **2006**, *51*, 427–556. [CrossRef]
3. Valiyev, R.Z.; Aleksandrov, I.A. *Bulk Nanostructure Metal Materials: Obtaining, Structure and Properties*; Akademkniga: Moscow, Russian, 2007; p. 398.
4. Pippan, R.; Scheriau, S.; Taylor, A.; Hafok, M.; Hohenwarter, A.; Bachmaier, A. Saturation of Fragmentation during Severe Plastic Deformation. *Annu. Rev. Mater. Res.* **2010**, *40*, 319–343. [CrossRef]
5. Sakai, T.; Belyakov, A.; Kaibyshev, R.; Miura, H.; Jonas, J.J. Dynamic and post-dynamic recrystallization under hot, cold and severe plastic deformation conditions. *Prog. Mater. Sci.* **2014**, *60*, 130–207. [CrossRef]
6. Valiev, R.; Estrin, Y.; Horita, Z.; Langdon, T.; Zehetbauer, M.; Zhu, Y. Fundamentals of superior properties in bulk nanoSPD materials. *Mater. Res. Lett.* **2016**, *4*, 1–21. [CrossRef]
7. Valiev, R.Z.; Estrin, Y.; Horita, Z.; Langdon, T.G.; Zehetbauer, M.J.; Zhu, Y. Producing bulk ultrafine-grained materials by severe plastic deformation: Ten years later. *JOM* **2016**, *68*, 1216–1226. [CrossRef]
8. Bagherpour, E.; Pardis, N.; Reihanian, M.; Ebrahimi, R. An overview on severe plastic deformation: Research status, techniques classification, microstructure evolution, and applications. *Int. J. Adv. Manuf. Technol.* **2019**, *100*, 1647–1694. [CrossRef]
9. Lopatin, Y.G.; Chuvildeev, V.N.; Nokhrin, A.V.; Kopylov, V.I.; Pirozhnikova, O.E.; Sakharov, N.V.; Piskunov, A.V.; Melyokhin, N.V. The effect of temperature in severe plastic deformation on grain refinement limit in metals and alloys. *Vestn. Lobachevsky State Univ. Nizhni Novgorod* **2010**, *5*, 132–137.
10. Prokofyev, E.A. Structure and properties of ultra-low grain TiNi alloy obtained by intensive plastic deformation. *Vestn. UGATU* **2006**, *8*, 169–174.
11. Yurchenko, L.I.; Dyupin, A.P.; Gunderov, D.V.; Valiev, R.Z.; Kuranova, N.N.; Pushin, V.G.; Uksusnikov, A.N. Mechanical Properties and Structure of High-Strength Nanostructured Nickel-Titanium Alloys Subjected to ECAP and Rolling. *Faz. Perekh. Uporyad. Sost. Novye Mater.* **2006**. Available online: http://www.ptosnm.ru (accessed on 4 October 2006).
12. Khmelevskaya, I.Y.; Prokoshkin, S.D.; Trubitsyna, I.B.; Belousov, M.N.; Dobatkin, S.V.; Tatyanin, E.V.; Korotitskiy, A.V.; Brailovski, V.; Stolyarov, V.V.; Prokofiev, E.A. Structure and properties of Ti–Ni-based alloys after equal-channel angular pressing and high-pressure torsion. *Mater. Sci. Eng. A* **2008**, *481–482*, 119–122. [CrossRef]
13. Kreitcberg, A.; Brailovski, V.; Prokoshkin, S.; Gunderov, D.; Khomutov, M.; Inaekyan, K. Effect of the grain/subgrain size on the strain-rate sensitivity and deformability of Ti–50 at%Ni alloy. *Mater. Sci. Eng. A* **2015**, *622*, 21–29. [CrossRef]
14. Shahmir, H.; Nili-Ahmadabadi, M.; Mansouri-Arani, M.; Langdon, T.G. The processing of NiTi shape memory alloys by equal-channel angular pressing at room temperature. *Mater. Sci. Eng. A* **2013**, *576*, 178–184. [CrossRef]
15. Lotkov, A.I.; Grishkov, V.N.; Kashin, O.A.; Baturin, A.A.; Zhapova, D.Y.; Timkin, V.N. Mechanisms of microstructure evolution in TiNi-based alloys under warm deformation and its effect on martensite transformations. In *Shape Memory Alloys: Properties, Technologies, Opportunities*; Resnina, N., Rubanic, V., Eds.; Materials Science Foundations; Trans. Tech. Publ. Ltd.: Pfäffikon, Switzerland, 2015; Volume 81–82, pp. 245–259. [CrossRef]
16. Lotkov, A.I.; Grishkov, V.N.; Dudarev, E.F.; Koval, Y.N.; Girsova, N.V.; Kashin, O.A.; Tabachenko, A.N.; Firstov, G.S.; Timkin, V.N.; Zhapova, D.Y. Ultrafine Structure and Martensitic Transformation in Titanium Nickelide after Warm abc Pressing. *Inorg. Mater. Appl. Res.* **2011**, *2*, 548–555. [CrossRef]
17. Lotkov, A.I.; Grishkov, V.N.; Baturin, A.A.; Dudarev, E.F.; Zhapova, D.Y.; Timkin, V.N. The effect of warm deformation by abc-pressing method on mechanical properties of titanium nickelide. *Lett. Mater.* **2015**, *5*, 170–174. [CrossRef]
18. Lotkov, A.I.; Grishkov, V.N.; Dudarev, Y.F.; Girsova, N.V.; Tabachenko, A.N. Formation of ultrafine grain structure, martensitic transformations and unelastic properties of titanium nickelide after abc-pressing. *Vopr. Materialoved.* **2008**, *1*, 161–165.
19. Lotkov, A.; Grishkov, V.; Zhapova, D.; Timkin, V.; Baturin, A.; Kashin, O. Superelasticity and shape memory effect after warm abc-pressing of TiNi-based alloy. *Mater. Today Proc.* **2017**, *4*, 4814–4818. [CrossRef]

20. Lotkov, A.I.; Kashin, O.A.; Grishkov, V.N.; Krukovskii, K.V. The influence of degree of deformation under isothermal abc pressing on evolution of structure and temperature of phase transformations of alloy based on titanium nickelide. *Inorg. Mater. Appl. Res.* **2015**, *6*, 96–104. [CrossRef]
21. Zhapova, D.Y.; Lotkov, A.I.; Grishkov, V.N.; Timkin, V.N.; Rodionov, I.S.; Kolevatov, A.S.; Belosludtseva, A.A. Inelastic properties of titanium nickelide after warm abc-pressing. *Izv. Vyssh. Uchebn. Zaved. Fiz.* **2016**, *59*, 60–63.
22. Kashin, O.; Krukovskii, K.; Lotkov, A.; Grishkov, V. Effect of True Strains in Isothermal abc Pressing on Mechanical Properties of Ti$_{49.8}$Ni$_{50.2}$ Alloy. *Metals* **2020**, *10*, 1313. [CrossRef]
23. Gunderov, D.; Lukyanov, A.; Prokofiev, E.; Kilmametov, A.; Pushin, V.; Valiev, R. Mechanical properties and martensitic transformations in nanocrystalline Ti$_{49.4}$Ni$_{50.6}$ alloy produced by high-pressure torsion. *Mater. Sci. Eng. A* **2009**, *503*, 75–77. [CrossRef]
24. Gunderov, D.; Lukyanov, A.; Prokofiev, E.; Pushin, V.G. Mechanical properties of the nanocrystalline Ti$_{49.4}$Ni$_{50.6}$ alloy produced by High Pressure Torsion. *Eur. Phys. J. Spec. Top.* **2008**, *158*, 53–58. [CrossRef]
25. Kockar, B.; Karaman, I.; Kulkarni, A.; Chumlyakov, Y.; Kireeva, I.V. Effect of severe ausforming via equal channel angular extrusion on the shape memory response of a NiTi alloy. *J. Nucl. Mater.* **2007**, *361*, 298–305. [CrossRef]
26. Potapova, A.A.; Stolyarov, V.V. Deformability and structural features of shape memory TiNi alloys processed by rolling with current. *Mater. Sci. Eng. A* **2013**, *579*, 114–117. [CrossRef]
27. Shahmir, H.; Nili-Ahmadabadi, M.; Huang, Y.; Jung, J.M.; Kim, H.S.; Langdon, T.G. Shape memory characteristics of a nanocrystalline TiNi alloy processed by HPT followed by post-deformation annealing. *Mater. Sci. Eng. A* **2018**, *734*, 445–452. [CrossRef]
28. Shahmir, H.; Nili-Ahmadabadi, M.; Huang, Y.; Jung, J.M.; Kim, H.S.; Langdon, T.G. Shape memory effect in nanocrystalline NiTi alloy processed by high-pressure torsion. *Mater. Sci. Eng. A* **2015**, *626*, 203–206. [CrossRef]
29. Shamsolhodaei, A.; Zarei-Hanzaki, A.; Moghaddam, M. Structural and functional properties of a semi equiatomic NiTi shape memory alloy processed by multi-axial forging. *Mater. Sci. Eng. A* **2017**, *700*, 1–9. [CrossRef]
30. Shri, D.N.A.; Tsuchiya, K.; Yamamoto, A. Surface characterization of TiNi deformed by high-pressure torsion. *Appl. Surf. Sci.* **2014**, *289*, 338–344. [CrossRef]
31. Tong, Y.X.; Chen, F.; Guo, B.; Tian, B.; Li, L.; Zheng, Y.F.; Gunderov, D.V.; Valiev, R.Z. Superelasticity and its stability of an ultrafine-grained Ti$_{49.2}$Ni$_{50.8}$ shape memory alloy processed by equal channel angular pressing. *Mater. Sci. Eng. A* **2013**, *587*, 61–64. [CrossRef]
32. Tong, Y.X.; Hu, K.P.; Chen, F.; Tian, B.; Li, L.; Zheng, Y.F. Multiple-stage transformation behavior of Ti49.2Ni50.8 alloy with different initial microstructure processed by equal channel angular pressing. *Intermetallics* **2017**, *85*, 163–169. [CrossRef]
33. Valiev, R.; Gunderov, D.; Prokofiev, E.; Pushin, V.; Zhu, Y. Nanostructuring of TiNi alloy by SPD processing for advanced properties. *Mater. Trans.* **2008**, *49*, 97–101. [CrossRef]
34. Zhang, X.; Song, J.; Huang, C.; Xia, B.; Chen, B.; Sun, X.; Xie, C. Microstructures evolution and phase transformation behaviors of Ni-rich TiNi shape memory alloys after equal channel angular extrusion. *J. Alloys Compd.* **2011**, *509*, 3006–3012. [CrossRef]
35. Gunderov, D.V.; Pushin, V.G.; Valiev, R.Z.; Prokofiev, E.A. Investigation of the nature of the high strength and ductility of the ultrafine-grained TiNi alloy obtained by equal-channel angular compression. *Deform. Fract. Mater.* **2007**, *10*, 13–21.
36. Kuranova, N.N.; Gunderov, D.V.; Ukusnikov, A.N.; Luk'yanov, A.V.; Yurchenko, L.I.; Prokof'ev, E.A.; Pushin, V.G.; Valiev, R.Z. Effect of heat treatment on the structural and phase transformations and mechanical properties of TiNi alloys subjected to severe plastic deformation. *Phys. Met. Metallogr.* **2009**, *108*, 556–568. [CrossRef]
37. Prokoshkin, S.D.; Khmelevskaya, I.Y.; Dobatkin, S.V.; Trubitsyna, I.B.; Tat'yanin, E.V.; Stolyarov, V.V.; Prokof'ev, E.A. Structure evolution under severe plastic deformation of shape memory alloys based on titanium nickelide. *Phys. Met. Metallogr.* **2004**, *97*, 619–625.
38. Stolyarov, V.V.; Prokofiev, E.A.; Prokoshkin, S.D.; Dobatkin, S.V.; Trubitsyna, I.B.; Khmelevskaya, I.Y.; Pushin, V.G.; Valiev, R.Z. Structure, mechanical properties, and shape memory effect in TiNi alloys exposed to equal channel angular pressing. *Phys. Met. Metallogr.* **2005**, *100*, 91–102.
39. Churakova, A.A.; Gunderov, D.V.; Lukyanov, A.V.; Lebedev, Y.A. Effect of thermal cycling in the range of phase transformation B2–B19' on the microstructure and mechanical properties of UFG alloy Ti$_{49.8}$Ni$_{50.2}$. *Lett. Mater.* **2013**, *3*, 166–168. [CrossRef]
40. Kudoh, Y.; Tokonami, M.; Miyazaki, S.; Otsuka, K. Cristal structure of the martensite in Ti+49.2 at. % Ni alloy analyzed by the single crystal X-ray diffraction method. *Acta Metall.* **1985**, *33*, 2049–2056. [CrossRef]
41. Sitepu, H. Use of synchrotron diffraction data for describing crystal structures and crystallographic phase analysis of R-phase NiTi shape memory alloy. *Textures Micrjstruct.* **2003**, *35*, 185–195. [CrossRef]
42. Huang, J.Y.; Zhu, Y.T.; Jiang, H.; Lowe, T.C. Microstructures and dislocation configurations in nanostructured Cu processed by repetitive corrugation and straightening. *Acta Mater.* **2001**, *49*, 1497–1505. [CrossRef]
43. Panin, V.E.; Likhachev, V.A.; Grinyaev, Y.V. *Structural Levels of Deformation in Solids*; Nauka: Novosibirsk, Russia, 1985; p. 255. (In Russian)

metals

Article

Structure and Multistage Martensite Transformation in Nanocrystalline Ti-50.9Ni Alloy

Tamara M. Poletika [1,*], Svetlana L. Girsova [1], Aleksander I. Lotkov [1], Andrej N. Kudryachov [2] and Natalya V. Girsova [1]

[1] Institute of Strength Physics and Materials Science SB RAS, 634055 Tomsk, Russia; girs@ispms.tsc.ru (S.L.G.); lotkov@ispms.tsc.ru (A.I.L.); girsova@ispms.tsc.ru (N.V.G.)
[2] Angioline Research, 630559 Novosibirsk, Russia; kudryashovan@gmail.com
* Correspondence: poletm@ispms.tsc.ru; Tel.: +7-909-548-7297

Abstract: An electron microscopic study of the evolution of the size, morphology, and spatial distribution of coherent Ti_3Ni_4 particles with a change in the aging temperature in a nanocrystalline (NC) Ti-50.9 at % Ni alloy with an inhomogeneous grain–subgrain B2-austenitic nanostructure has been carried out. It was found that with an increase in the aging temperature, along with a change in the size and shape of Ti_3Ni_4 nanoparticles, their spatial distribution changes from location at dislocations to precipitates at subboundaries. Research has shown that the presence of different types of internal interfaces in the nanostructure contributes to the heterogeneous distribution of coherent Ti_3Ni_4 nanoparticles in the volume of the B2 matrix, which is associated with the precipitation of particles in the region of low-angle subboundaries and the suppression of the Ti_3Ni_4 precipitation in nanograins with high-angle boundaries. The difference in the structural-phase state of nanograins and subgrains regions is the main reason for the implementation of the anomalous R-phase transformation effect in the sequence of multistage martensitic transformations $B2\leftrightarrow R\leftrightarrow B19'$.

Keywords: nanocrystalline TiNi alloy; annealing; grain–subgrain structure; Ti_3Ni_4 particles; martensite transformations

Citation: Poletika, T.M.; Girsova, S.L.; Lotkov, A.I.; Kudryachov, A.N.; Girsova, N.V. Structure and Multistage Martensite Transformation in Nanocrystalline Ti-50.9Ni Alloy. *Metals* **2021**, *11*, 1262. https://doi.org/10.3390/met11081262

Academic Editor: Elena Pereloma

Received: 25 June 2021
Accepted: 9 August 2021
Published: 10 August 2021

Publisher's Note: MDPI stays neutral with regard to jurisdictional claims in published maps and institutional affiliations.

1. Introduction

Nanocrystalline TiNi alloys with unique properties of shape memory effect and superelasticity, due to the reversible transformation between B2-austenite and $B19'$ martensite, have high strength and functional stability and are widely used in medicine [1,2]. Their representatives in biomedicine are Ni-rich TiNi alloys, which are aged with the formation of coherent Ti_3Ni_4 particles [1–3]. Such precipitates in TiNi polycrystals can change the path of martensitic transformations from $B2\rightarrow B19'$ to the sequence of transformations through the transition R-phase: $B2\rightarrow R\rightarrow B19'$ [3–5]. This is mainly due to inhomogeneous Ti_3Ni_4 particles distribution in B2-austenite grains and the associated difference in the Ni content between B2-austenite grain volumes and boundaries [6,7]. However, in NC TiNi alloys with a large volume fraction of interfaces, the cause of multistage transformations is unclear, and the formation of coherent Ti_3Ni_4 particles as well as their spatial distribution is poorly understood. This is largely due to the lack of data on the localization of coherent Ti_3Ni_4 particles in the nanostructure of TiNi alloys due to the complexity of their detection and certification in NC material. In addition, the study of NC TiNi alloys is complicated by the necessity of accounting for grain refinement and associated increase in critical martensite shear stresses [8], possible suppression of B2-austenite solid solution decomposition in nanograins [9], and high defect density in the nanostructure after severe deformation [10]. Another important factor is that the nanostructure of TiNi alloys after cold deformation is highly inhomogeneous. For example, a mixed grain–subgrain B2 structure is typical of textured TiNi after severe cold deformation [11]. Such a structure is observed in semi-finished products (wires, thin-walled tubes, etc.) used for manufacturing various medical

Metals **2021**, *11*, 1262. https://doi.org/10.3390/met11081262

devices. The purpose of this work is to better understand the effect of the size and spatial distribution of Ti_3Ni_4 particles in grain–subgrain nanostructure of an aged Ti-50.9 at % Ni alloy on the multistage character of martensite transformations. This knowledge can be used to fine-tune the interval of martensitic transformations in NC Ni-rich TiNi alloy tubing in the process of medical implants manufacturing.

2. Materials and Methods

The test material was a commercial nanocrystalline TiNi alloy with a Ni content of 50.9 at % (Vascotube GmbH, Birkenfeld, Germany). The tubular samples cut from microtubes for medical stents were annealed at 300 °C (low-temperature aging) and 400 °C (region of the highest rate of Ti_3Ni_4 precipitation) for 1 h with further water quenching at room temperature. These aging treatments will allow the precipitation of Ti_3Ni_4 particles differing in size and morphology. The start and finish temperatures of direct and reverse $B2\leftrightarrow R\leftrightarrow B19'$ transformations were determined by electrical resistivity measurements. In all specimens the sequence of direct and reverse martensite transformations was $B2\leftrightarrow R\leftrightarrow B19'$. The critical start temperatures of $B2\rightarrow R$ (T_R) as well as the start and finish temperatures of $R\leftrightarrow B19'$ (M_S, M_F, A_S, A_F) determined by resistivity measurements are presented in Table 1. The features of $B2\leftrightarrow R\leftrightarrow B19'$ transformations were analyzed on a NETZSCH DSC 404 F1 differential scanning calorimeter with a heating and cooling rate of 10 K/min.

Table 1. Martensite transformation temperatures in nanocrystalline Ti-50.9Ni alloy.

Temperature	T_R, °C	M_S, °C	M_F, °C	A_S, °C	A_F, °C
Initial	7	−74	−140	−45	−26
300 °C	22	−50	−130	−37	−14
400 °C	33	−30	−115	−2	15

The structure of the specimens cut along the microtube axis was studied on a JEOL JEM 2100 transmission electron microscope. The preparation of thin foils included grinding, mechanical polishing to a thickness of 0.1 mm, and ion thinning on a JEOL EM 09100IS ion slicer. The average size of grains and subgrains was determined by measuring the average diameter of no less 100 grains on bright- and dark-field images. A comparative analysis of the size of the Ti_3Ni_4 precipitations in samples heat-treated at various temperatures was carried out mainly on dark-field images.

3. Results

3.1. Microstructure Characterization

The initial alloy has an inhomogeneous hierarchical B2-austenite nanostructure with elements of two scales: conglomerates of slightly misoriented nanosubgrains of micron sizes and nanograins with high-angle boundaries, which provides a wide range of internal interfaces (Figure 1). There are two types of nanograins: (i) dislocation-containing nanograins formed under severe deformation and further annealing and (ii) dislocation-free nanograins crystallized from amorphous state [11]. The average size of dislocation-free nanograins with high-angle boundaries is 70 nm. In subgrains of size 40–60 nm, a high dislocation density is observed. Such subgrains, having imperfect low-angle boundaries azimuthally misoriented by less than 3°, form regions of up to 800 nm (Figure 1b). On selected area electron diffraction (SAED) patterns, the grain–subgrain structure is represented mostly by circular point reflections. In addition to B2-austenite reflections, there are individual reflections identifiable as R-phase and Ti_3Ni_4-phase reflections.

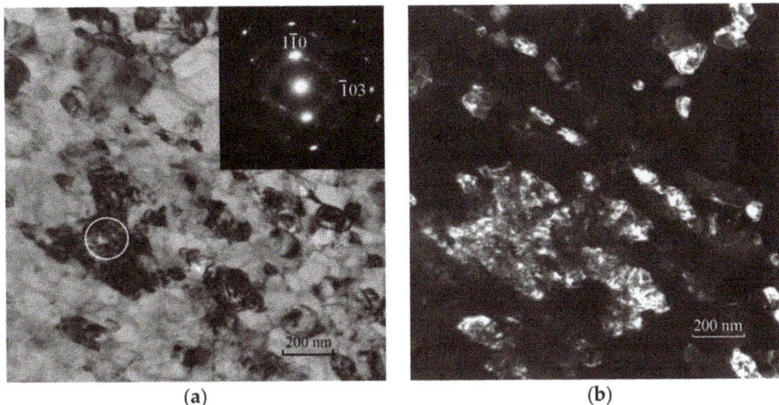

Figure 1. Bright-field image of mixed grain–subgrain structure with a submicron subgrain region, corresponding SAED pattern from the circled region, zone axis [331] (**a**), and dark-field image (**b**) in B2-austenite reflection ($1\bar{1}0$).

Figure 2 shows the structure of the alloy after annealing at 300 °C. Such low-temperature aging involves recovery and polygonization processes, dislocation density reduction, and dislocation rectification of low-angle boundaries, but the size of grains and subgrains remains, on average, unchanged. In the structure, coherent Ti_3Ni_4 particles smaller than 10 nm precipitate (Figure 2a,b), which is characteristic of the initial stage of B2 phase decomposition in TiNi during low-temperature aging [12]. Such particles precipitate mostly inside subgrains and at dislocations, showing weak reflections in positions 1/7 along the B2 lattice directions <321> (Figure 2c) and a "coffee bone" contrast induced by elastic distortion fields (Figure 2b). For comparison, dislocation-free grains are shown (Figure 2e,f), which are usually several tens of nanometers in size. No dislocations were found in these nanograins at admissible tilt angles. In addition, such grains did not show a clear TEM diffraction of Ti_3Ni_4 precipitates. Figure 2c shows that on the SAED patterns, the diffraction spots occurring at 1/3 <110> B2 are observed, indicating the formation of an R-phase. This is consistent with the data in Table 1, which show that the B2- and R-phases coexist at room temperature. As can be seen on the dark-field image in a ($\bar{1}0\bar{2}$) reflex (Figure 2d), disperse and imperfect morphology of R-martensite is observed.

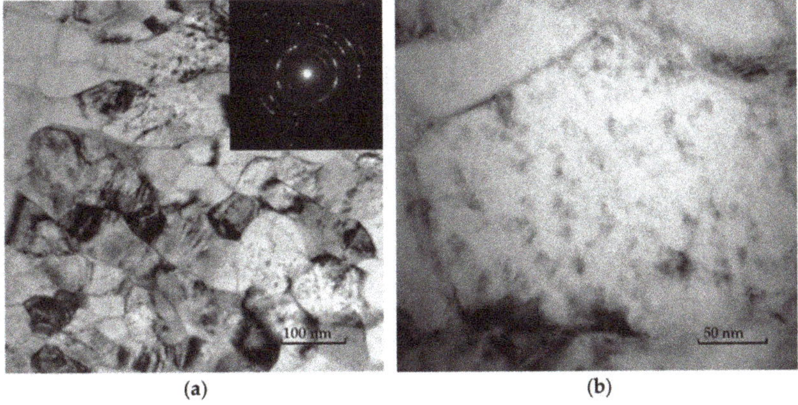

Figure 2. *Cont.*

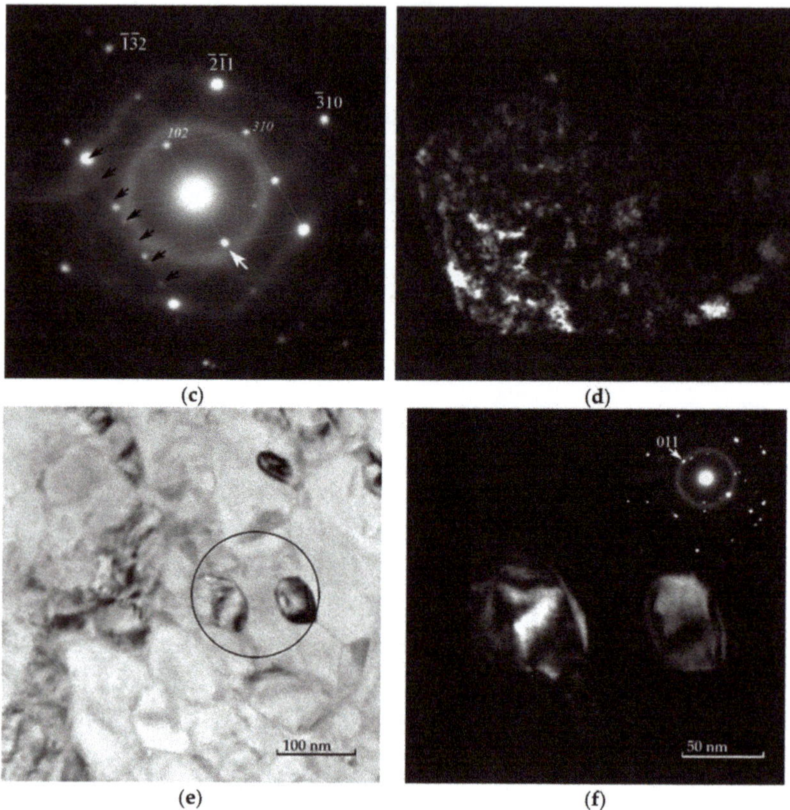

Figure 2. Structure of Ti-50.9Ni after annealing at 300 °C: (**a**) bright-field image of subgrains and corresponding SAED pattern, arcs correspond to the positions of reflections of B2, (**b**) bright-field image of large grain with "coffee bone" contrast indicative of the presence of coherent Ti_3Ni_4 particles, (**c**) corresponding SAED pattern showing weak reflections of Ti_3Ni_4 particles in positions 1/7 along the B2 lattice directions <321> (dark arrows), zone axis [135], the R-phase reciprocal lattice section is indicated by white lines, (**d**) R-phase dark-field image in a $(\overline{1}0\overline{2})$ reflection indicated by white arrows on (**c**), (**e,f**) bright-field image of dislocation-free grains and corresponding dark-field in (011) reflection of B2 on SAED pattern.

The precipitation of Ti_3Ni_4 particles at dislocations during low-temperature annealing is due to the low volumetric diffusion of atoms and the predominance of diffusion of atoms along dislocation tubes. It should be noted that this type of precipitate is found in dislocation-containing nanograins and is not found in dislocation-free nanograins. Such Ti_3Ni_4 nanoparticles are rarely observed in grain boundary–subboundary regions because the low aging temperature (300 °C) does not allow Ni atoms to diffuse from internal grain–subgrain volumes to their boundaries. The fine Ti_3Ni_4 particles decorate the few dislocations present in nanograins. Our results agree with data on polycrystalline TiNi alloys whose structure after low-temperature annealing (<330 °C) contains a large amount of coherent spherical nanoparticles of diameter <10 nm [12,13].

Annealing at 400 °C decreases the dislocation density and slightly widens the size range of polygonized nanograins, i.e., no recrystallization occurs. At the stage of recovery, dislocations inside subgrains are redistributed such that some of them are annihilated and others escape to low-angle subboundaries. As a result, the subboundaries change their shape from linear after annealing at 300 °C (Figure 2a,b) to convex after annealing at 400 °C (Figure 3). Their bright-field images reveal a diffuse diffraction contrast (Figure 3) which can be indicative of their nonequilibrium state [14,15] due to substantial B2 matrix distortion in elastic stress fields induced by excess dislocations and by coherent Ti_3Ni_4 particles formed at subgrain boundaries [15].

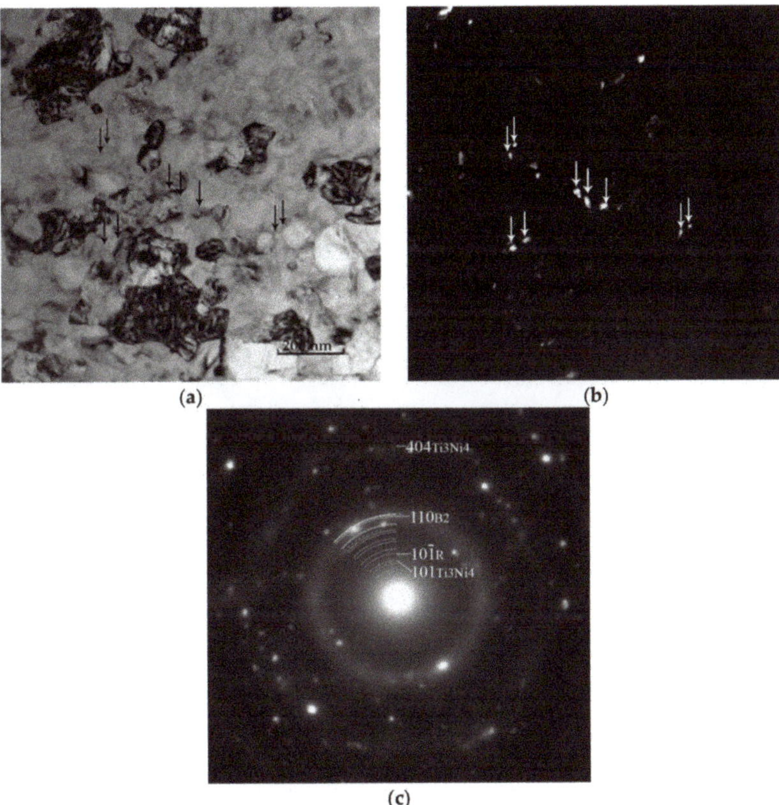

Figure 3. Subgrain structure of Ti-50.9Ni with Ti_3Ni_4 particles after annealing at 400 °C: (**a**) bright-field images with arrows for particle lines near low-angle boundaries, (**b**) dark field in reflection {101} of Ti_3Ni_4 particles, (**c**) SAED pattern, solid and dash arcs correspond to positions of reflections of R-and Ti_3Ni_4–phases.

Under annealing at 400 °C, the alloy undergoes intense aging due to intense Ti_3Ni_4 precipitation at 400–450 °C [16]. As can be seen from Figure 3c, its microdiffraction patterns reveal numerous point reflections of R martensite and Ti_3Ni_4 particles precipitated mainly in nonequilibrium subboundary regions.

The shape of Ti_3Ni_4 particles is close to lenticular with a lateral dimension of up to 5 nm and length of up to 20 nm. Their tendency to line up in rows in the planes {111} is observed. In Figure 3a, one can identify lines of Ti_3Ni_4 particles adjacent to subboundaries. Such an arrangement points to their autocatalytic character, which compensates the nucleation and growth energy of Ti_3Ni_4 [5]. The morphology of Ti_3Ni_4 particles is such that they can lose their lenticular shape caused by the lattice mismatch between Ti_3Ni_4 and B2 matrix [5]. Apparently, this is due to their asymmetric growth under the action of high inhomogeneous stress fields near subboundaries [5,17]. In our analysis, no lenticular Ti_3Ni_4 particles and lines were found in nanograins with high-angle boundaries, and this agrees with the conclusion that in TiNi aged at 400 °C, the B2 phase decomposition with the formation of Ti₃Ni₄ particles in nanograins smaller than 150 nm is suppressed [9].

At room temperature, the alloy aged at 400 °C has a two-phase state: B2-austenite and R-phase (Table 1). Considering that the grain and subgrain sizes remain unchanged, we think that the intense formation of Ti_3Ni_4 effectively decreases the Ni concentration in the B2 matrix and increases the temperature of B2↔R transformations (T_R). The R-phase is dispersed into nanodomains, and its morphology is mostly imperfect (Figure 4b), which is due to the effect of elastic stress fields created by coherent Ti_3Ni_4 particles and dislocations. This phase is present in substructural regions and is not present in nanograins free of dislocations and particles. The transformation to the R-phase can cover the whole volume of individual subgrains and their groups (Figure 4).

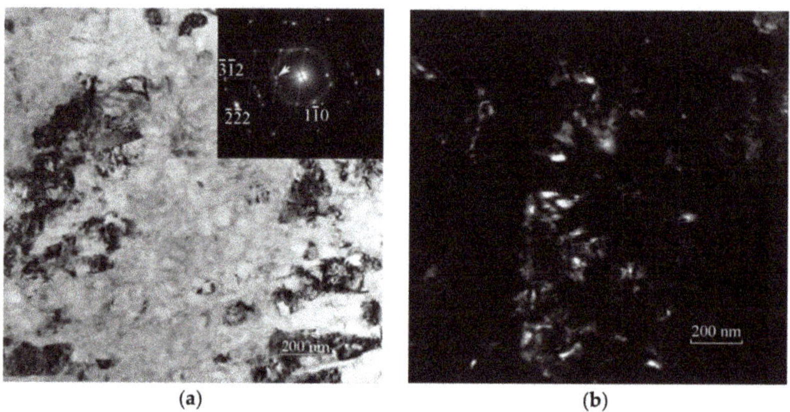

(a)   (b)

Figure 4. R-phase in substructure: (**a**) bright-field image with complex contrast created by Ti_3Ni_4 particles, dislocations, and R-phase; corresponding microdiffraction pattern, zone axis [112], reflections from B2-phase are indexed, R-phase reciprocal lattice section are indicated by white lines (**b**) corresponding dark-field image in R-phase reflection ($\bar{3}10$) are indicated by white arrow on (**a**) with visible phase dispersion.

3.2. Multistage Martensite Transformations

Figure 5 shows calorimetric curves for NC Ti-50.9Ni alloy in the initial state and aged at 300 °C and 400 °C. DSC peaks were interpreted using the results of measuring the temperatures of martensitic transformations by the thermoresistometry method, presented in Table 1, as well as the above results of structural studies of aged alloy samples by the TEM method. On cooling, the initial NC alloy reveals two diffuse exothermal peaks due to direct B2→R and B2→ B19′ transformations characteristic of aged TiNi alloys with defect B2 structure [18]. On heating, it shows a single broad endothermal peak corresponding to B19′→R→B2 transformations. The latent heats ΔH of the B2→R and the R→B19′ transformations were 3.8 and 2 J g⁻¹, respectively. The R→B19′ transformation is significantly suppressed, showing a low and broad exothermal peak. Thus, we can assume that the sample exhibited a single-stage B2→R transformation on cooling and a single stage R→B2 transformation on heating. It can be seen that after aging at 300 °C, the peaks on the DSC curves represent B2→R and B2→ B19′ on cooling and B19′→R and R→B2 on heating, which are the normal two-stage martensitic transformations for aged Ti–Ni alloys. The corresponding value of $\Delta H_{A\rightarrow R}$ was 5.4 J g⁻¹ and that of $\Delta H_{R\rightarrow A}$ was 5.3 J g⁻¹. These values are in the range of 1–6 J g⁻¹ for B2↔R transformations [19]. The diffuse nature of these peaks does not allow a sufficient accuracy the temperature hysteresis, but the estimate gives a value of about 6 °C, which is the typical characteristic of R-phase transformation. Thus, it can be deduced that these DSC peaks represent B2→R and R→B2, respectively, which is also close to the data presented in Table 1.

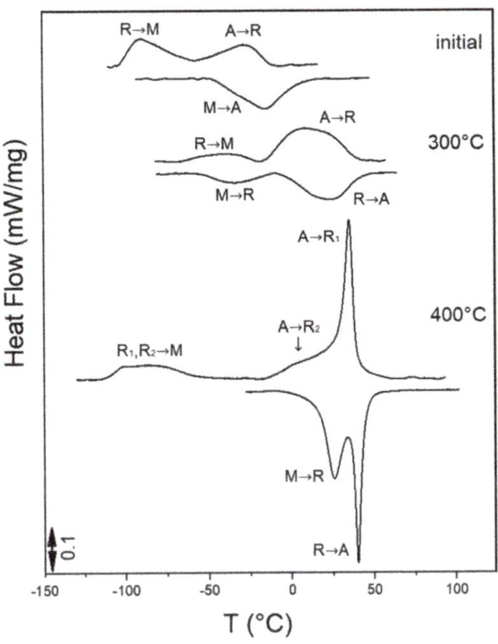

Figure 5. Calorimetric curves for cooling (upper traces) and heating (lower traces) of Ti-50.9Ni in initial state and annealed at 300 °C and 400 °C for 1 h.

Another form of calorimetric curves is observed for the alloy aged at 400 °C (Figure 5). On cooling, the B2→R transformation occurred over a broad temperature range. As can be seen from Figure 5, the narrow B2→R peak has a shoulder on the left, which can be interpreted as at least one additional peak. The bases of the peaks merge, giving no way to separate additional peaks with sufficient accuracy. The combined latent heat of this peak is 6.8 J g^{-1}. As shown above, aging at 400 °C leads to intense Ti_3Ni_4 particles precipitation that shifts the B2→R transformation to higher temperatures. As a result, the R-phase is stabilized and the R→B19′ is significantly suppressed, showing a very broad exothermal peak. The blurring of the exothermal peak corresponding to R→B19′ is due to the particle distribution inhomogeneity and to the extended temperature range of this transformation as R_1, R_2→B19′. The latent heat of this peak is 3.3 J g^{-1}. We cannot exclude the implementation in a number of nanograins of the direct transformation B2→B19′. On heating, two merging peaks corresponding to successive B19′→R and R→B19′ transformations are observed (Figure 5). Though merging, the main peaks for the direct and reverse B2↔R transformations on cooling and on heating are rather narrow and well identifiable, which makes it possible to determine the temperature hysteresis as 5.6 °C. The small hysteresis confirms the correspondence of these peaks to B2↔R transformation. The results obtained for NC Ti-50.9Ni alloy after aging under conditions of intense Ti_3Ni_4 precipitation are consistent with the Table 1 and TEM data, according to which T_R is 33 °C and the alloy is in the B2 + R state.

Thus, we can assume that the broad B2→R peak upon cooling contains two peaks corresponding to the B2→R_1 and B2→R_2 transformations. The TEM analysis results allow us to interpret these peaks as follows: (i) first, the transformation B2→R_1, which is characterized by quite high and narrow peaks developing in the substructure prevailing in the volume of the grain–subgrain nanostructure; (ii) then, there is a temperature-extended transformation of B2→R in groups of nanograins, which is designated as B2→R2.

4. Discussion

Aging at different temperatures is found to provide coherent precipitation of Ti$_3$Ni$_4$ particles differing in size, morphology, and spatial distribution in grain–subgrain structure and cause different variations to the transformation behavior of NC Ti-50.9Ni alloy. After aging at 300 °C, coherent particles precipitate mainly in the subgrain structure at dislocations. At the same time, Ti$_3$Ni$_4$ particles are not observed in dislocation-free nanograins but precipitate in nanograins, including those with a size less than 100 nm in the case of the presence of dislocations in them. Thus, the initial non-uniform grain-subgrain structure of the NC Ti-50.9Ni alloy causes an inhomogeneous distribution of both Ti$_3$Ni$_4$ precipitates and dislocations in the nanostructure. We believe that the reason for this precipitation behavior during aging at 300 °C lies in the features of the initial NC structure formation after severe cold deformation.

As noted above, two types of nanograins can be distinguished in the NC TiNi grain-subgrain structure: dislocation-containing nanograins formed under deformation and dislocation-free nanograins crystallized from amorphous state [10,11]. Unfortunately, it is impossible to determine with a sufficient degree of reliability in each case the type of observed nanograins without a special study. Reasoning from our analysis, the absence of Ti$_3$Ni$_4$ particles in dislocation-free nanograins suggests that their precipitation during low-temperature aging needs such defects as dislocations [12,13]. This conclusion agrees with data [19] showing that TiNi nanograins crystallized from an amorphous state are unfavorable for Ti$_3$Ni$_4$ precipitation in terms of the energy requirements energy due to the release of stored energy and in terms of chemical composition. In addition, the R-phase which usually nucleates at the interfaces between the particle and the matrix [2,6,20] is not detected in dislocation-free nanograins, which is consistent with the absence of coherent Ti$_3$Ni$_4$ particles in such nanograins.

It was found that with an increase in the aging temperature to 400 °C and the diffusion mobility of Ni atoms, the size of Ti$_3$Ni$_4$ nanoparticles increases, their shape changes from spherical to lenticular, and the spatial distribution in the nanostructure changes from location at dislocations to precipitation at subboundaries. No aging effects were found in nanograins. It is believed that the suppression of the deposition of coherent particles is provided by geometric constraints on the side of high-angle boundaries, which prevent the formation of arrays of self-accommodated lenticular particles inside nanocrystalline grains [5,6]. Thus, the presence of different types of internal boundaries (low-angle and high-angle) in the grain–subgrain TiNi nanostructure is the main factor affecting the process of heterogeneous nucleation of coherent Ti$_3$Ni$_4$ particles at temperatures of intense decomposition of the B2 solid solution. Such precipitates emerge in the region of low-angle subboundaries and are not found in nanograins with high-angle boundaries.

Note that the precipitation of Ti$_3$Ni$_4$ particles and the corresponding appearance of the R-phase upon annealing at 400 °C are accompanied by a shift in the characteristic temperatures of martensitic transformation (Table 1). It is consistent with the general regularities of the evolution of the structural-phase state with a change in aging temperature that are obtained for polycrystalline Ni-rich TiNi alloys [2–7,18–20]. At the same time, there are some features of the multistage martensitic transformations in the NC Ti–50.9Ni alloy due to the inhomogeneous Ti$_3$Ni$_4$ particles precipitation.

Various hypotheses of the mechanisms of this phenomenon have been proposed [4–7,20]. In polycrystalline TiNi-based alloys, two types of inhomogeneity can change the sequence of martensite transformations: (i) local inhomogeneities of Ni concentrations and internal stresses near coherent Ti$_3$Ni$_4$ particles and (ii) microscopic heterogeneity due to the different nature of the decomposition of the B2-solid solution TiNi between the region of grain boundaries and their internal volume [4–7,18,20]. The most probable cause of multistage transformations is the presence of zones differing in size and particle distribution [18]. The inhomogeneity of the distribution of Ti$_3$Ni$_4$ particles within the grains provides a difference in the concentration of Ni in the B2-austenite within the boundaries and in the internal volume of the grains. As a result, the B2↔R transformation first occurs at the grain boundary with a low Ni content, and then inside the grains with a higher Ni content, which gives rise to the abnormal three-stage martensitic transformation [4–7].

Evidently, the grain size determines the ratio between the volume fractions of grain boundaries and grain volumes and can thus influence the Ti_3Ni_4 distribution and martensite transformations in aged specimens [7,20–22]. Note that according to [23], the size of nickel-depleted zones near the particle–matrix interface varies from 20 to 100 nm, depending on the particle size, which corresponds to the observed range of nanograin sizes and allows considering only local Ni concentration inhomogeneities. Thus, it is unclear what kind of inhomogeneity is responsible for the abnormal three-peak behavior of NC TiNi alloy under intensive aging at 400 °C. At the same time, our analysis identifies inhomogeneities associated with different Ti_3Ni_4 precipitation behavior in substructural regions and groups of nanograins. After aging at 400 °C, the B2 matrix of the Ti-50.9Ni NC alloy exhibits a difference in the Ni concentration in the substructural regions and groups of nanograins, and the scale of this inhomogeneity is much larger than the scale of local stresses and the Ni concentration near Ti_3Ni_4 particles. In this case, an anomalous R-phase transformation effect is possible: higher temperature transformations $B2 \leftrightarrow R_1$ in substructural regions with Ti_3Ni_4 particles and lower temperature transformations $B2 \leftrightarrow R_2$ in nanograins free of Ti_3Ni_4. This is consistent with the expectation of Ni content depletion in the matrix as Ni-rich precipitation progresses. We believe that the joint occurrence of the two transformation streams is due to the heterogeneity of the grain–subgrain nanostructure of Ti-50.9Ni NC alloy. The study of martensitic transformations by the DSC method made it possible to confirm these assumptions (Figure 5).

5. Conclusions

Electron microscopy analysis of the size and morphology of coherent Ti_3Ni_4 precipitates in grain–subgrain NC Ti-50.9Ni alloy aged at 300 °C and 400 °C is performed.

It was shown that on aging at 300 °C, spherical Ti_3Ni_4 particles smaller than 10 nm precipitate at dislocations in the alloy substructure. Such particles are found in dislocation-containing nanograins and are not found in dislocation-free nanograins.

On aging at 400 °C, the Ti_3Ni_4 particles increase in size, changing their shape from spherical to near lenticular and their distribution within the nanostructure from location at dislocations to precipitation at subboundaries. The presence of different types of internal boundaries in Ti-50.9Ni is the main factor that influences the inhomogeneous nucleation of coherent Ti_3Ni_4 particles in the temperature range of their intense precipitation at 400 °C. Such particles nucleate at low-angle sub-boundaries, while decomposition of B2-austenite is not observed in nanograins with high-angle boundaries.

The Ti_3Ni_4 distribution inhomogeneity in NC Ti-50.9Ni provides different structural-phase states in the grain–subgrain nanostructure and accordingly different Ni concentrations in subgrains and nanograins regions. This is one of the driving factors responsible for the occurrence of the anomalous R-phase transformation effect in the sequence of multistage martensitic transformations $B2 \leftrightarrow R \leftrightarrow B19'$.

The results reported can help in choosing appropriate thermal and thermomechanical treatment modes for Ni-rich TiNi alloys to improve their performance as cardiovascular implant materials.

Author Contributions: Conceptualization, T.M.P. and A.I.L.; methodology, T.M.P. and S.L.G.; investigation, S.L.G. and N.V.G.; writing—original draft preparation, T.M.P.; writing—review and editing, T.M.P. and. S.L.G.; project administration, A.I.L. and. A.N.K. All authors have read and agreed to the published version of the manuscript.

Funding: This research was performed under State Assignment for ISPMS SB RAS (project No. FWRW-2021-0004).

Institutional Review Board Statement: Not applicable.

Informed Consent Statement: Not applicable.

Data Availability Statement: Data are available from the corresponding author on reasonable request.

Conflicts of Interest: The authors declare no conflict of interest.

References

1. Elahinia, M.H.; Hashemi, M.; Tabesh, M.; Bhaduri, S.B. Manufacturing and processing of NiTi implants: A review. *Prog. Mater. Sci.* **2012**, *57*, 911–946. [CrossRef]
2. Otsuka, K.; Ren, X. Physical metallurgy of TiNi-based shape memory alloys. *Prog. Mater. Sci.* **2005**, *50*, 511–678. [CrossRef]
3. Zheng, Y.; Jiang, F.; Li, L.; Yang, H.; Liu, Y. Effect of ageing treatment on the transformation behaviour of Ti–50.9 at.% Ni alloy. *Acta Mater.* **2010**, *58*, 3444–3458. [CrossRef]
4. Fan, G.L.; Chen, W.; Yang, S.; Zhu, J.H.; Ren, X.B.; Otsuka, K. Origin of abnormal multi-stage martensitic transformation behavior in aged Ni-rich TiNi shape memory alloys. *Acta Mater.* **2004**, *52*, 4351–4362. [CrossRef]
5. Khalil-Allafi, J.; Dlouhý, A.; Eggeler, G. Ni$_4$Ti$_3$-precipitation during aging of NiTi shape memory alloys and its influence on martensitic phase transformations. *Acta Mater.* **2002**, *50*, 4255–4274. [CrossRef]
6. Khalil-Allafi, J.; Ren, X.; Eggeler, G. The mechanism of multistage martensitic transformations in aged Ni-rich NiTi shape memory alloys. *Acta Mater.* **2002**, *50*, 793–803. [CrossRef]
7. Wang, X.; Kustov, S.; Li, K.; Schryvers, D.; Verlinden, B.; Van Humbeeck, J. Effect of nanoprecipitates on the transformation behavior and functional properties of a Ti-50.8 at.% Ni alloy with micron-sized grains. *Acta Mater.* **2015**, *82*, 224–233. [CrossRef]
8. Waitz, T.; Kazykhanov, V.; Karnthaler, H.P. Martensitic phase transformations in nanocrystalline NiTi studied by TEM. *Acta Mater.* **2004**, *52*, 137–147. [CrossRef]
9. Prokofiev, E.A.; Burow, A.; Payton, E.; Bhaduri, S. Suppression of Ni$_4$Ti$_3$ precipitation by grain size refinement in Ni-rich NiTi shape memory alloys. *Mem. Alloys Adv. Eng. Mater.* **2010**, *12*, 747–753. [CrossRef]
10. Prokoshkin, S.; Brailovski, V.; Dubinskiy, S.; Inaekyan, K.; Kreitcberg, A. Gradation of nanostructures in cold-rolled and annealed Ti–Ni shape memory alloys. *Shap. Mem. Superelasticity* **2016**, *2*, 12–17. [CrossRef]
11. Prokoshkin, S.; Dubinskiy, S.; Brailovski. Features of a nanosubgrained structure in deformed and annealed Ti-Ni SMA: A Brief Review. *Shap. Mem. Superelasticity* **2019**, *5*, 336–345. [CrossRef]
12. Nishida, M.; Wayman, C.M.; Honma, T. Precipitation processes in near-equiatomic TiNi shape memory alloys. *Metall. Trans.* **1986**, *A17*, 1505–1555. [CrossRef]
13. Kim, J.I.; Miyazaki, S. Effect of nano-scaled precipitates on shape memory behavior of Ti-50.9 at. % Ni alloy. *Acta Mater.* **2005**, *53*, 4545–4554. [CrossRef]
14. Valiev, R.Z.; Islamgaliev, R.K.; Alexandrov, I.V. Bulk nanostructured materials from severe plastic deformation. *Prog. Mater. Sci.* **2000**, *45*, 911–946. [CrossRef]
15. Sauvage, X.; Wilde, G.; Divinski, S.V.; Horitac, Z.; Valiev, R.Z. Grain boundaries in ultrafine grained materials processed by severe plastic deformation and related phenomena. *Mater. Sci. Eng. A* **2012**, *540*, 1–12. [CrossRef]
16. Pelton, A.R.; Russell, S.M.; DiCello, J. The physical metallurgy of nitinol for medical applications. *JOM* **2003**, *55*, 33–37. [CrossRef]
17. Cao, S.; Ke, C.B.; Zhang, X.P.; Schryvers, D. Morphological characterization and distribution of autocatalytic-grown Ni$_4$Ti$_3$ precipitates in a NiTi single crystal. *J. Alloys Compd.* **2013**, *577*, 215–218. [CrossRef]
18. Karbakhsh Ravari, B.; Farjami, S.; Nishida, M. Effects of Ni concentration and aging conditions on multistage martensitic transformation in aged Ni-rich Ti–Ni alloys. *Acta Mater.* **2014**, *69*, 17–29. [CrossRef]
19. Hu, L.; Jiang, S.; Zhang, Y. Role of severe plastic deformation in suppressing formation of R-phase and Ni$_4$Ti$_3$ precipitate of NiTi shape memory alloy. *Metals* **2017**, *7*, 145. [CrossRef]
20. Dlouhy, A.; Khalil-Allafi, J.; Eggeler, G. Multiple-step martensitic transformations in Ni-rich NiTi alloys—An in situ transmission electron microscopy investigation. *Phil. Mag.* **2003**, *83*, 339–363. [CrossRef]
21. Zhou, Y.; Zhang, J.; Fan, G.; Ding, X.; Sun, J.; Ren, X.; Otsuka, K. Origin of 2-stage R-phase transformation in low-temperature aged Ni-rich Ti–Ni alloys. *Acta Mater.* **2005**, *53*, 5365–5377. [CrossRef]
22. Wang, X.; Verlinden, B.; Van Humbeeck, J. Effect of post-deformation annealing on the R-phase transformation temperatures in NiTi shape memory alloys. *Intermetallics* **2015**, *62*, 43–49. [CrossRef]
23. Yang, Z.; Tirry, W.; Schryvers, D. Analytical TEM investigations on concentration gradients surrounding Ni$_4$Ti$_3$ precipitates in NiTi shape memory material. *Scr. Mater.* **2005**, *52*, 1129–1134. [CrossRef]

Article

Cyclic Stress Response Behavior of Near β Titanium Alloy and Deformation Mechanism Associated with Precipitated Phase

Siqian Zhang [1], Haoyu Zhang [1,*], Junhong Hao [1], Jing Liu [1], Jie Sun [2] and Lijia Chen [1]

[1] School of Materials Science and Engineering, Shenyang University of Technology, Shenyang 110870, China; sqzhang@alum.imr.ac.cn (S.Z.); wangronnie2016@163.com (J.H.); zwtsysmm@163.com (J.L.); chenlijia@sut.edu.cn (L.C.)

[2] State Key Laboratory of Rolling and Automation, Northeastern University, Shenyang 110004, China; sunjie@ral.neu.edu.cn

* Correspondence: zhanghaoyu@sut.edu.cn; Tel.: +86-186-2408-7566

Received: 25 September 2020; Accepted: 4 November 2020; Published: 6 November 2020

Abstract: The cyclic stress response behavior of Ti-3Al-8V-6Cr-4Mo-4Zr alloy with three different microstructures has been systematically studied. The cyclic stress response was highly related to the applied strain amplitude and precipitated phase. At low strain amplitude, the plastic deformation was mainly restricted to soft α phase, and a significant cyclic saturation stage was shown until fracture for all three alloys. At high strain amplitude, three alloys all displayed an initial striking cyclic softening. However, the softening mechanism was obviously difference. Interestingly, a significant cyclic saturation stage was noticed after an initial cyclic softening for alloy aging for 12 h, which could be attributed to the deformation of {332}<113> twin and precipitation of α″ martensite.

Keywords: cyclic stress response; cyclic softening; cyclic saturation; {332}<113> twinning; stress induced α″ martensite

---

1. Introduction

Near β titanium alloys are extensively used in industry due to their high specific strength, adequate ductility, fracture toughness, biocompatibility and excellent corrosion resistance [1–3]. It is generally known that low cycle fatigue behavior is of utmost importance in the selection of engineering materials, and cyclic stress–strain data have been connected with fatigue crack initiation and fatigue crack propagation rate [4–7]. Therefore, in order to ensure engineering applications, the cyclic stress response behavior must be comprehensively characterized.

As titanium alloys are cyclically strained during fatigue, the response stresses generally show an increase or a decrease as the number of cycles increases, which is termed cyclic hardening or cyclic softening, respectively. In practical applications, the cyclic stress response behavior of titanium alloys tends to be remarkably complex, which is highly dependent on the test temperature, applied plastic strain amplitude, strain rate, microstructure and so on [8–11]. For example, it may display an initial cyclic hardening, then an obvious softening, and finally a saturation stage is reached. Or, an obvious cyclic saturation stage is noticed after an initial cyclic softening until fracture.

Generally speaking, the cyclic stress response behavior is influenced by the competitive effect of the back stress and friction stress during fatigue deformation, which is closely related to dislocation motion [12]. For two phase titanium alloys, the softening behavior is due to a decrease in the kinematic component of stress at room temperature. It is controlled by dislocation configuration in the α phase changing from the heterogeneous distribution to the gradual homogenization [13]. On the contrary, at high temperature, the observed cyclic softening behavior is related to the second phase shearing process because of the homogeneously distributed dislocations and the occurrence of cross-slip. During fatigue deformation, the interaction between the dislocation-dislocation or mobile dislocation and the precipitation phase results in hardening [14].

For near β titanium alloys, different heat treatment processes often result in obviously different α/β morphology [15–17]. However, little information is available on the intrinsic connection between the cyclic deformation behavior and the transition of precipitation phases. In addition, it should be emphasized that the formation of α'' is often induced during fatigue deformation [18,19]. Additionally, for titanium alloys, twinning has great effect on maintaining the homogeneous plastic deformation [20–24]. Therefore, it is necessary to study the effect of α'' and twins on cyclic stress response behavior.

2. Materials and Methods

A 300 mm diameter ingot of the Ti-3Al-8V-6Cr-4Mo-4Zr alloy was produced by vacuum arc melting using pure Ti, V-Al alloy, Cr, Zr and Mo as raw materials. The ingot was forged at 850 °C to 55 mm diameter cylindrical bars, and then hot rolled at 800 °C to 16 mm in diameter rods. The chemical compositions results are given in Table 1. According to the previous studies [25], after solution treated at 800 °C for 0.5 h (AC)+aging at 500 °C, the alloys have a good resistance/ductility combination. So, samples were heat treated at 800 °C for 0.5 h and cooled in air. Then some of these specimens were aged at 500 °C for 4, 8, 12 and 24 h, respectively, and cooled in air.

Table 1. Chemical composition of alloys (wt.%).

Alloy	Al	V	Cr	Mo	Zr	Ti
Ti-3Al-8V-6Cr-4Mo-4Zr	2.9	7.8	6.1	3.9	4.0	Bal.

Uniaxial tensile properties were tested at room temperature (~25 °C) and atmosphere using a rectangular specimen with cross section 2.0 mm × 3.0 mm and a gage length 13 mm. The LCF specimens with a gage length 20 mm and diameter 5 mm were processed, ground and polished. The LCF tests under total strain controlled were conducted at room temperature (~25 °C) and atmosphere using an MTS landmark 370.10 servohydraulic test system (MTS, Eden Prairie, Minnesota, USA) with the strain ratios (R) of −1 and a frequency of 0.5 Hz.

The TEM slices were mechanically thinned down to a thickness of approximately 50 μm. Discs of 3 mm in diameter were punched out of the thin sheets and electro-polished with 60 mL perchloric acid, 85 mL n-butanol and 150 mL methanol at temperature −33 °C. Microstructural observations were conducted with SEM (Zeiss Gemini 500, Heidenheim, Germany) and TEM (JEOL JEM-2100, Tokyo, Japan).

3. Results

3.1. Microstructure before Fatigue Deformation

Figure 1 shows TEM microstructure of the alloy with three different heat treatment processes. It can be found that α phase was precipitated from β grains after heat treatment. With increasing the ageing time, the quantity of α phase decreased (volume fraction), but the size increased gradually. Moreover, the distance between α phase was also increasing.

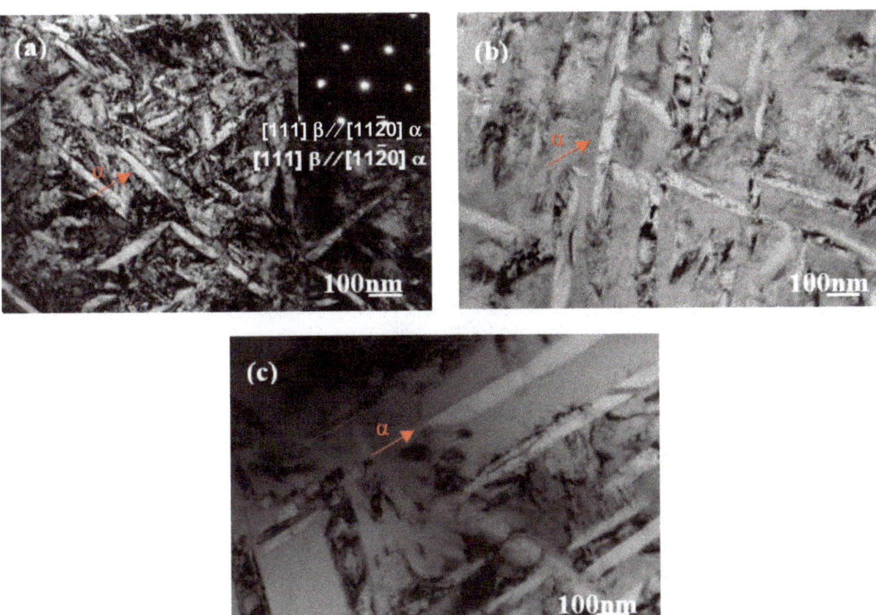

Figure 1. Microstructures of the alloy in different heat treatment conditions, (**a**) 800 °C/30 min + 500 °C/4 h (**b**) 800 °C/30 min + 500 °C/12 h (**c**) 800 °C/30 min + 500 °C/24 h.

3.2. Tensile Properties

The mechanical properties of the alloy with different heat treatment microstructures were evaluated first by tensile tests, and the results are shown in Table 2. As the aging time increased, both the ultimate tensile strength (σ_b) and yield strength ($\sigma_{0.2}$) increased significantly and then decreased rapidly, but the ductility almost stayed the same. Therefore, we can confirm that the alloy possessed the best tensile property after 12 h aging.

Table 2. Tensile properties of the alloy at room temperature.

Heat Treatment States	σ_b/MPa	$\sigma_{0.2}$/MPa	φ(Elongation at Maximum Tensile Strength)	ψ(Reduction of Area)
800 °C/30 min + 500 °C/4 h	1205.2	1137.8	12.98%	16.94%
800 °C/30 min + 500 °C/12 h	1452.1	1436.9	12.75%	15.85%
800 °C/30 min + 500 °C/24 h	1115.6	1088.6	12.67%	15.18%

3.3. Strain Amplitude in Fatigue

It can be seen from Figure 2 that the three curves (total, elastic and plastic strains) are linear on log–log scale. There is an intersection of elastic and plastic strain-life curves, called as the transition fatigue life 2Nt, the life at which elastic and plastic regions of strain are equal. For the alloy after aging, at the lower cycle region when 2Nf ≤ 2Nt, the plastic strain plays a main role and fatigue properties are dominated by strength, while at the higher cycle region when 2Nf ≥ 2Nt, the elastic strain plays a main role and fatigue properties are dominated by ductility.

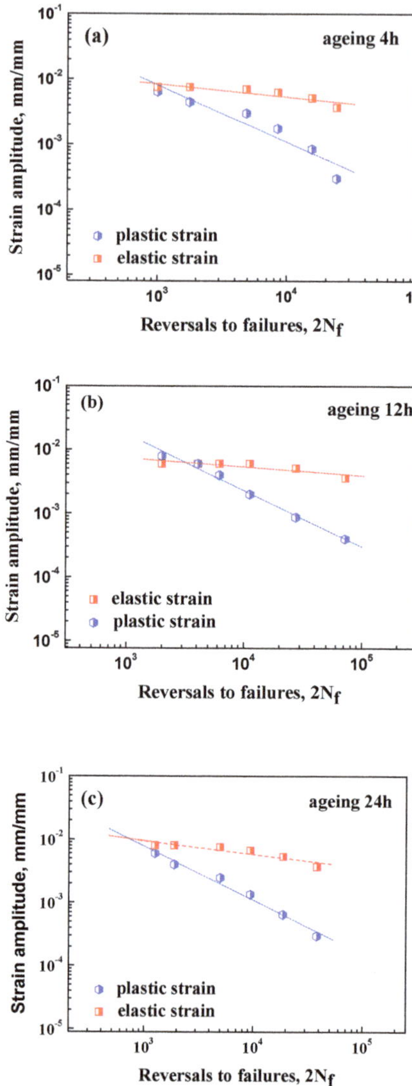

Figure 2. Strain amplitude versus reversals to failure curves for Ti-3Al-8V-6Cr-4Mo-4Zr alloys in different state. (**a**) Aging after 4 h, (**b**) aging after 12 h, (**c**) aging after 24 h.

3.4. Cyclic Stress Response Behaviour

Strain-controlled low cycle fatigue behaviors of the alloys with different heat treatment microstructures were studied at R = −1 and the frequency of 0.5 Hz at room temperature and atmosphere. The cyclic stress response curves of alloy with different aging times are showed in Figure 3. It can be found that fatigue life for each alloy decreased as the strain amplitude increases. The cyclic stress response behavior was highly dependent on aging time and strain amplitude.

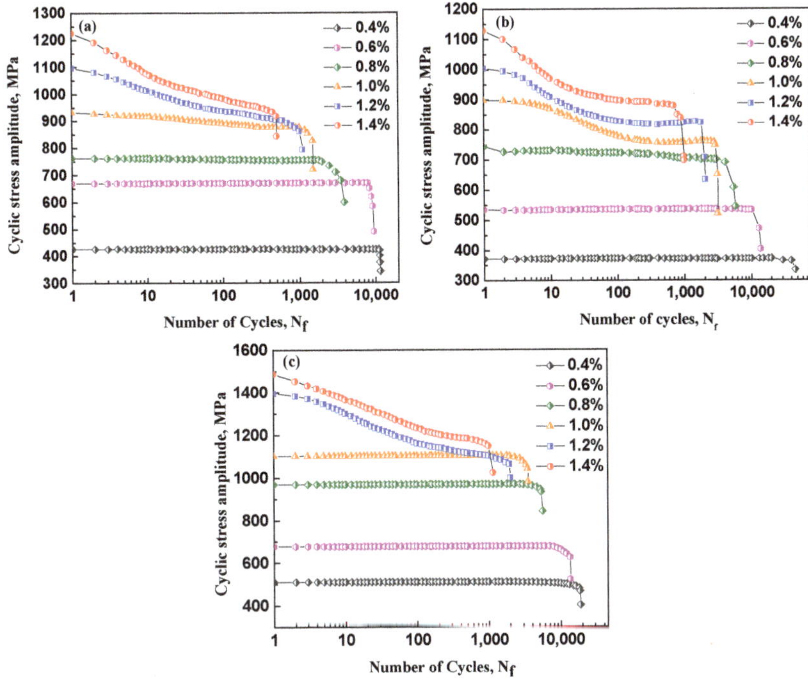

Figure 3. Cyclic stress response curves of the alloy after aging treatment. (**a**) 4 h, (**b**) 12 h (**c**) 24 h.

(a) At low total strain amplitude (alloys aging for 4 h and 24 h: $\Delta\varepsilon/2 \leq 1.0\%$, alloy aging for 12 h: $\Delta\varepsilon/2 \leq 0.8\%$), cyclic saturation was exhibited, followed by a short stage of softening.

(b) At high total strain amplitude (alloys aging for 4 h and 24 h: $\Delta\varepsilon/2 > 1.0\%$, alloy aging for 12 h: $\Delta\varepsilon/2 > 0.8\%$), the alloy generally exhibited a rapid cyclic softening. It is worth mentioning that, for the alloy aging for 12 h at $\Delta\varepsilon/2 \geq 1.0\%$, cyclic saturation stage after the initial softening was reached and extended through the last part of the fatigue life.

3.5. Microstructure after Fatigue Deformation

Microstructures corresponding to the different heat treatment alloy after the low cycle fatigue (LCF) were characterized by TEM analysis. Figure 4 is a TEM microstructure of alloy after LCF under 0.4% total strain amplitude. From the picture, it can be noticed that the dislocation density increased obviously after LCF deformation, compared to heat treatment microstructure. Figure 5 shows the TEM microstructure of alloy after LCF under 1.4% total strain amplitude. Compared with the microstructure under 0.4% total strain amplitude, the dislocation density increased further, especially for alloy aging for 4 h (Figure 5a). In addition, through a lower magnification overview of the microstructure (Figure 5b), it can be also found that the alloy aging for 12 h consists of nano-size twins which were locally distributed. The selected-area taken from the red circle region marked in Figure 5b is presented in Figure 5c, from which we can determined that the ~5 nm twin plate is a typical {332}<113> twin pattern. Interestingly, some extra reflections which can be inferred as the α'' martensite structure were also visible (Figure 5d), in addition to the formation of {332}<113> twin. It can be seen that the stress induced the formation of α'' martensite is located on the one side of the twin boundary. Moreover, a detwinning process also took place after LCF (Figure 5e). Some parts of the twins decreased obviously as the size of α'' martensite was increased. Interestingly, the twins might transform back into the matrix, thus significantly decreasing the area of the twin interface i.e., decreasing of the total interfacial energy, which might be the driving force of the process of detwinning. The {332}<113> twins partly transformed into the matrix and the α'' martensite at the shared interface took place of the detwinning area (Figure 5e). These observations indicated that the formation of stress induced α'' martensite might have undergone a complex process, and the nature of the phenomenon needs to be investigated deeply later. As to the alloy aging for 4 h and 24 h under 1.4% total strain amplitude, With the increase of dislocation density, it is easy to produce multi-system slip and reduce the effective stress, thus showing the trend of cyclic softening(Figure 5a,f).

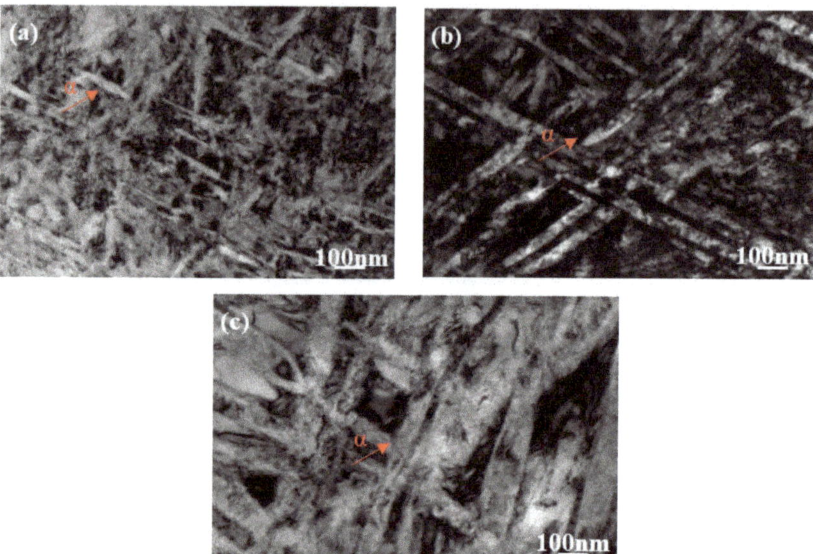

Figure 4. TEM images of the alloys with different aging treatment after fatigue tests under 0.4% total strain range. (**a**) 4 h, (**b**) 12 h and (**c**) 24 h.

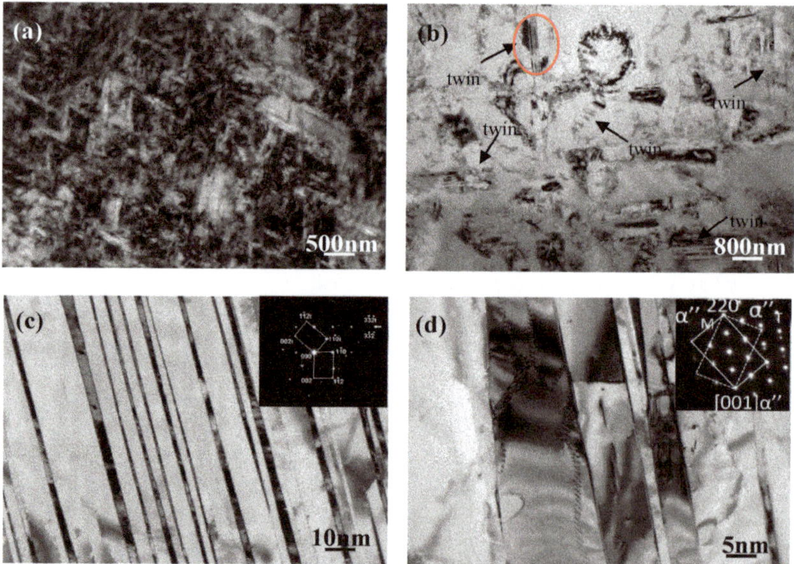

Figure 5. *Cont.*

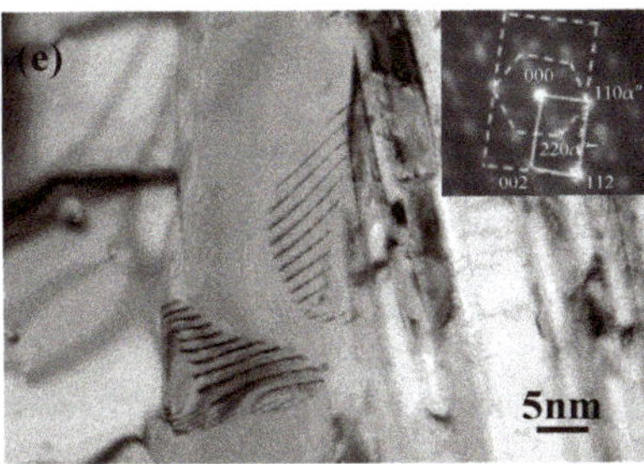

Figure 5. TEM images of the alloys with different aging treatment after fatigue tests under 1.4% total strain range. (**a**) 4 h, (**b–e**) 12 h and (**f**) 24 h.

4. Discussion

According to the above results, we can infer that the cyclic stress response behavior of near β titanium alloy was highly dependent on the applied plastic strain amplitude and precipitated phase.

At low strain amplitude, an apparently cyclic saturation stage was exhibited until fracture for all three alloys. It appeared that the plastic deformation was restricted to the soft α phase, with the hard β-phase staying in its elastic domain. Therefore, it can be considered that the dislocations movement in matrix was negligible due to the large quantity of α/β interfaces which hinder gliding. It is interesting, for alloys aged for 24 h possess a perfect cyclic stability up to a total strain amplitude of $\Delta\varepsilon/2 = 1.0\%$ and stress amplitude up to 1100 MPa. It could be because the distance between α phase was relatively large, and α″ martensite and twins were not easily precipitated for alloys aged for 24 h.

At high strain amplitude, three alloys all displayed an initial striking cyclic softening. However, the softening mechanism may be obviously different. For alloys aging for 4 h and 24 h, the softening behavior was closely related to mobile dislocation. It was well known that prismatic slip could be activated at high cyclic strain amplitude. The continuous softening was controlled by the gradual homogenization of the initially heterogeneous distribution of the dislocations in the α phase in association with cross-slip. The cross-slip activation at room temperature may be related to the presence of additional elements in the α phase and to a high stacking fault energy associated with the prismatic slip [11]. In addition, for alloys aging for 12 h, the formation of the twin due to strain incompatibility might also have effect on the initial softening response of the samples during cycling. It was noticed that the microstructure evolution and cyclic stress response were connected with the {332}<113> twins in this study. As the twins formed and grew, the trend of the cyclic stress curve, shown in Figure 4, might be well explained by softening the mechanism in terms of a crystallographic orientation of a deforming crystal structure. Earlier reports indicated that deformation twinning can induce the structural softening in a twinned region, since texture evolution was associated with the lattice reorientation [26]. TEM studies have indicated that slip precedes twin formation [27].

Interestingly, for alloy aging for 12 h, a significantly cyclic saturation stage was noticed after an initial cyclic softening. This would imply that there was a balance between the counteracting effects of the softening and hardening mechanism. According to the TEM microstructure, we can infer that the precipitation of α″ martensite was the main reason for the hardening. As the cycling proceeds, the formation of α″ pined the dislocations and further increased the stress. It can be said that the irregular twin boundary assists the transformation of α″ martensite through lattice reorientation [28,29]. From Figure 4, it can be found that some of the twins partly transformed into the matrix and α″ martensite at the shared interface took the place of the detwinning area due to the decrease in strain energy. During the LCF process, the residual stress near the twin region might be partially released and the lattice strain around the twin would gradually relax. Furthermore, the twins make opposite contributions to the macroscopic strains. As deformation progresses, the stress can be released by local compatible deformation. Under an external applied stress, a crystal with a twinned part has to make a positive contribution with respect to the macroscopic strain field to be compatible with its surroundings, which gives rise to the occurrence of detwinning [30]. Then, the SIM at the Shared interface took place of the detwinning area. The cyclic saturation stress in the alloy aging for 12 h ($\Delta\varepsilon/2 > 0.8\%$) which associates with the increased strain hardening caused by SIM can be clarified by dynamic Hall–Petch theory [31]. The formation of α'' martensite rather adjoined with the twin boundaries obviously decreased the effective dislocation glide distance, and β grains could be segmented into smaller zones by the deformation of the SIM phase and the phase boundaries acted as impenetrable barriers for the dislocation glide [32]. Figure 4 shows that many dislocations slip into α'' martensite and form an immovable dislocation array, and the movable dislocation decreases with the cyclic stress, which provides support for the enhanced strain hardening, and the cooperation between α'' martensite and twins causes the formation of cyclic stress stability.

5. Conclusions

In the present work, the cyclic stress response behavior of Ti-3Al-8V-6Cr-4Mo-4Zr alloy with three different microstructures has been studied. The following main conclusions can be drawn in this work:

(1) The alloy aging for 12 h possessed the optimal α/β morphology and exhibited the best tensile and LCF properties.

(2) At low strain amplitude, an obvious cyclic saturation stage was revealed until fracture for all three alloys.

(3) At high strain amplitude, three alloys all displayed an initial striking cyclic softening. However, for alloy aging for 12 h, a cyclic saturation stage was obtained after an initial cyclic softening.

(4) The deformation of {332}<113> twin and precipitation of α'' martensite were found to have a crucial influence on the cyclic stress response behavior.

Author Contributions: Contributed materials and performed the experiments, J.L., H.Z., J.H. and L.C.; designed the experiments, J.S.; analyzed the data and wrote the paper, S.Z. All authors have read and agreed to the published version of the manuscript.

Funding: This research was funded by Liaoning provincial Natural Science Foundation projects (20180550998), National Natural Science Foundation of China (51501117) and (52071219), and Foundation of state key laboratory of rolling and automation, northeastern university (2018RALKFKT010).

Conflicts of Interest: The authors declare no conflict of interest.

References

1. Koizumi, H.; Takeuchi, Y.; Imai, H.; Kawai, T.; Yoneyama, T. Application of titanium and titanium alloys to fixed dental prostheses. *J. Prosthodont. Res.* **2019**, *63*, 266–270. [CrossRef] [PubMed]

2. Hong, S.H.; Hwang, Y.J.; Park, S.W.; Park, C.H.; Yeom, J.-T.; Park, J.M.; Kim, K. Low-cost beta titanium cast alloys with good tensile properties developed with addition of commercial material. *J. Alloys Compd.* **2019**, *793*, 271–276. [CrossRef]

3. Bin Asim, U.; Siddiq, A.; Kartal, M.E. A CPFEM based study to understand the void growth in high strength dual-phase titanium alloy (Ti-10V-2Fe-3Al). *Int. J. Plast.* **2019**, *122*, 188–211. [CrossRef]

4. Srivatsan, T.; Soboyejo, W.; Lederich, R. The cyclic fatigue and fracture behavior of a titanium alloy metal matrix composite. *Eng. Fract. Mech.* **1995**, *52*, 467–491. [CrossRef]

5. Grosdidier, T.; Combress, Y.; Gautier, E.; Philippe, M.J. Effect of microstructure variations on the formation of deformation-induced martensite and associated tensile properties in a b metastable Ti alloy. *Metall. Mater. Trans. A* **2000**, *31A*, 1095–1106. [CrossRef]

6. Hua, K.; Zhang, Y.; Gan, W.; Kou, H.; Beausir, B.; Li, J.; Esling, C. Hot deformation behavior originated from dislocation activity and β to α phase transformation in a metastable β titanium alloy. *Int. J. Plast.* **2019**, *119*, 200–214. [CrossRef]

7. Sun, Q.Y.; Song, S.J.; Zhu, R.H.; Gu, H.C. Toughening of titanium alloys by twinning and martensite transformation. *J. Mater. Sci.* **2002**, *37*, 2543–2547. [CrossRef]

8. Gouthama, N.S.; Singh, V. Low cycle fatigue behaviour of Ti alloy Timetal 834 at 873K. *Int. J. Fatigue* **2007**, *29*, 843–851. [CrossRef]

9. Zhang, Y.; Chen, Z.; Qu, S.J.; Feng, A.; Mi, G.; Shen, J.; Huang, X.; Chen, D. Microstructure and cyclic deformation behavior of a 3D-printed Ti–6Al–4V alloy. *J. Alloys Compd.* **2020**, *825*, 153971. [CrossRef]

10. Ibrahim, A.M.H.; Balog, M.; Krizik, P.; Novy, F.; Cetin, Y.; Svec, P.; Bajana, O.; Drienovsky, M. Partially biodegradable Ti-based composites for biomedical applications subjected to intense and cyclic loading. *J. Alloys Compd.* **2020**, *839*, 155663. [CrossRef]

11. Luquiau, D.; Feaugas, X.; Clavel, M. Cyclic softenimg of the Ti-10V-2Fe-3Al titanium alloy. *Mater. Sci. Eng. A.* **1997**, *224*, 146–156. [CrossRef]

12. Cottrell, A.H. *Dislocations and Plastic Flow in Crystals*; Oxford University Press: Oxford, London, 1953.

13. Béranger, A.; Feaugas, X.; Clavel, M. Low cycle fatigue behavior of an α + β titanium alloy: Ti6246. *Mater. Sci. Eng. A* **1993**, *172*, 31–41. [CrossRef]

14. Giugliano, D.; Cho, N.-K.; Chen, H.; Gentile, L. Cyclic plasticity and creep-cyclic plasticity behaviours of the SiC/Ti-6242 particulate reinforced titanium matrix composites under thermo-mechanical loadings. *Compos. Struct.* **2019**, *218*, 204–216. [CrossRef]

15. Santos, P.F.; Niinomi, M.; Cho, K.; Liu, H.; Nakai, M.; Narushima, T.; Ueda, K.; Itoh, Y. Effects of Mo Addition on the Mechanical Properties and Microstructures of Ti-Mn Alloys Fabricated by Metal Injection Molding for Biomedical Applications. *Mater. Trans.* **2017**, *58*, 271–279. [CrossRef]

16. Du, Z.; Ma, Y.; Liu, F.; Zhao, X.; Chen, Y.; Li, G.; Liu, G.; Chen, Y. Improving mechanical properties of near beta titanium alloy by high-low duplex aging. *Mater. Sci. Eng. A* **2019**, *754*, 702–707. [CrossRef]

17. Bertrand, E.; Castany, P.; Péron, I.; Gloriant, T. Twinning system selection in a metastable β-titanium alloy by Schmid factor analysis. *Scr. Mater.* **2011**, *64*, 1110–1113. [CrossRef]

18. Lee, B.-S.; Im, Y.-D.; Kim, H.G.; Kim, K.; Kim, W.-Y.; Lim, S.-H. Stress Induced α" Martensitic Transformation Mechanism in Deformation Twinning of Metastable β-Type Ti-27Nb-0.5Ge Alloy under Tension. *Mater. Trans.* **2016**, *57*, 1868–1871. [CrossRef]

19. Blackburn, M.J.; Feeny, J.A. Stress-induced transformations in Ti-Mo alloys. *J. Inst. Met.* **1971**, *99*, 132–134.

20. Lai, M.J.; Tasan, C.C.; Raabe, D. On the mechanism of {332} twinning in metastable β titanium alloys. *Acta Mater.* **2016**, *111*, 173–186. [CrossRef]

21. Gao, Y.; Zheng, Y.; Fraser, H.; Wang, Y. Intrinsic coupling between twinning plasticity and transformation plasticity in metastable β Ti-alloys: A symmetry and pathway analysis. *Acta Mater.* **2020**, *196*, 488–504. [CrossRef]

22. Gong, M.Y.; Xu, S.; Capolungo, L.; Tomé, C.N.; Wang, J. Interactions between <a>dislocations and three-dimensional {1122} twin in Ti. *Acta Mater.* **2020**, *195*, 597–610. [CrossRef]

23. Chen, N.; Aldareguia, J.M.M.; Kou, H.C.; Qiang, F.M.; Wu, Z.H.; Li, J.S. Reversion martensitic phase transformation induced {3 3 2}<1 1 3> twinning in metastable β-Ti alloys. *Mater. Lett.* **2020**, *272*, 127883. [CrossRef]

24. Ren, L.; Xiao, W.; Kent, D.; Wan, M.; Ma, C.; Zhou, L. Simultaneously enhanced strength and ductility in a metastable β-Ti alloy by stress-induced hierarchical twin structure. *Scr. Mater.* **2020**, *184*, 6–11. [CrossRef]

25. Cao, S.; Zhou, X.; Lim, C.V.S.; Boyer, R.R.; Williams, J.C.; Wu, X. A strong and ductile Ti-3Al-8V-6Cr-4Mo-4Zr (Beta-C) alloy achieved by introducing trace carbon addition and cold work. *Scr. Mater.* **2020**, *178*, 124–128. [CrossRef]

26. Shin, S.; Zhu, C.Y.; Vecchio, K.S. Observations on {332}<113> twinning- induced softening in Ti-Nb Gum Metal. *Mater. Sci. Eng. A.* **2018**, *724*, 189–198. [CrossRef]

27. De Cooman, B.C.; Estrin, Y.; Kim, S.K. Twinning-induced plasticity (TWIP) steels. *Acta Mater.* **2018**, *142*, 283–362. [CrossRef]

28. Chen, B.; Sun, W. Transitional structure of {332}<113> β twin boundary in a deformed metastable β-type Ti-Nb-based alloy, revealed by atomic resolution electron microscopy. *Scr. Mater.* **2018**, *150*, 115–119. [CrossRef]

29. Castany, P.; Yang, Y.; Bertrand, E.; Gloriant, T. Reversion of a parent {130}<310>α" martensitic Twinning system at the origin of {332}<113> β twins observed in metastable β titanium alloys. *Phys. Rev. Lett.* **2016**, *117*, 245501. [CrossRef] [PubMed]

30. Qu, L.; Yang, Y.; Lu, Y.F.; Feng, L.; Ju, J.H.; Ge, P.; Zhou, W.; Han, D.; Ping, D.H. A detwinning process of {332}<113> twins in beta titanium alloys. *Scr. Mater.* **2013**, *69*, 389–392. [CrossRef]

31. Idrissi, H.; Renard, K.; Schryvers, D.; Jacques, P. On the relationship between the twin internal structure and the work-hardening rate of TWIP steels. *Scr. Mater.* **2010**, *63*, 961–964. [CrossRef]

32. Cho, K.; Morioka, R.; Harjo, S.; Kawasaki, T.; Yasuda, H.Y. Study on formation mechanism of {332}<113> deformation twinning in metastable β-type Ti alloy focusing on stress-induced α″ martensite phase. *Scr. Mater.* **2020**, *177*, 106–111. [CrossRef]

Publisher's Note: MDPI stays neutral with regard to jurisdictional claims in published maps and institutional affiliations.

 *metals* 

Article

Recovery of Scratch Grooves in Ti-6Al-4V Alloy Caused by Reversible Phase Transformations

Artur R. Shugurov [1,*], **Alexey V. Panin** [1,2], **Andrey I. Dmitriev** [1,3] **and Anton Yu. Nikonov** [1,3]

[1]   ISPMS Institute of Strength Physics and Material Science SB RAS, Akademicheskii pr. 2/4, 634055 Tomsk, Russia; pav@ispms.ru (A.V.P.); dmitr@ispms.ru (A.I.D.); anickonoff@ispms.ru (A.Y.N.)
[2]   TPU National Research Tomsk Polytechnic University, School of Nuclear Science & Engineering, Lenin av. 30, 634050 Tomsk, Russia
[3]   TSU Tomsk State University, Department of Metal Physics, Lenin av. 36, 634050 Tomsk, Russia
*   Correspondence: shugurov@ispms.ru; Tel.: +7-3822-286-823

Received: 18 September 2020; Accepted: 2 October 2020; Published: 5 October 2020

Abstract: The deformation behaviors of Ti-6Al-4V alloy samples with lamellar and bimodal microstructures under scratch testing were studied experimentally and using molecular dynamics simulation. It was found that the scratch depth in the sample with a bimodal microstructure was twice as shallow as that measured in the sample with a lamellar microstructure. This effect is attributed to the higher hardness of the sample with a bimodal microstructure and the larger amount of elastic recovery of scratch grooves in this sample. On the basis of the results of molecular dynamics simulation, a mechanism was proposed, which associates the recovery of the scratch grooves with the inhomogeneous vanadium distribution in the β-areas. The calculations showed that at a vanadium content typical for Ti-6Al-4V alloy, both the body-centered cubic (BCC) and hexagonal close-packed (HCP) structures can be more energetically favorable depending on the atomic volume. Therefore, compressive or tensile stresses induced by the indenter could facilitate $\beta \rightarrow \alpha$ and $\alpha \rightarrow \beta$ phase transformations, respectively, in the vanadium-depleted domains of the β-areas, which contribute to the recovery of the Ti-6Al-4V alloy subjected to scratching.

Keywords: Ti-6Al-4V; scratch testing; molecular dynamics; microstructure; phase transformations

---

1. Introduction

Due to the remarkable combination of high specific strength, strong resistance to creep, excellent corrosion resistance and low heat conductivity, titanium and its alloys have been extensively used in a wide range of applications in aerospace, biomedical, chemical, marine, automotive and many other industries [1]. In most applications, high-strength $\alpha + \beta$ two-phase titanium alloys composed of HCP-α phase and BCC-β phase are generally utilized, among which Ti-6Al-4V is the most widely used and commercially available alloy. However, the poor wear resistance of titanium alloys, including Ti-6Al-4V, is still the main shortcoming that restricts their application, particularly in areas involving friction and wear [2,3].

Classical theories of wear suppose that the wear resistance of materials under abrasive [4] and adhesive [5] wear conditions is primarily defined by their hardness, and harder materials are usually more wear-resistant. However, as early as in the middle of the last century an improvement of the wear resistance of materials was also attributed to an increase in their ultimate elastic strain, i.e., to an increase in the contribution of the elastic strain to the total strain [6]. Therefore, it is currently commonly believed that the combination of high hardness H with low Young's modulus E, i.e., high H/E ratio, provides a better wear resistance to materials [7,8].

Since the mechanical properties of materials are highly dependent on their microstructure, the obvious way to obtain a material with a high H/E ratio is the optimization of its microstructure by varying processing parameters. In the case of Ti-6Al-4V alloy, a large variety of microstructures, including lamellar, martensite and equiaxed globular microstructures, as well as bimodal (duplex) microstructures composed of equiaxed α grains and transformed β areas, can be obtained depending on the thermomechanical processing routes [9,10]. While the effect of the microstructure of Ti-6Al-4V alloy on its mechanical properties, including hardness and Young's modulus, has been extensively studied, there is a lack of information about the effect of the Ti-6Al-4V microstructure on its elastic recovery, which should be pronounced at high H/E ratios. At the same time, recent studies have indicated that elastic recovery substantially affects the ploughing of metals, which is the main mechanism of abrasive wear [11,12].

Scratch testing is a promising technique for gaining insight into the mechanisms underlying plastic ploughing and the elastic recovery of metals. This technique makes it possible to study the deformation of multiphase materials, taking into account local features of their microstructure, i.e., to investigate the deformation of grains with different crystallographic orientations [13,14], to demonstrate the effects of internal interfaces [15,16], individual phases and inclusions [14,17] on deformation behavior, as well as to reveal the development of deformation phase transformations in the materials [18,19]. In addition to the experimental studies, molecular dynamics (MD) simulation is widely used for the investigation of nucleation and the development of plastic deformation in materials subjected to scratch testing. In particular, defect-free single crystals with face-centered cubic (FCC) [20–22] and BCC [23,24] crystal lattices were the focus of MD simulation of scratching. Alhafez et al. [25] used MD simulation to study the defect generation and plasticity in HCP crystals, with the basal and prismatic planes as the surfaces subjected to scratching by a hard tip. The authors' MD simulation showed that in the course of the scratching of Ti crystals, there is extensive defect-generation in the surface layers of the materials that is followed by the formation of disordered nanodomains [16,26]. The improvement of computer performance has promoted MD simulations of scratch testing of nanosized polycrystals [27,28], amorphous materials [29] and layered materials [30].

Despite a large amount of experimental and MD studies dealing with scratching different materials, there are only a few works considering the effect of the elastic recovery of a scratch groove on the mechanical and tribological properties of materials and mechanisms responsible for their enhanced recovery. Fan et al. [31] showed that the elastic recovery of fused silica under scratching has a significant effect on its hardness and friction coefficient. It has been supposed that reversible phase transformations could be responsible for the substantial recovery of scratch grooves in ultrasonically treated commercially pure titanium and Ti-6Al-4V [32]. More recently, it has been found that the development of direct and reverse $\alpha'' \rightarrow \alpha \rightarrow \alpha''$ martensitic transformations in the uppermost surface layer of Ti-6Al-4V samples subjected to low-energy, high-current pulsed electron beam treatment resulted in the enhancement of the elastic recovery of scratch grooves [33]. However, the effect of the microstructure of Ti-6Al-4V alloy on its ploughing behavior and elastic recovery has not been properly addressed. The objective of this work is to study, experimentally and using MD simulations, the mechanisms underlying the elastic recovery of scratch grooves in Ti-6Al-4V alloys with lamellar and bimodal microstructures.

2. Experimental Details

Two polycrystalline samples of Ti-6Al-4V alloy subjected to different processing routes were studied: a sample subjected to vacuum annealing for 2 h at a temperature of 980 °C (sample 1) and a sample obtained by hot compression at 950 °C with subsequent annealing at 900 °C for 1 h (sample 2). The rectangular samples 10×10 mm^2 in size and 2 mm in height were cut from Ti-6Al-4V alloy billets. The samples were mechanically ground and subsequently electropolished to obtain smooth surfaces. The microstructure characterization of the samples was performed with an Axiovert 40 Mat optical microscope (Carl Zeiss, Göttingen, Germany). The phase compositions and structures of the coatings were investigated using X-ray diffraction (XRD) with a Shimadzu XRD-7000 X-ray diffractometer (Shimadzu Corporation, Kyoto, Japan). The investigations were performed using CuK$_\alpha$ radiation

(λ = 1.5406 Å) in the Bragg–Brentano geometry. The measurements of the mechanical properties of the samples as well as their scratch testing were carried out using a NanoTest system (Micro Materials Ltd., Wrexham, UK). The nanoindentation measurements were performed in a load-controlled mode with a Berkovich diamond tip at a maximum load of 50 mN. The hardness H, the Young's modulus E and elastic recovery R were determined using the Oliver–Pharr method.

The scratch tests were performed using a conical diamond with a tip radius of 25 µm. The scratching was carried out with a constant velocity of 10 µm/s. Scratching tracks 600 µm long were applied to all samples. In the experiment, an initial surface profile of the tested samples was detected by pre-scanning with a very low load of 0.1 µN (no wear occurs at this load). During scratching, the surface profile could be sensed and recorded by the depth sensing system. After scratching, the surface profile of the samples was scanned again to record the deformation recovery. In the second step (scratching), the normal load applied to the indenter was sharply increased to maximum loads of 50, 100 and 200 mN, and maintained at a constant level thereafter. Five scratches were performed for each sample. After scratching, the surface topography of the samples in the vicinity of the scratch tracks was scanned using a Solver HV atomic force microscope (AFM, NT-MDT Co., Moscow, Russia) operating in a contact mode. A V-shaped silicon nitride cantilever (type SNL-10, Bruker Co., Berlin, Germany) was used with a nominal spring constant of 0.35 N/m and a tip radius of 12 nm. A series of 10 cross-sectional profiles of the scratch tracks were made and averaged to determine the residual scratch depth for each titanium sample.

3. Model Description

The model crystallite of HCP-Ti (α-phase) with the inclusion of BCC-Ti (β-phase) had the shape of a parallelepiped, with dimensions of 50 × 26 × 12 nm along the X, Y and Z directions, respectively. The inclusion of BCC-Ti was modeled as an interlayer between two grains (grain 1 and grain 2) of HCP-Ti, as shown in Figure 1. The dimensions along the X axis were 20 nm for the grains and 10 nm for the inclusion. Taking into account the experimental data, the crystallographic orientations of the HCP grains along the X, Y and Z axes were chosen as $[\bar{1}6\bar{5}0]$, $[5\bar{2}3\bar{7}]$ and $[\bar{3}12\bar{2}]$ (grain 1) and $[\bar{1}2\bar{3}2]$, $[3\bar{4}1\bar{3}]$ and $[3\bar{1}2\bar{2}]$ (grain 2), correspondingly. In the case of the inclusion of BCC-Ti, the [100], [010] and [001] directions were oriented along the X, Y and Z axes, respectively. Two different Ti crystallites were studied to reveal the effect of vanadium, which is a stabilizer of the β-phase in Ti-6Al-4V alloy, on the mechanical behavior of the sample during scratching. In the first case, the inclusion of BCC-Ti contained 13 at. % of vanadium, which corresponds to a content of 4 wt. % in the model sample, which is typical for Ti-6Al-4V alloy. In the second case, a Ti crystallite with a vanadium-free inclusion was investigated. Thus, in both cases a Ti crystallite was modeled with incoherent boundaries between the grains and inclusion, for which the equilibrium configuration was obtained by the relaxation of the sample for 10 ns before scratching.

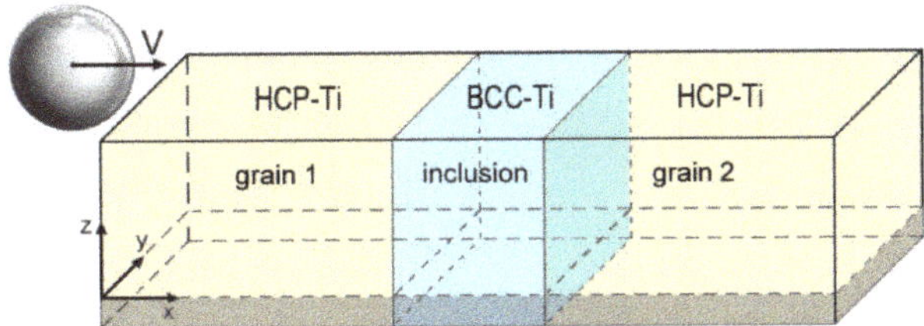

Figure 1. Molecular dynamic model of the scratch test for a HCP-Ti crystallite with inclusion of BCC-Ti.

The scratching was realized through the movement of a spherical indenter with a radius R of 6.5 nm. The indenter was moved along the X axis at a fixed depth of 3.5 nm and with a constant scratching speed of 15 m/s. Thus, atoms whose distance was less than the equilibrium radius R

were acted upon by a force directed from the force center, equal to $F = -k(R - r)^2$, where k is the coefficient reflecting tip stiffness and r is the distance between the centers of the indenter and the atom. The coefficient k was chosen to be equal to 10 eV/Å^3, similar to the earlier works [16,25,26]. The vertical immersion of the indenter into the crystallite was not considered. The lateral "incursion" of a previously immersed indenter on the initially defect-free crystallite was modeled, as shown in Figure 1. A 1.5 nm thick bottom layer (shown grey in Figure 1) simulated a fixed substrate, while other surfaces of the sample were considered free. The total number of atoms in the simulated crystallite exceeded one million. The interaction between Ti atoms was described by a potential [34] constructed using the embedded atom method. According to [34], this potential is suitable for the simulation of phase transitions and BCC-Ti. A potential obtained within the frame of the modified embedded atom method was used to describe the interaction between titanium and vanadium atoms [35]. The model sample was considered as an NVT ensemble that maintains the number of atoms N, the occupied volume V and the temperature of the system T. The initial temperature (100 K) of the simulation was achieved by using the velocity rescaling method during the MD simulation process from the balance between kinetic and thermal energies. The low temperature was used to avoid large thermal fluctuations, in order to provide a clear visualization of the structural defects. All MD calculations were implemented using the LAMMPS (large-scale atomic/molecular massively parallel simulator) molecular dynamics code [36]. To analyze the structure, the Dislocation Extraction Algorithm (DXA) and Common Neighbor Analysis (CNA) algorithms, implemented in the open visualization tool OVITO, were used [37]. The dislocation structure that evolved in HCP-Ti was analyzed only. The potential energy per atom in BCC and HCP Ti crystallites was calculated within the frame of the modified embedded atom method, using the following expression:

$$E = \frac{1}{N} \sum_i \left\{ F_i(\overline{\rho}_i) + \frac{1}{2} \sum_{i \neq j} \phi_{ij}(r_{ij}) \right\} \tag{1}$$

where F_i is the embedding function that represents the energy required to place atom i into the electron cloud, $\overline{\rho}_i$ is the electron charge density at the location of atom i, ϕ_{ij} is a pair-wise potential function, and r_{ij} is the distance between atoms i and j.

4. Experimental Results

The typical microstructures observed in the Ti-6Al-4V samples studied are shown in Figure 2. It is seen that sample 1 exhibits a fully lamellar morphology consisting of colonies of α-phase lamellae within large-body primary β-phase grains (Figure 2a). Sample 2 is characterized by a bimodal microstructure, which contains globular alpha grains and alpha/beta colonies (Figure 2b).

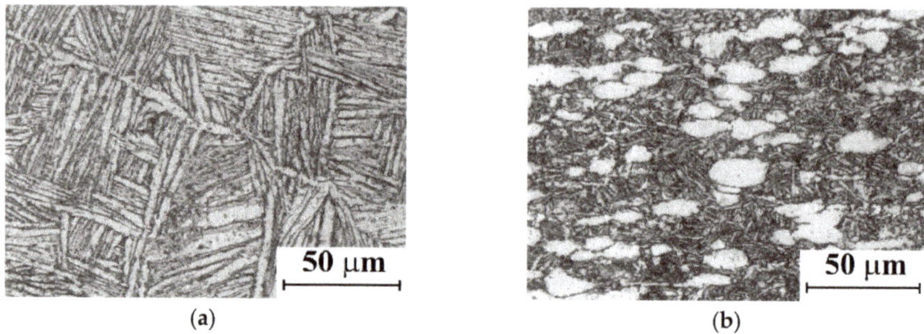

Figure 2. Lamellar (**a**) and bimodal (**b**) microstructures of Ti-6Al-4V alloy samples.

The XRD patterns of the samples primarily contain α-Ti peaks, which is typical for Ti-6Al-4V (Figure 3). The XRD pattern of sample 1 shows no preferred orientation of the α-phase (Figure 3, curve 1), while sample 2 demonstrates a pronounced crystallographic texture along the [002] direction (Figure 3, curve 2). In addition, the presence of small peaks of β-Ti phase is revealed in the XRD patterns, which indicates that there is a small amount of the β-phase retained in the final microstructure. It can be seen from Table 1 that the volume fraction of the β-phase in sample 2 is nearly 1.7 times lower than in sample 1. The lattice constants of the β-phase in samples 1 and 2 are 0.323 and 0.319 nm, respectively, which is less than the equilibrium value (0.332 nm). Since it is known that an increase in vanadium content results in decreasing the lattice constant of the β-phase in Ti-6Al-4V alloy [38], it can be inferred that the sample with a bimodal structure is characterized by higher vanadium content in the β-areas.

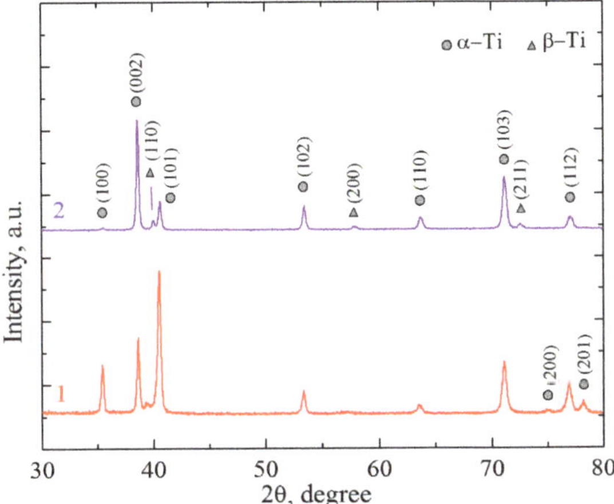

Figure 3. X-ray diffraction patterns of Ti-6Al-4V alloys with lamellar (1) and bimodal (2) microstructures.

The mechanical characteristics of the Ti-6Al-4V samples determined by instrumented indentation are listed in Table 1. It can be seen that sample 1 is characterized by a lower hardness than sample 2. In contrast, the Ti-6Al-4V sample with a bimodal microstructure has a lower Young's modulus. As a result, the *H/E* ratio and elastic recovery E_r are higher in sample 2 than in sample 1.

Table 1. Volume fraction of β-phase and mechanical properties of Ti-6Al-4V alloy samples.

Sample	Volume Fraction of β-phase, %	*H*, GPa	*E*, GPa	*H/E*	E_r, %
1	7.9	3.85 ± 0.42	154 ± 17	0.025	9
2	4.7	4.51 ± 0.44	146 ± 14	0.031	11

Scratch testing of the Ti-6Al-4V samples showed that they were subjected to ductile ploughing, which is a typical deformation mechanism for the scratching of metals. A combination of normal and tangential loads applied to the indenter resulted in the plastic edging of the material from the scratch grooves to their flanks, which led to the formation of pile-ups along the scratches (see Figure 4). It was found that deeper grooves form during scratching in sample 1, while the scratch grooves in sample 2 are rather shallow. The difference in the residual groove depths grows with increasing the applied load. As can be seen from the AFM images obtained in the vicinity of the scratch grooves and the corresponding surface profiles shown in Figure 4, the residual scratch depth at a load of 200 mN is close to 500 nm in sample 1, whereas it is less than half of this value in sample 2. The cross-sectional areas of the scratch grooves are 3.47 and 1.23 μm², respectively. Moreover, a comparison of the longitudinal surface profiles scanned before, during and after scratching shown in Figure 5 indicates that sample 2 is characterized not only by the smaller penetration depth of the indenter during scratching, but also by the higher recovery of the scratch groove. Indeed, while the groove in sample 1 is recovered after scratching, from ~800 to ~400 nm, i.e., by 50%, the groove recovery in sample 2 increases up to ~65% (from ~600 to ~200 nm). A large amount of elastic recovery of the material in sample 2 results in the formation of rather wide but shallow scratch grooves (see Figure 4d). As a result, the ratio of the cross-sectional area of the pile-ups at the groove flanks (1.02 and 0.68 μm^2 in samples 1 and 2, respectively) to the cross-sectional area of the grooves increases from 0.29 (sample 1) to 0.55 (sample 2).

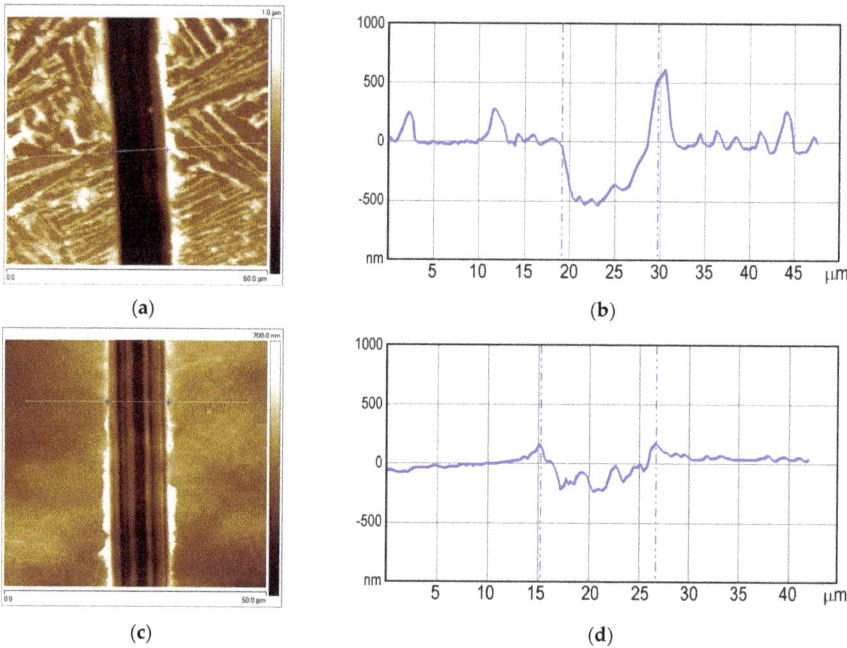

Figure 4. AFM images (**a,c**) and corresponding cross-sectional surface profiles (**b,d**) of scratch grooves formed at the applied load of 200 mN in Ti-6Al-4V alloy samples with lamellar (**a,b**) and bimodal (**c,d**) microstructures.

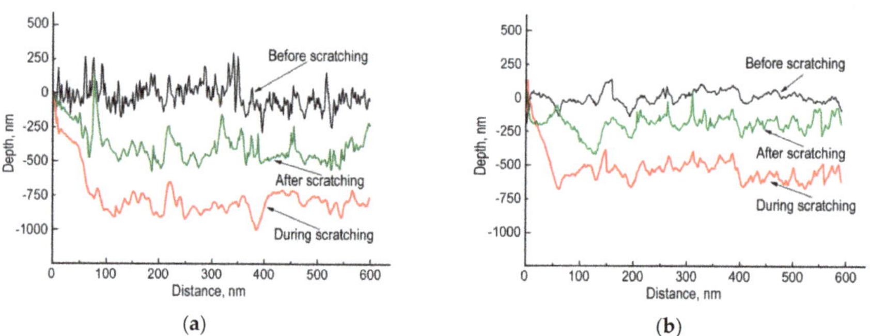

Figure 5. Surface profiles in Ti-6Al-4V samples 1 (**a**) and 2 (**b**) before, during and after scratching with an applied load of 200 mN.

The above described experiments revealed substantial difference in the elastic recovery of scratch grooves and imprints made by nanoindentation in the samples studied. The elastic recovery of the scratch grooves in the Ti-6Al-4V samples with lamellar and bimodal microstructures (50% and

65% respectively) is considerably higher than the recovery after nanoindentation (9% and 11%, correspondingly). Moreover, the difference between the E_r values for the samples reaches 15% in the case of scratching and drops to 2% for nanoindentation. These differences indicate that additional mechanisms of reversible deformation are activated upon the scratching of the Ti-6Al-4V alloy, which promotes material recovery under unloading. In particular, reversible deformation phase transformations can underlie this effect. Therefore, a molecular dynamics simulation of the evolution of the atomic structure in α-Ti crystallites with β-Ti inclusions during scratching was performed in order to gain a deeper insight into the development of reversible phase transformations in Ti-6Al-4V alloy.

5. Simulation Results

Figure 6 exhibits the evolution of the atomic structure in a Ti crystallite with a β-phase inclusion containing 13 at. % vanadium during scratching. Only BCC atoms are shown in Figure 6 in order to clearly visualize the observed changes in the atomic structure. Figure 6a,b correspond to the indenter positions in the middle of grain 1 and the BCC inclusion, respectively. A decrease in the inclusion volume, resulting from the rearrangement of a significant part of the BCC atoms into the HCP lattice of α-Ti, is clearly seen from the comparison of Figure 6a,b. After the indenter moved into grain 2, the BCC structure of the inclusion was largely recovered, as shown in Figure 6c.

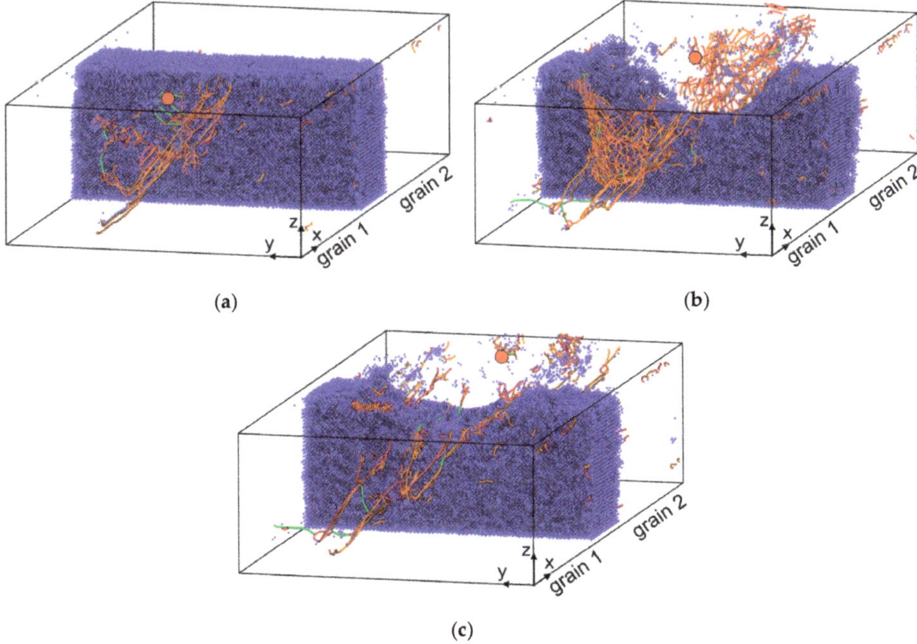

(a)　　　　　　　　　　(b)

(c)

Figure 6. Snapshots displaying the structure of the model Ti crystallite with inclusion of β-Ti phase containing 13 at. % of vanadium after 10 (**a**) 25 (**b**) and 40 nm (**c**) of scratching. BCC atoms are shown as blue dots. Colored curves denote the dislocation core. Dislocations with the Burgers vectors $\mathbf{b_1} = 1/3 \langle 1\bar{1}00 \rangle$ and $\mathbf{b_2} = 1/3 \langle 1\bar{2}10 \rangle$ are shown in orange and green, respectively. Other dislocation lines are shown in red. Only the dislocations generated in HCP-Ti are shown. The large red dot marks the position of the indenter's center.

Figure 7 demonstrates the evolution of the volume fraction of BCC atoms in the β-Ti inclusions during scratching of the Ti crystallites. It is seen that in the case of the inclusion, where the BCC structure is stabilized by vanadium, the α→β phase transformation leads to decreasing the fraction of

BCC atoms down to ~0.7 (Figure 7, curve 1). The drop in the fraction of BCC atoms starts long before the indenter reaches the interphase boundary between grain 1 and the β-Ti inclusion. The minimum of the dependence corresponds to the point when the indenter is in the middle of the inclusion. A nearly full recovery of the fraction of BCC atoms is observed after scratching.

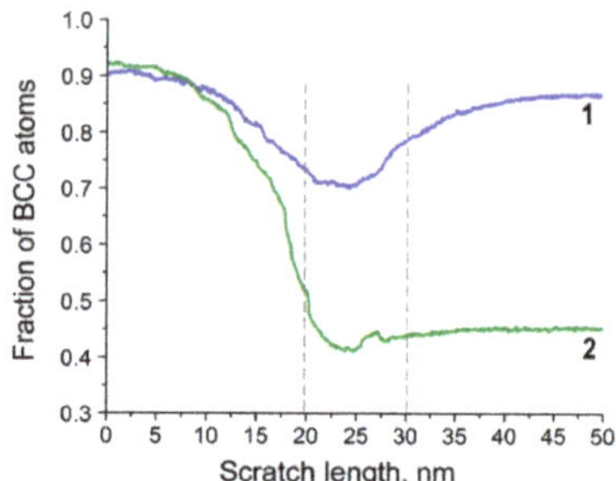

Figure 7. Volume fraction of BCC atoms in the β-Ti inclusion with 13 at. % of vanadium (curve 1) and the vanadium-free β-Ti inclusion (curve 2) as a function of the scratch length. Dashed lines indicate the boundaries of the inclusions.

The detailed studies of the evolution of atomic structure in the β-Ti inclusion containing 13 at. % of vanadium revealed that the development of phase transformations in the inclusion is concerned with its concentration inhomogeneity, i.e., non-uniform distribution of V atoms in the BCC lattice of titanium. Figure 8a exhibits the initial atomic structure of the analyzed fragment of the β-Ti inclusion (denoted as "A"), illustrating the configuration of the Ti and V atoms and the schematic showing the location of this fragment in the simulated crystallite. The dashed ovals in Figure 8a indicate the vanadium-enriched local regions (domains) of the fragment. Figure 8b–d demonstrate the evolution of the atomic structure of the fragment during scratching. CNA analysis was used to visualize atoms belonging to different local configurations of the crystal lattice. It is seen from the comparison of Figure 8a,b that the atomic rearrangement from the BCC lattice into the HCP lattice begins in the vanadium-depleted domains. In contrast, the vanadium-enriched domains largely maintain their BCC structure even when the indenter is in the middle part of the inclusion, which is accompanied by large lattice distortions (Figure 8c). The latter results in the formation of regions with a disordered atomic structure, the crystal lattice of which cannot be identified using CNA. Evidently, the atomic disorder depends on the arising of dislocations generated in the contact area of the indenter with the sample. When the indenter passes through the inclusion, the majority of the HCP atoms rearrange again into the BCC lattice, whereas the regions with a disordered lattice structure keep their structure (Figure 8d).

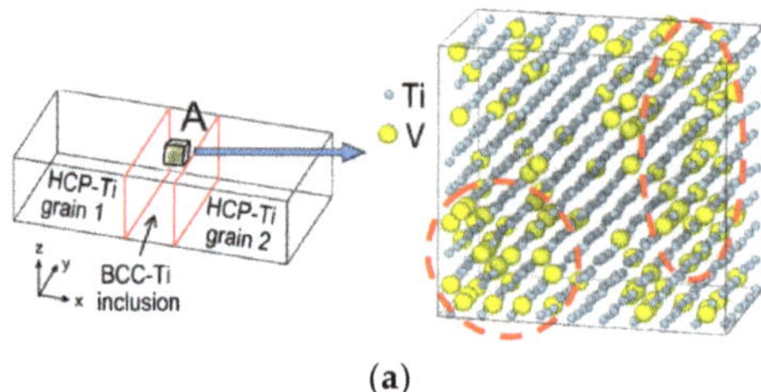

(a)

Figure 8. *Cont.*

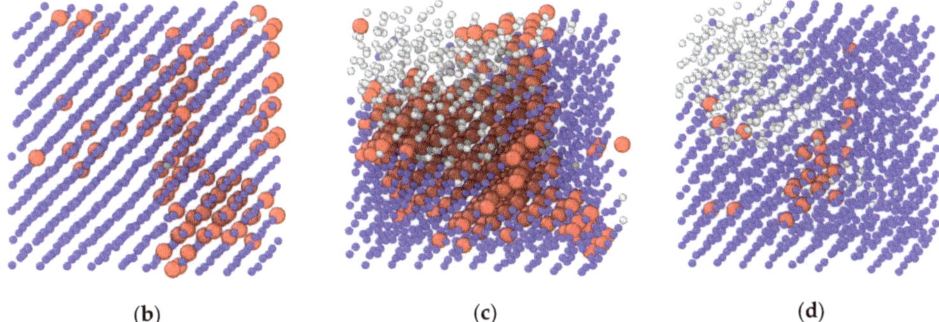

(b)   (c)   (d)

Figure 8. Location and atomic structure of fragment "A" in the β-Ti inclusion containing 13 at. % of vanadium (**a**). The dashed ovals indicate vanadium-enriched domains. Evolution of the atomic structure of the fragment at scratch lengths of 2 (**b**), 25 (**c**) and 35 nm (**d**). According to CNA [37], the HCP atoms are shown red, BCC atoms are shown in blue, and atoms belonging to unidentified lattice structures are shown in grey. For clarity, HCP atoms are shown as larger spheres.

Figure 9 shows the variation in the fractions of atoms with different local configurations of a crystal lattice in fragment "A" during scratching. It is seen that the decrease in the fraction of BCC atoms visible in Figure 7 is concerned with their rearrangement into the HCP configuration. The moving of the indenter into the β-Ti inclusion is accompanied by increasing the fraction of atoms the configuration of which is not identified because of strong lattice distortions and local atomic disordering. It should be noted that the increase in the fraction of the atoms with an unidentified lattice structure primarily occurs at the expense of decreasing the fraction of HCP atoms. After passing the indenter through the middle of the inclusion (at a scratch length of 25 nm), the restoration of the fraction of BCC atoms, and the corresponding reduction in the fractions of HCP atoms and atoms with unidentified lattice structures, are observed. Finally, when the indenter moves into grain 2, the fraction of HCP atoms in the studied fragment of the inclusion decreases by more than 90%, while the fraction of atoms with unidentified lattice structures decreases by only 20%. Thus, the restoration of the initial BCC structure primarily occurs by means of the reverse phase transformation of the HCP lattice formed under the influence of the indenter.

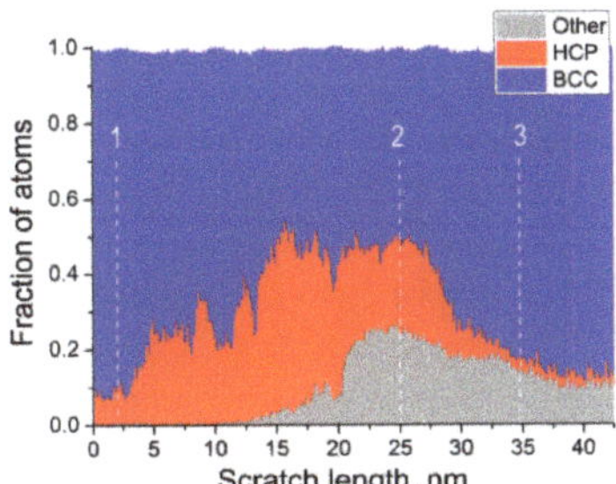

Figure 9. Fractions of atoms with different configurations of crystal lattice in fragment "A" of the β-Ti inclusion as a function of the scratch length. Vertical dashed lines 1, 2 and 3 mark positions of the indenter corresponding to the atomic structures shown in Figure 8b–d, respectively.

The key role of the domain structure of the β-Ti inclusions in the development of reverse α→β phase transformations is clearly revealed in the simulation of scratching of a Ti crystallite with a vanadium-free BCC inclusion. As shown in Figure 10, in this case, a greater fraction of BCC atoms rearranges into the HCP lattice as compared with the inclusion containing vanadium (see Figure 6).

As a result, after passing the indenter through the vanadium-free β-Ti inclusion, the initial BCC structure maintains only in its bottom layers adjacent to the substrate with the rigidly fixed BCC lattice. It is also seen from Figure 7 (curve 2) that the minimum fraction of BCC atoms in the vanadium-free β-Ti inclusion drops to 0.4, and only slightly increases (up to 0.45) after passing the indenter.

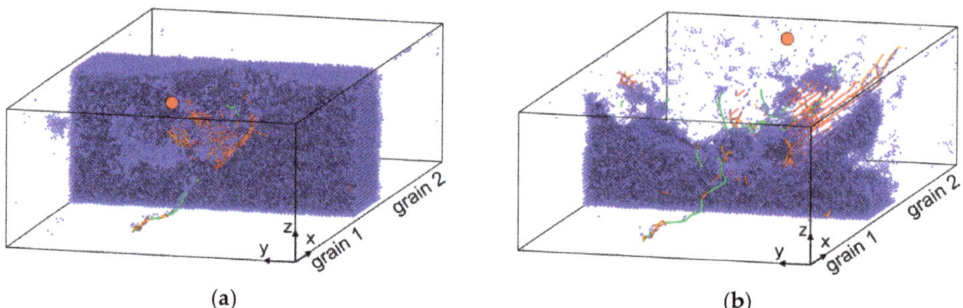

(a)　　　　　　　　　　　　　　(b)

Figure 10. Snapshots displaying the resulting structure of the model Ti crystallite with the vanadium-free β-Ti inclusion after 10 (**a**) and 37 nm (**b**) of scratching. BCC atoms are shown as blue dots. Colored curves denote the dislocations cores. Dislocations with the Burgers vectors $\mathbf{b_1} = 1/3 \langle 1\bar{1}00 \rangle$ and $\mathbf{b_2} = 1/3 \langle 1\bar{2}10 \rangle$ are shown in orange and green, respectively. Other dislocation lines are shown in red. Only the dislocations generated in HCP-Ti are shown. The large red dot marks the position of the indenter.

6. Discussion

The results of scratch testing indicate that the Ti-6Al-4V sample with a bimodal microstructure is characterized by a substantially smaller residual scratch depth than the sample with a lamellar microstructure. This is in good agreement with the previous studies [10,38], which concluded that among the various morphologies, the bimodal microstructure possesses the best combination of strength and ductility. The origin of these improved mechanical properties is usually attributed to the fact that dislocation gliding inside the α-grains was blocked at the interfaces between them and the transformed β-areas, i.e., the β-areas with a finer morphology acted as a strengthening component for the mechanical properties of the bimodal microstructure [39]. On the other hand, recent studies argued that plastic strain can localize preferentially within the transformed β-areas, and then propagate into the surrounding α-grains [40]. Strain gradients from the interface to the interior of α-grains were supposed to increase the number of geometrically necessary dislocations near the interface, which contribute to the enhanced work-hardening rate of the bimodal microstructure. Evidently, both the mechanisms can contribute to the higher hardness of the Ti-6Al-4V sample with a bimodal microstructure, which results in the smaller penetration depth of the indenter during scratching. The mechanisms should become more pronounced with increasing the load applied to the indenter during scratching, because this results in the enlargement of the zone of local plastic strains and involves a greater number of interfaces between the α-grains and the β-areas in the deformation process. Therefore, the difference in the residual scratch groove depths in the samples with lamellar and bimodal microstructures grows with increasing the applied load.

Another origin of the smaller residual scratch depth in the sample with a bimodal microstructure is its higher recovery after scratching. This effect can be partially explained by increasing the contribution of elastic deformation to the total deformation of the material that is confirmed by the nanoindentation results, which demonstrate that the Ti-6Al-4V sample with a bimodal microstructure is characterized by higher values of *H/E* and E_r than the sample with a lamellar microstructure. However, the values of the recovery of the samples under scratching (15%) are substantially different from those under nanoindentation (2%). This indicates that, in contrast to the recovery of indenter imprints in the case of nanoindentation, the recovery of scratch grooves is caused not only by the elastic deformation of the Ti-6Al-4V samples, but also by the development of additional mechanisms of reversible deformation.

One of the mechanisms contributing to the formation and recovery of the scratch grooves is the reversible phase transformations developing during scratching. In particular, the enhanced elastic recovery of scratch grooves in Ti-6Al-4V samples subjected to electron beam treatment has been attributed to transformations of the more close-packed orthorhombic α''-Ti phase into the HCP α-Ti phase, and vice versa [33]. The analysis of the total energy per atom, performed using MD calculations for Ti crystallites with BCC and HCP lattices, showed that the possibility of $\beta \leftrightarrow \alpha$ phase transformations substantially depends on the V content. Figure 11 shows the total energy per atom as a function of atomic volume in BCC and HCP Ti crystallites containing 5, 13 and 19 at. % of vanadium. It is seen from the Figure that at a low vanadium content (5 at. %), the HCP structure is more energetically favorable, while at the high content (19 at. %) the preferable configuration is BCC. At the same time, at the medium vanadium content (13 at. %), which is typical for the β-phase in Ti-6Al-4V alloy, the BCC and HCP strictures are characterized by close values of the total energy. Therefore, in the latter case both the BCC and HCP structures can be more favorable, depending on the applied external load.

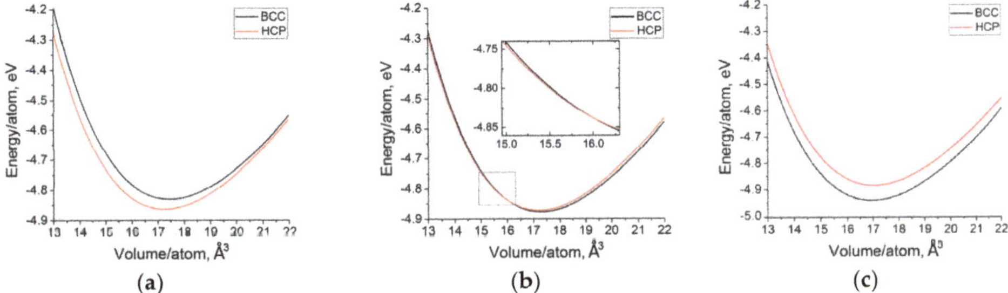

Figure 11. Energy per atom in the β-Ti inclusions with 5 (**a**), 13 (**b**) and 19 at. % of vanadium (**c**) as a function of atomic volume.

It has been shown that the indenter movement results in arising compressive stresses before and underneath it, as well as tensile stresses behind it [41]. Since the BCC β-phase of titanium is characterized by a lower packing density (0.68) than the HCP α-phase (0.74), it can be supposed that the compressive stresses arising ahead of and under the indenter facilitate the transformation of the β-phase into the more close-packed α-phase that results in deeper penetration of the indenter into the sample. It is clearly seen from Figure 11b that, although the minimum energy of the Ti crystallite with 13 at. % vanadium content in the β-phase, which lies in the range typical for Ti-6Al-4V alloy [42], corresponds to the BCC lattice, the HCP structure becomes more favorable with decreasing atomic volume.

The MD simulations showed that this $\beta \rightarrow \alpha$ phase transformation is possible only in the vanadium-depleted domains of the β-Ti inclusion, because vanadium is a β-phase stabilizer. Evidently, this effect should be more pronounced in the Ti-6Al-4V sample with lamellar microstructure, which is characterized by lower V content in the β-phase, and it is an additional factor that provides deeper scratch grooves in this sample (Figure 5a).

Phase transformations also contribute to the recovery of scratch grooves after scratching, which is substantially more pronounced in the sample with a bimodal microstructure. As can be seen from Figure 11b, with an increasing atomic volume, the BCC lattice structure becomes more favorable again in the Ti crystallite with a 13 at. % vanadium content. Therefore, when the indenter moves away, the relaxation of the compressive stresses and the arising of tensile stresses behind it favor the development of the reverse $\alpha \rightarrow \beta$ phase transformations in the vanadium-depleted domains of the β-Ti inclusion. This transformation behind the moving indenter allows a reduction in the energy of such domains, which are characterized by strong lattice distortions and, consequently, by high deformation energy. Moreover, the $\alpha \rightarrow \beta$ phase transformations in the HCP domains, surrounded by areas where the BCC lattice is stabilized by vanadium atoms, is accompanied by the annihilation of interphase boundaries that also favors a reduction in the total energy of the Ti crystallites. Apparently, the dimensions of the vanadium-depleted domains should decrease with increasing V content in the β-Ti inclusions. The decrease in the domain size leads to the growth of the volume fraction of the boundaries between the HCP and BCC domains, and consequently, their annihilation contributes more to reducing the total energy of the β-Ti inclusions. Therefore, the recovery of scratch grooves is noticeably more pronounced in the Ti-6Al-4V sample with a bimodal microstructure, which is characterized by the higher vanadium content in the β-phase.

7. Conclusions

The investigation of Ti-6Al-4V alloys with lamellar and bimodal morphologies subjected to scratch testing revealed the strong dependence of scratch groove parameters (maximum and residual scratch depths, elastic recovery, volume of pile-up generated on the scratch flanks, etc.) on the microstructure. The bimodal microstructure was found to provide a scratch depth twice as shallow as that measured in the sample with a lamellar microstructure. This effect is assumed to be caused by the following: (i) the higher hardness of the Ti-6Al-4V sample with a bimodal microstructure resulting from the smaller grain size, and an enhanced work-hardening rate induced by the accumulation of dislocations at the interfaces between α-grains and β-areas; (ii) the greater degree of elastic recovery of scratch grooves in this sample due to the higher H/E ratio and the development of reversible $\beta \rightarrow \alpha \rightarrow \beta$ phase transformations in the β-areas.

According to the performed MD simulations, an important condition for the reversibility of the $\beta \rightarrow \alpha \rightarrow \beta$ phase transformations is the presence of vanadium in the BCC crystal lattice of titanium. The calculations showed that with a medium vanadium content in the β phase (13 at. %), both the BCC and HCP structures can be more energetically favorable, depending on the atomic volume. Therefore, compressive or tensile stresses induced by the indenter can facilitate direct $\beta \rightarrow \alpha$ or reverse $\alpha \rightarrow \beta$ phase transformations, respectively. In contrast, at low (5 at. %) and high (19 at. %) vanadium contents, the HCP and BCC structures, correspondingly, are more preferable regardless of the atomic volume. As a result, the lamellar microstructure, which is characterized by the larger volume fraction of the β-phase and, consequently, the lower V content in the β-areas, does not favor reverse $\alpha \rightarrow \beta$ phase transformations, which results in the lower recovery of the scratch grooves. The results obtained indicate the need to consider the effect of the microstructure of $\alpha + \beta$ titanium alloys on the ploughing mechanisms under abrasive wear, and can be useful for elaborating the methods to increase their abrasion resistance.

Author Contributions: Conceptualization, A.R.S. and A.I.D.; methodology, A.R.S.; software, A.I.D. and A.Y.N.; investigation, A.R.S., A.V.P. and A.Y.N.; writing—original draft preparation, A.R.S. and A.I.D.; writing—review and editing, A.R.S. and A.V.P.; visualization, A.Y.N.; project administration, A.R.S.; funding acquisition, A.R.S. All authors have read and agreed to the published version of the manuscript.

Funding: The work was performed according to the Government research assignment for ISPMS SB RAS, Projects III.23.1.1 and III.23.2.4. The authors also gratefully acknowledge the financial support granted by the Russian Foundation for Basic Research and the Government of the Tomsk region of the Russian Federation, grant № 18-48-700009.

Acknowledgments: The supercomputer of Tomsk State University was used in the framework of the TSU competitiveness improvement program.

Conflicts of Interest: The authors declare no conflict of interest.

References

1. Froes, F.H. *Titanium: Physical Metallurgy, Processing and Application*; ASM International: Materials Park, OH, USA, 2015.
2. Budinski, K.G. Tribological properties of titanium alloys. *Wear* **1991**, *151*, 203–217. [CrossRef]
3. Revankar, G.D.; Shetty, R.; Rao, S.S.; Gaitonde, V.N. Wear resistance enhancement of titanium alloy (Ti–6Al–4V) by ball burnishing process. *J. Mater. Res. Technol.* **2017**, *6*, 13–32. [CrossRef]

4.   Rabinowicz, E. *Friction and Wear of Materials*; Wiley-Interscience: New York, NY, USA, 1965.

5.   Archard, J.F. Contact and rubbing of flat surfaces. *J. Appl. Phys.* **1953**, *24*, 981–988. [CrossRef]

6.   Oberle, T.L. Properties influencing wear of metals. *J. Metals.* **1951**, *3*, 438–439. [CrossRef]

7.   Leyland, A.; Matthews, A. On the significance of the H/E ratio in wear control: A nanocomposite coating approach to optimised tribological behavior. *Wear* **2000**, *246*, 1–11. [CrossRef]

8.   Attar, H.; Bermingham, M.J.; Ehtemam-Haghighi, S.; Dehghan-Manshadi, A.; Kent, D.; Dargusch, M.S. Evaluation of the mechanical and wear properties of titanium produced by three different additive manufacturing methods for biomedical application. *Mater. Sci. Eng. A* **2019**, *760*, 339–345. [CrossRef]

9.   Lutjering, G. Influence of processing on microstructure and mechanical properties of (α+β) titanium alloys. *Mater. Sci. Eng. A* **1998**, *243*, 32–45. [CrossRef]

10.   Mironov, S.; Sato, Y.S.; Kokawa, H.; Hirano, S.; Pilchak, A.L.; Semiatin, S.L. Microstructural characterization of friction-stir processed Ti-6Al-4V. *Metals* **2020**, *10*, 976. [CrossRef]

11.   Schibicheski Kurelo, B.C.E.; de Oliveira, W.R.; Serbena, F.C.; de Souza, G.B. Surface mechanics and wear resistance of supermartensitic stainless steel nitrided by plasma immersion ion implantation. *Surf. Coat. Technol.* **2018**, *353*, 199–209. [CrossRef]

12.   Tsybenko, H.; Xia, W.; Dehm, G.; Brinckmann, S. On the commensuration of plastic plowing at the microscale. *Tribol. Int.* **2020**, *151*, 106477. [CrossRef]

13.   Shugurov, A.R.; Panin, A.V.; Dmitriev, A.I.; Nikonov, A.Y. The effect of the crystallographic grain orientation of polycrystalline Ti on plastic ploughing under scratch testing. *Wear* **2018**, *408–409*, 214–221. [CrossRef]

14.   Kimm, J.; Sander, M.; Pöhl, F.; Theisen, W. Micromechanical characterization of hard phases by means of instrumented indentation and scratch testing. *Mater. Sci. Eng. A* **2019**, *768*, 138480. [CrossRef]

15.   Machado, P.C.; Pereira, J.I.; Penagos, J.J.; Yonamine, T.; Sinatora, A. The effect of in-service work hardening and crystallographic orientation on the micro-scratch wear of Hadfield steel. *Wear* **2017**, *376*, 1064–1073 [CrossRef]

16.   Dmitriev, A.I.; Nikonov, A.Y.; Shugurov, A.R.; Panin, A.V. The role of grain boundaries in rotational deformation in polycrystalline titanium under scratch testing. *Phys. Mesomech.* **2019**, *22*, 365–374. [CrossRef]

17.   Liu, J.; Zeng, Q.; Xu, S. The state-of-art in characterizing the micro/nano-structure and mechanical properties of cement-based materials via scratch test. *Constr. Build. Mater.* **2020**, *254*, 119255. [CrossRef]

18.   Chavoshi, S.Z.; Gallo, S.C.; Dong, H.; Luo, X. High temperature nanoscratching of single crystal silicon under reduced oxygen condition. *Mater. Sci. Eng. A* **2017**, *684*, 385–393. [CrossRef]

19.   Wang, B.; Melkote, S.N.; Saraogi, S.; Wang, P. Effect of scratching speed on phase transformations in high-speed scratching of monocrystalline silicon. *Mater. Sci. Eng. A* **2020**, *772*, 138836. [CrossRef]

20.   Zhang, P.; Zhao, H.; Shi, C.; Zhang, L.; Huang, H.; Ren, L. Influence of double-tip scratch and single-tip scratch on nano-scratching process via molecular dynamics simulation. *Appl. Surf. Sci.* **2013**, *280*, 751–756. [CrossRef]

21.   Zhu, P.; Hu, Y.; Fang, F.; Wang, H. Multiscale simulations of nanoindentation and nanoscratch of single crystal copper. *Appl. Surf. Sci.* **2012**, *258*, 4624–4631. [CrossRef]

22.   Ren, J.; Hao, M.; Lv, M.; Wang, S.; Zhu, B. Molecular dynamics research on ultra-high-speed grinding mechanism of monocrystalline nickel. *Appl. Surf. Sci.* **2018**, *455*, 629–634. [CrossRef]

23.   Gao, Y.; Ruestes, C.J.; Urbassek, H.M. Nanoindentation and nanoscratching of iron: Atomistic simulation of dislocation generation and reactions. *Comput. Mater. Sci.* **2014**, *90*, 232–240. [CrossRef]

24.   Wu, C.-D.; Fang, T.-H.; Lin, J.-F. Atomic-scale simulations of material behaviors and tribology properties for FCC and BCC metal films. *Mater. Lett.* **2012**, *80*, 59–62. [CrossRef]

25.   Alabd Alhafez, I.; Urbassek, H.M. Scratching of hcp metals: A molecular-dynamics study. *Comput. Mater. Sci.* **2016**, *113*, 187–197. [CrossRef]

26.   Dmitriev, A.I.; Nikonov, A.Y.; Shugurov, A.R.; Panin, A.V. Numerical study of atomic scale deformation mechanisms of Ti grains with different crystallographic orientation subjected to scratch testing. *Appl. Surf. Sci.* **2019**, *471*, 318–327. [CrossRef]

27.   Liu, Y.; Li, B.; Kong, L. A molecular dynamics investigation into nanoscale scratching mechanism of polycrystalline silicon carbide. *Comput. Mater. Sci.* **2018**, *148*, 76–86. [CrossRef]

28.   Gao, Y.; Urbassek, H.M. Scratching of nanocrystalline metals: A molecular dynamics study of Fe. *Appl. Surf. Sci.* **2016**, *389*, 688–695. [CrossRef]

29. Guo, X.; Zhai, C.; Kang, R.; Jin, Z. The mechanical properties of the scratched surface for silica glass by molecular dynamics simulation. *J. Non-Cryst. Solids.* **2015**, *420*, 1–6. [CrossRef]
30. Yang, W.; Ayoub, G.; Salehinia, I.; Mansoor, B.; Zbib, H. Multiaxial tension/compression asymmetry of Ti/TiN nano laminates: MD investigation. *Acta Mater.* **2017**, *135*, 348–360. [CrossRef]
31. Fan, X.; Li, C.; Sun, L.; Sun, H.; Jiang, Z. Hardness and friction coefficient of fused silica under scratching considering elastic recovery. *Ceram. Int.* **2020**, *46*, 8200–8208. [CrossRef]
32. Panin, V.E.; Panin, A.V.; Pochivalov, Y.I.; Elsukova, T.F.; Shugurov, A.R. Scale invariance of structural transformations in plastically deformed nanostructured solids. *Phys. Mesomech.* **2017**, *20*, 55–68. [CrossRef]
33. Sinyakova, E.A.; Panin, A.V.; Perevalova, O.B.; Shugurov, A.R.; Kalashnikov, M.P.; Teresov, A.D. The effect of phase transformations on the recovery of pulsed electron beam irradiated Ti-6Al-4V titanium alloy during scratching. *J. Alloys Compd.* **2019**, *795*, 275–283. [CrossRef]
34. Mendelev, M.I.; Underwood, T.L.; Ackland, G.J. Development of an interatomic potential for the simulation of defects, plasticity, and phase transformations in titanium. *J. Chem. Phys.* **2016**, *145*, 154102. [CrossRef] [PubMed]
35. Maisel, S.B.; Ko, W.-S.; Zhang, J.-L.; Grabowski, B.; Neugebauer, J. Thermomechanical response of NiTi shape-memory nanoprecipitates in TiV alloys. *Phys. Rev. Mater.* **2017**, *1*, 033610. [CrossRef]
36. Plimpton, S. Fast Parallel Algorithms for Short-Range Molecular Dynamics. *J. Comput. Phys.* **1995**, *117*, 1–19. [CrossRef]
37. Stukowski, A.; Bulatov, V.V.; Arsenlis, A. Automated identification and indexing of dislocations in crystal interfaces. *Model. Simul. Mater. Sci. Eng.* **2012**, *20*, 085007. [CrossRef]
38. Elmer, J.W.; Palmer, T.A.; Babu, S.S.; Specht, E.D. In situ observations of lattice expansion and transformation rates of α and β phases in Ti–6Al–4V. *Mater. Sci. Eng. A* **2005**, *391*, 104–113. [CrossRef]
39. Bridier, F.; Villechaise, P.; Mendez, J. Slip and fatigue crack formation processes in an α/β titanium alloy in relation to crystallographic texture on different scales. *Acta Mater.* **2008**, *56*, 3951–3962. [CrossRef]
40. Chong, Y.; Bhattacharjee, T.; Park, M.-H.; Shibata, A.; Tsuji, N. Factors determining room temperature mechanical properties of bimodal microstructures in Ti-6Al-4V alloy. *Mater. Sci. Eng. A* **2018**, *730*, 217–222. [CrossRef]
41. Holmberg, K.; Laukkanen, A.; Ronkainen, H.; Wallin, K.; Varjus, S.; Koskinen, J. Tribological contact analysis of a rigid ball sliding on a hard coated surface Part I: Modelling stresses and strains. *Surf. Coat. Technol.* **2006**, *200*, 3793–3809. [CrossRef]
42. Lee, Y.T.; Welsch, G. Young's modulus and damping of Ti6Al4V alloy as a function of heat treatment and oxygen concentration. *Mater. Sci. Eng. A* **1990**, *128*, 77–89. [CrossRef]

Article

Microstructural Features and Surface Hardening of Ultrafine-Grained Ti-6Al-4V Alloy through Plasma Electrolytic Polishing and Nitrogen Ion Implantation

Marina K. Smyslova, Roman R. Valiev, Anatoliy M. Smyslov, Iuliia M. Modina, Vil D. Sitdikov and Irina P. Semenova *

Institute of Physics of Advanced Materials, Ufa State Aviation Technical University, 12 K. Marx st., 450008 Ufa, Russia; smyslovam@yandex.ru (M.K.S.); rovaliev@gmail.com (R.R.V.); smyslovmail@gmail.com (A.M.S.); modina_yulia@mail.ru (I.M.M.); svil@ugatu.su (V.D.S.)
* Correspondence: semenova-ip@mail.ru; Tel.: +7-347-273-3422

Abstract: This work studies a near-surface layer microstructure in Ti-6Al-4V alloy samples subjected to plasma electrolytic polishing (PEP) and subsequent high-energy ion implantation with nitrogen (II). Samples with a conventional coarse-grained (CG) structure with an average α-phase size of 8 μm and an ultrafine-grained (UFG) structure (α-phase size up to 0.35 μm) produced by equal channel angular pressing were used in the studies. Features of phase composition and substructure in the thin surface layers are shown after sequential processing by PEP and II of both substrates with CG and UFG structures. Irrespective of a substrate structure, the so-called "long-range effect" was observed, which manifested itself in enhanced microhardness to a depth of surface layer up to 40 μm, exceeding the penetration distance of an implanted ion he. The effect of a UFG structure on depth and degree of surface hardening after PEP and ion-implantation is discussed.

Keywords: titanium alloys; ultra-fine grained structure; plasma electrolytic polishing; ion implantation; modified layer; substructure; microhardness; long-range effect

Citation: Smyslova, M.K.; Valiev, R.R.; Smyslov, A.M.; Modina, I.M.; Sitdikov, V.D.; Semenova, I.P. Microstructural Features and Surface Hardening of Ultrafine-Grained Ti-6Al-4V Alloy through Plasma Electrolytic Polishing and Nitrogen Ion Implantation. *Metals* **2021**, *11*, 696. https://doi.org/10.3390/met11050696

Academic Editor: Francesca Borgioli

Received: 26 March 2021
Accepted: 20 April 2021
Published: 23 April 2021

Publisher's Note: MDPI stays neutral with regard to jurisdictional claims in published maps and institutional affiliations.

1. Introduction

In recent decades, large scientific capacity has been accumulated in the studies on the surface modification of structural materials, including steels and titanium alloys, to improve their service performance. Among the most widely used methods are plasma electrolytic polishing (PEP) and methods associated with the impact of concentrated energy flows, which include ion implantation (II) [1–5]. Parts treated with PEP have a surface roughness Ra of up to 0.01 μm, while microscopic marks and a defective layer with foreign inclusions are removed off the surface. A high quality of a polished surface improves the corrosion resistance and fatigue strength of materials [4]. The physics of ion implantation consists of the introduction of an alloying element into the surface layer of a part as a result of its bombardment with high kinetic energy ions [2,3]. The so-called "long-range effect" occurs, one of the manifestations of which is the formation of a complex dislocation substructure at a depth that is tens of times greater than the penetration depth of a doping ion [6]. The surface modification in conventional coarse-grained (CG) Ti-6Al-4V alloy by nitrogen ion implantation improves its tribological [7] and corrosion properties [5] and its creep rupture strength at 600 °C [8].

In recent years, ultrafine-grained (UFG) metals and alloys processed by severe plastic deformation techniques have attracted great interest [9–12]. Formation of a bulk UFG structure in titanium alloys makes it possible to significantly improve their service and technological properties (strength, fatigue resistance, superplasticity, etc.) [12]. Integration of these technological methods associated with the creation of a UFG state in the material volume and surface modification to improve performance of titanium alloys is an urgent task for modern mechanical engineering.

127

At the same time, there arise a number of unsolved problems connected with the effec of these modification methods on the surface of a substrate with a UFG structure. Since oxidation processes in UFG titanium alloys occur much more intensively, and thin bu rather dense oxide films form, this can, for example, complicate the electrochemical process during PEP and, as a result, affect the physical and chemical state of the surface. During high-energy ion implantation, a complex dislocation substructure is formed in the surface layer, and ions embedded in the crystal lattice of a metal substrate form solid solution. or new chemical compounds [2]. In contrast to conventional CG materials, nucleation and accumulation of new dislocations in ultrafine grains with high dislocation density are hindered [11]. This can probably affect the mechanisms of modified layer formation in the surface of UFG titanium alloy and, therefore, its service characteristics.

The work is aimed at studying the effect of a Ti-6Al-4V substrate UFG structure on structural and phase changes in the alloy surface layers after plasma electrolytic polishing and ion implantation with nitrogen.

2. Experimental Procedure

2.1. Materials and Sample Preparation Procedure

The investigations were carried out on Ti-6Al-4V alloy. The chemical composition is presented in Table 1.

Table 1. Chemical composition of Ti-6Al-4V alloy (in wt.%).

Ti	Fe	C	Al	O	V	N	H	Si	Zr
basis	0.18	0.007	6.6	0.17	4.9	0.01	0.002	0.033	0.02

Equal channel angular pressing (ECAP) was used to form the UFG structure in the Ti-6Al-4V alloy. The microstructure of as-received billets was predominantly equi-axed with an average grain size of the α-phase of 15 µm, which is typical of hot-rolled rods. Rods with a diameter of 20 mm and 100 mm in length were preliminarily heat-treated to produce a mixed duplex (globular–lamellar) microstructure according to the following regime: quenching at T = 960 °C (by ~30 °C below the β-transus temperature) for 1 h followed by tempering at T = 675 °C for 4 h. Then, the rods were deformed in a die-se: with an angle of the channels intersection at 120° at T = 650 °C along the Bc route, 6 passes with a total accumulated strain ε~3 [13].

Disc-shaped samples with a thickness of 5 mm were cut out of the rods (Figure 1a) Before ion implantation, the sample surface was subjected to mechanical polishing and plasma electrolytic polishing to a mirror finish. The details of PEP processing are given in [14]. The ion implantation of the sample surface was carried out on an experimental se: with a broadband ion source. Before implantation, the samples were washed with gasoline in an ultrasound bath and wiped with ethanol. Based on the results of earlier studies [15] on Ti-6Al-4V alloy (Russian analog VT6), the following treatment modes were selected the dose of ion implantation D = 2 × 10^{17} ion/cm^2; the accelerating voltage U_{acc} = 25 kV the residual pressure in the chamber p = 5.5 . . . 6.5 Pa; the heating temperature of samples in the chamber did not exceed T = 250 °C. After implantation the samples were cooled in vacuum for 2–3 min.

2.2. Structure Studies in the Bulk and Surface Layer of Samples

The microstructure was studied by scanning electron microscopy (SEM) using a JEOL JSM 6390 microscope (JEOL Ltd., Tokyo, Japan) and transmission electron microscopy (TEM, JEOL Ltd., Tokyo, Japan) on a JEOL JEM 2100 with an accelerating voltage of 200 kV Thin foils for TEM were prepared by spark cutting plates with a thickness of 0.8 mm in accordance with the scheme displayed in Figure 1a. Then, the foils were mechanically thinned till a thickness of ~100 µm and electro-polished using a TenuPol-5 facility with a solution of 5% perchloric acid, 35% butanol, and 60% methanol at a polishing temperature in the range from −20 to −35 °C. Only the non-polished side of the foils was thinned (Figure 1b). The foils allowed for revealing structural changes in the near-surface layer at a depth of no more than 200 nm.

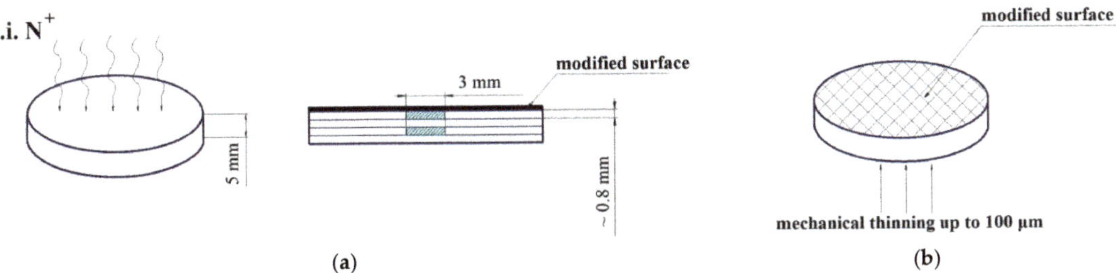

Figure 1. Scheme of sample processing and preparation for the studies: disc-shaped samples for plasma electrolytic polishing and ion implantation (**a**); scheme of mechanical thinning of foils for microstructure studies by TEM (**b**).

2.3. X-ray Structure Analysis

The study on the surface layer structural-phase and chemical composition was carried out by X-ray diffraction analysis (XRD). To determine the size of coherent scattering domains (CSD) and dislocation density (ρ), X-ray diffraction patterns were obtained on a Rigaku Ultima IV diffractometer (Rigaku Corporation, Tokyo, Japan), which implements a focusing method (Bragg–Brentano goniometer scheme). All X-ray diffraction patterns were obtained using CuKα radiation at a voltage and current of 40 kV and 40 mA, respectively, on the X-ray tube. The diffraction patterns were measured within the scattering angle 2θ from 30° to 150° with a step of 0.02° and exposure of 4 s per point. A graphite monochromator (Rigaku Corporation, Tokyo, Japan) was used on a reflected beam. The WPPM approach implemented in the PM2K software (version 2.1, University of Trento, Trento, Italy) was used to analyze the diffraction patterns in order to calculate the distribution of CSD over size, density of edge and screw dislocations, and effective radius of dislocations [16,17]. Such parameters as sample plane displacement, lattice parameter a, dislocation density ρ, volume fraction of edge dislocations m_{ixp}, effective dislocation radius R_e, and the shape and size of crystallites D were varied to calculate diffraction patterns according to [16,17]. The average crystallite size D_{ave} was set in the form of a sphere with lognormal distribution $D_{ave} = \exp(\mu + \sigma^2/2)$, where μ and σ are the average value and its deviation, respectively. The instrumental broadening of diffraction lines, i.e., the parameters U, V, W, a, b, and c of the Cagliotti function were determined by processing the X-ray diffraction patterns of LaB$_6$ obtained under the same conditions in which the titanium alloy was studied. The diffraction patterns were calculated as a result of 40 iterations with the initial parameters equal to the following: $a = 0.2956$ nm, $c = 0.4687$ nm, $\rho = 1.0 \times 10^{14}$ m^{-2}, $m_{ixp} = 0.5$, $R_e = 3.0$ nm, $\mu = 4.0$, $\sigma = 0.1$, $W = 1.93000000 \times 10^3$, $V = 6.27346000 \times 10^4$, $U = 2.03000000 \times 10^3$, $a = 2.38030000 \times 10^1$, $b = 9.93000000 \times 10^3$, and $c = 0$.

The qualitative phase analysis was performed using a PDF-2 X-ray database in the PDXL software package (version 1.8.1, Rigaku Corporation, Tokyo, Japan). To enhance the quantitative phase analysis accuracy, the volume fraction f of precipitates was calculated by the Rietveld method [18]; the pseudo-Voigt function was applied to describe the shape of a peak profile, with account of the asymmetry of a peak, subtracting the background radiation by the Sonneveld–Visser method [19]. All this made it possible to calculate precipitates with a volume fraction of less than 0.5%.

2.4. Microhardness

The surface microhardness was measured using a Struers Duramin microhardness tester (Struers A/S, Ballerup, Denmark) with a load of 25 g for 10 s. The microhardness was measured on the surface of the studied samples, as well as on the inclined microspecimens prepared according to the scheme shown in Figure 2. The investigated surface was inclined to the initial one at an angle of about 2°. At least 5 measurements were performed at each point at depth.

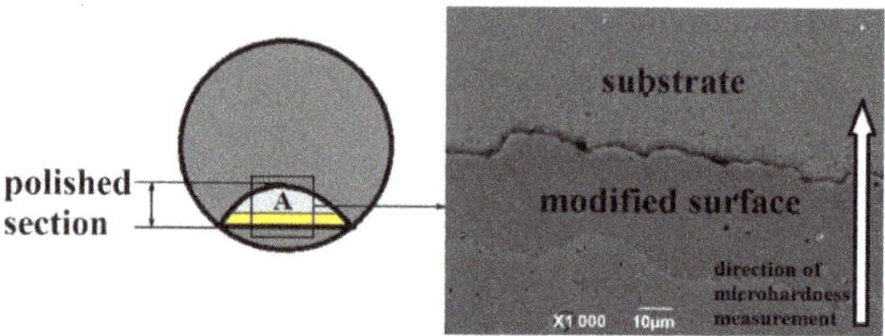

Figure 2. Scheme of microhardness measurement over the sample depth.

3. Results

3.1. Microstructure of VT6 in CG and UFG States

The initial microstructure of a heat-treated workpiece was a mixed globular–lamellar structure (Figure 3a) and consisted of a primary α-phase (the average size was about 5 µm, the fraction was about 30%) and areas with (α + β) thin-lamellar structure (Figure 3b). The thickness of plates was no more than 1 µm on average (Figure 3b). This type of structure ensures good technological ductility of the material and the most effective refinement of the alloy under severe plastic deformation by ECAP. This approach was described earlier in [20]. The results of the X-ray phase analysis revealed that the ratio of the α and β phases was approximately 85:15%.

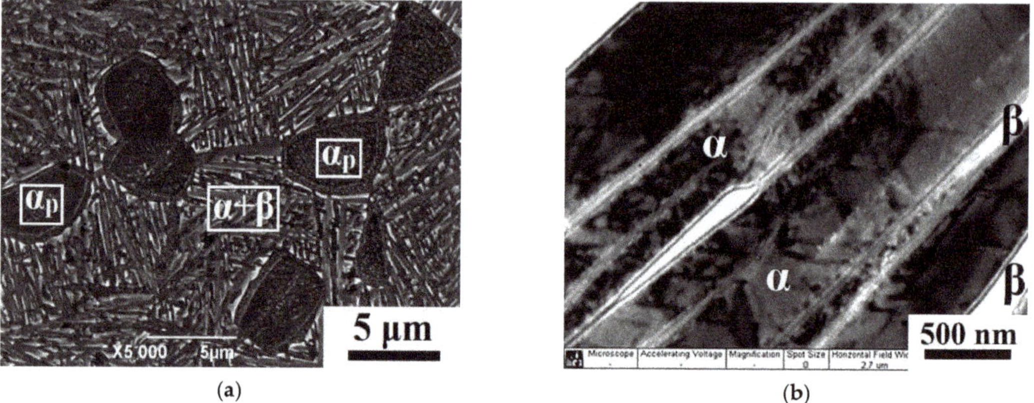

Figure 3. Microstructure of Ti-6Al-4V in the initial CG state: general view of structure, SEM (**a**); (α + β) area of lamellar structure, TEM (**b**).

Figure 4 shows the alloy microstructure after ECAP obtained by SEM and TEM. Weakly deformed grains of a primary α-phase with an average size of about 5 µm were observed in the microstructure. In the two-phase α + β areas, a mixture of α- and β phases in the form of ultrafine grains was found (Figure 4a). The TEM images show that the average grain/subgrain size was about 350 nm (Figure 4b). The interiors of the α grains were characterized by a high density of dislocations and a cellular-type dislocation substructure formed inside the weakly deformed primary α-phase (card No. 00-044-1294). The fraction of the β-phase, identified with card No. 00-044-1288, decreased from 15 to 6%, compared to the initial state, as a result of partial decomposition and dissolution $\beta_m \rightarrow \alpha + \beta$ initiated by severe plastic deformation [21]. The total dislocation density in the workpiece after SPD, as estimated by X-ray analysis, was about 12×10^{14} m^{-2}, which was significantly higher than the dislocation density in the initial state of the alloy (0.05×10^{14} m^{-2}) (see Table 2). The microhardness of the sample surface before and after ECAP was 3800 and 4500 MPa, respectively.

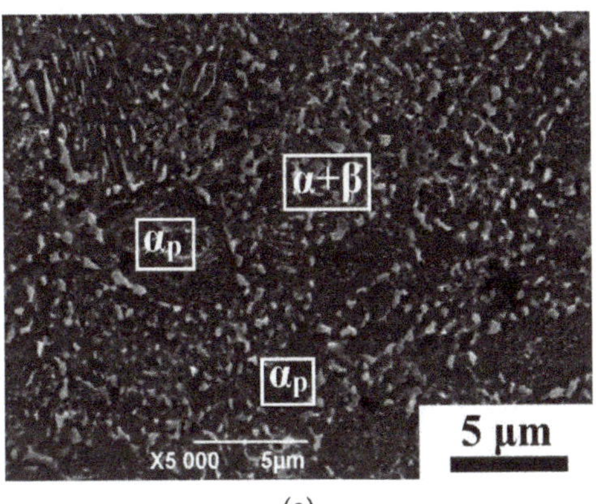

(a)

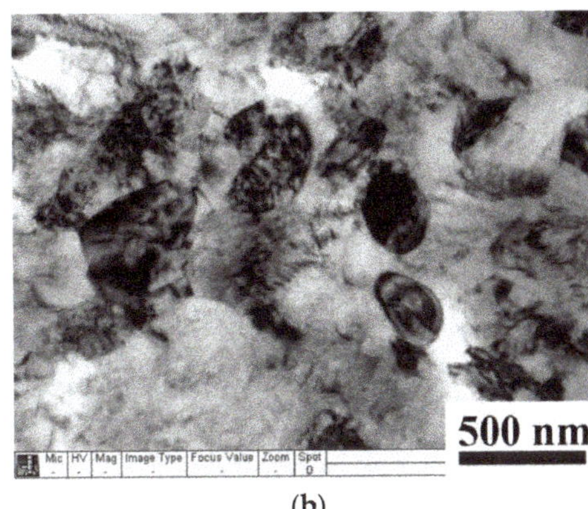

(b)

Figure 4. Microstructure of Ti-6Al-4V alloy in the UFG state: general view of the structure, SEM (**a**); (α + β) area of a mixture of ultrafine grains, TEM (**b**).

Table 2. Results of X-ray analysis.

No.	Surface Treatment	CSD, nm	Dislocations Density, $\rho \times 10^{14}$ m^{-2}	β-Ti, %	TiN, %	TiO2, %
1	CG annealed, no treatment	93	0.05	15.0	-	-
2	CG + PEP	-	2.0	-	-	-
3	CG + PEP + II	42	10.0	15.0	0.12	0.12
4	UFG no treatment	23	12.0	8.5	-	-
5	UFG + PEP	-	9.0	-	-	-
6	UFG + PEP + II	63	23.0	7.5	0.10	0.10

3.2. Microstructure of the Near-Surface Layer of Samples after PEP

The near-surface layer microstructure in the CG and UFG sample states after PEP treatment is shown in Figures 5 and 6.

The TEM images showed the formation of a reticular and inhomogeneously cellular dislocation substructure in the near-surface layer in coarse grains of a primary α-phase (Figure 5a). Dislocation pile-ups were observed at the interphase boundaries of primary α-phase grains and between plates (Figure 5b). This is consistent with the results of the X-ray diffraction analysis of the CG samples surface after PEP, which noted a significant increase in the dislocation density from 0.05 to 2.0×10^{14} m^{-2} (see Table 2). Microdischarges arising in the PEP process probably promote the generation of new dislocations, which accumulate on obstacles in the form of interphase boundaries.

The investigation of UFG samples after PEP revealed that the microstructure retained its pattern in the sample surface, but it became more inhomogeneous due to the redistribution and partial annihilation of dislocations inside and along the boundaries of ultrafine grains/subgrains (Figure 6a). The largest pile-ups of dislocations were observed in the grain and phase boundaries (Figure 6b). The TEM data agree with the X-ray analysis results with respect to the total dislocation density decreasing slightly from 12.0 to 9.0×10^{14} m^{-2}, in contrast to the CG substrate (see Table 2). The total value of dislocation density in the near-surface zone of the UFG substrate remains several times higher than that of the CG alloy (9.0 and 2.0 × 10^{14} m^{-2}, respectively).

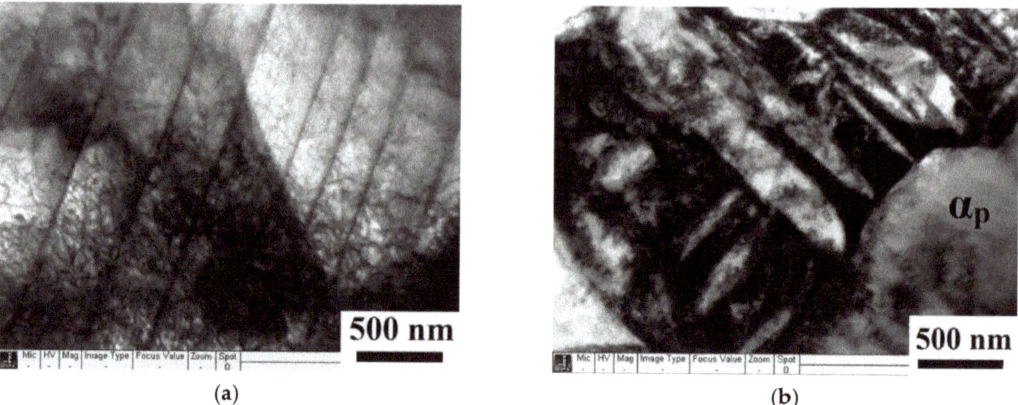

(a)   (b)

Figure 5. Microstructure of the near-surface layer of a Ti-6Al-4V sample with CG structure after PEP: dislocation substructure in the primary α_p (**a**); ($\alpha + \beta$) lamellar structure (**b**). TEM.

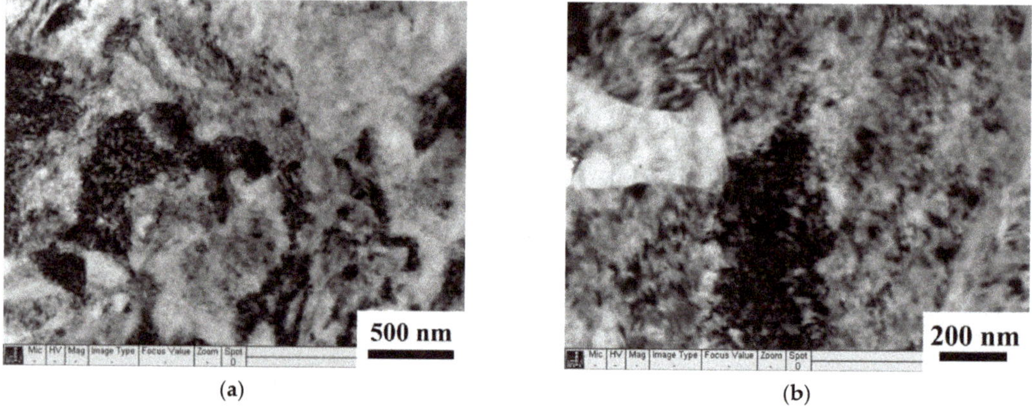

(a)   (b)

Figure 6. Microstructure of the near-surface layer of a Ti-6Al-4V sample with UFG structure after PEP: general view of the inhomogeneous structure (**a**); pile-ups of dislocations (**b**). TEM.

3.3. Microstructure of a Sample Modified Layer after Ion Implantation

Figures 7 and 8 show TEM images of the near-surface layer of the CG and UFG substrates of Ti-6Al-4V alloy after ion implantation with nitrogen. By comparing the layer microstructure after PEP (Section 3.1) and II, one can note that, due to the impact of ions with high kinetic energy, the structure in both CG and UFG substrates undergoes significant changes, which leads to the formation of a strong state of nonequilibrium with a large number of crystal defects as a result of radiation-stimulated diffusion [3].

In the CG substrate, strong disordering of the structure (Figure 7a) and the formation of blocks with high dislocation density are observed (Figure 7b). According to the results of X-ray structural analysis, a noticeable decrease in the CSD size (from 93 to 42 nm) was found, and the total dislocation density increased by another five times: from 2 to 10 × 10^{14} m^{-2} compared to the surface state after PEP (see Table 2).

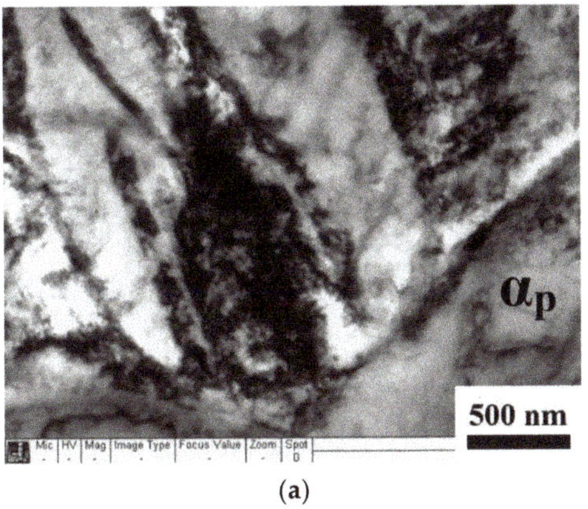

(a)

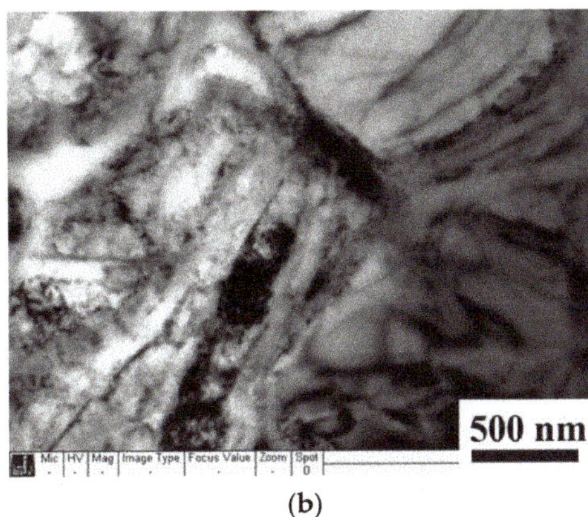

(b)

Figure 7. Microstructure of a near-surface layer of a CG Ti-6Al-4V sample after PEP + II: general view of structure disordering (**a**); blocks with high dislocation density (**b**). TEM.

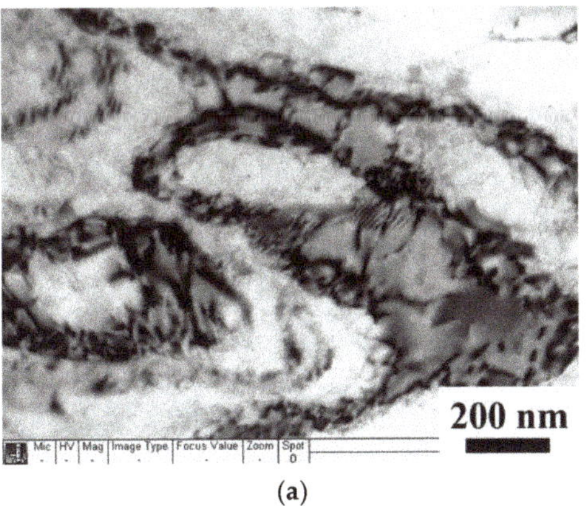

(a)

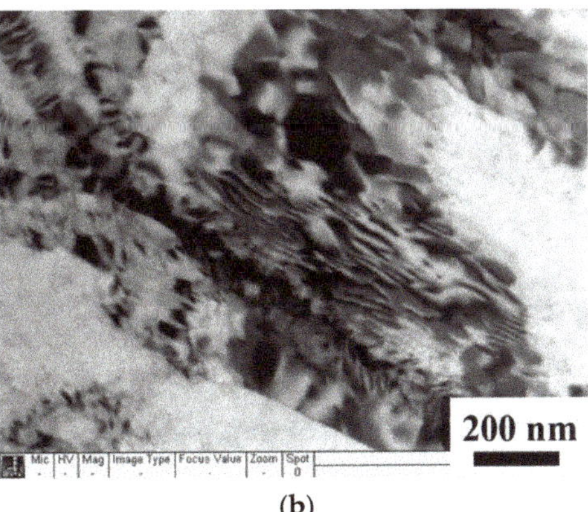

(b)

Figure 8. Microstructure of the near-surface layer of a Ti-6Al-4V sample with UFG structure after PEP + II: accumulations of crystal defects along subgrain boundaries (**a**); redistribution and partial annihilation of dislocations (**b**). TEM.

In the UFG substrate, accumulations of dislocations and other crystal defects were also observed, mainly along subgrain boundaries (Figure 8a). In separate areas (as well as in subgrains), one can observe a slight decrease in the dislocation density as a result of their redistribution and partial annihilation (Figure 8b). The dislocation density grew only by 2.5 times in comparison with a non-modified state (from 9.0 to 23.0 $\times$ 10^{14} m^{-2}), in contrast to the CG substrate, the dislocation density of which increased by five times after ion implantation (see Table 2). Apparently, after severe plastic deformation, the dislocation density had already reached critical values; therefore, ion bombardment of a UFG substrate surface led to their partial annihilation. This is also evidenced by an increase in the CSD size to 63 nm (see Table 2). The dislocation density in the ion-implanted surface of a UFG sample remains much higher than that in the surface of a CG alloy after ion implantation (23 and 10 $\times$ 10^{14} m^{-2}, respectively). This can be explained by the pinning of dislocations on nitride and oxide precipitates as a result of nitrogen ions' penetration into the crystal lattice [22]. This is confirmed by the results of X-ray structural analysis in Table 2. The volume fraction of nitride (card No. 00-038-1420) and oxide (card No. 00-021-1272) precipitates in both cases is approximately the same, and is in the range of 0.10–0.12%.

It should be noted that ion implantation with nitrogen did not lead to significan changes in the ratio of α- and β-phases in CG and UFG substrates (see Table 2).

3.4. Microhardness of a Sample Modified Layer after Ion Implantation

Figure 9 displays the results of microhardness measurements in the CG and UFG substrates subjected to implantation with nitrogen. The microhardness was measured on the inclined samples in accordance with Figure 2. One can see that the maximum value of microhardness in both substrates is approximately the same, and amounts to about 5300 MPa. At a depth of ~5 μm, the microhardness slightly decreases to 5000 MPa in UFG and to 4700 in CG samples. It should be noted that the earlier studies on an ion-implanted surface of a UFG alloy showed that the penetration depth of nitrogen ions did not exceed 6 μm [15].

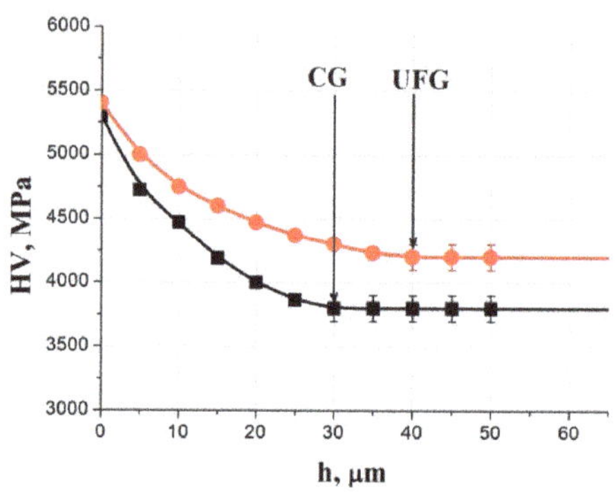

Figure 9. Microhardness behavior over the depth of UFG and CG substrates of a Ti-6Al-4V alloy after high-energy nitrogen ion implantation.

The microhardness in the surface layer is enhanced in comparison with that in the internal volume of samples in both CG and UFG states and retains practically up to a depth of 30 and 40 μm, respectively (Figure 9). The researchers attribute this to a "long-range effect" during ion implantation, when a change in the dislocation substructure in the surface was observed in deeper layers several times exceeding the implantation depth of ions [6,23].

4. Discussion

This work presented some features of a structural-phase state of a modified surface layer of samples made of ultrafine-grained Ti-6Al-4V alloy treated by plasma electrolytic polishing followed by ion implantation with nitrogen.

Plasma electrolytic polishing is usually performed to reduce the surface roughness of products. It is a preparatory operation for subsequent modification by ion implantation and/or the application of various coatings. As is known, the microstructure of a substrate affects the electrochemical activity of a surface. In particular, the transition of a metal to a UFG state, which is characterized by a high density of boundaries and defects, is accompanied by the enhanced activity of electrons and an increased diffusion coefficient in grain boundary areas [24]. The indicated effect of electrochemical activity enhancement can be explained by intensification in the heterogeneity of the surface substructure and its thermodynamically unstable state [25]. These differences in the electrochemical behavior of the CG and UFG alloys affected the dislocation substructure in the near-surface layers of the samples after treatment. First of all, PEP processing of the CG substrate surface led to its strengthening due to an increase in the total dislocation density from 0.05 to 2.0×10^{14} m^{-2}. On the contrary, partial softening and dislocation density reduction from 12.0 to 9.0×10^{14} m^{-2} were observed in the UFG substrate, although its value is several times higher than that of the CG alloy. The UFG material, unlike the CG one, already has a high degree of hardening and a state of less equilibrium. At energy pulses, probably from microdischarges during PEP, the generation of motion of unstable dislocation clusters occurs, leading to stress relaxation and some softening of the UFG sample surface. Thus, as applied to UFG titanium alloy, PEP can be an independent process for surface treatment, allowing one to achieve high surface quality characteristics that are not attainable by other processing methods [14].

Ion implantation is now widely used for steels and titanium alloys. It is known that the surface hardening of titanium alloys is achieved as a result of the interaction of implanted nitrogen ions with the resulting radiation defects of the crystal structure, as well as a significant solid solution hardening mechanism due to the formation of dispersed nitride phases [7,8]. A similar effect of nitrogen ion implantation in the surface was observed in both CG and UFG substrates. The total density of dislocations in the near-surface layer with CG structure increased by five times, to 10×10^{14} m^{-2}, and that with UFG structure grew only by 2.5 times (Table 2). In both samples, strong disordering of the structure was observed because of large distortions of the crystal lattice as a result of shock-wave action during ion implantation and the formation of a large number of radiation defects. A significant increase in the dislocation density in the near-surface zone of the CG and UFG samples affected the microhardness value (Figure 9). Its values were 5000 and 4700 MPa for UFG and CG specimens, respectively. In this case, no significant difference in the microhardnesses of CG and UFG structured surfaces was observed. This is probably due to the UFG surface having a critical dislocation density before ion implantation. Therefore, further bombardment led to their redistribution and partial annihilation. This fact is proven by the different increment in the dislocation density after ion implantation as compared to the no treatment state of CG (from 0.02 to 10×10^{14} m^{-2}) and UFG (from 12 to 23×10^{14} m^{-2}) substrates (Table 2)

It should be noted that a high microhardness in the UFG sample was approximately at a depth of up to 5 µm, which is consistent with the depth of nitrogen ions penetrating into the surface, which was determined using the Auger spectroscopy in [15]. The same depth of nitrogen ion penetration was observed in the surface of conventional titanium alloy [26]. The inhibition of a doping ion in the near-surface layer of the UFG substrate probably occurs similarly to that in the CG state, i.e., the processes in thin layers of ion penetration are insensitive to the substrate structure. The formation of highly dispersed segregations and nitride phases is confirmed by X-ray diffraction analysis, according to the results of which, an average of 0.10–0.12% nitride phases was found in the irradiated surface of the CG and UFG substrates. A similar result was obtained in our recent work [27].

Enhanced microhardness, as compared to that in the material bulk, was found in the ion-implanted surface of both CG and UFG substrates. In the UFG material it retains at a depth of up to 40 µm, and in the CG alloy it was observed at a depth of 30 µm (Figure 9). Such an unusual effect was demonstrated in earlier studies devoted to the ion implantation of titanium alloys [26]. The authors previously identified a superdeep change in the structure of the implanted surface, which manifested itself in the formation of a complex dislocation structure at a depth of up to 30 µm. A model of deep formation of a developed dislocation structure was proposed. During ion bombardment of the surface, vacancy dislocation loops are formed in the zone of atomic collision cascades as a result of diffusion-induced rearrangement. The dislocation loops grow and transform under the conditions of increasing concentration of point defects (vacancies), which also contributes to the mass transfer of impurity atoms. An increase in the depth of a hardened surface layer in the UFG alloy up to 40 µm is apparently due to a more active redistribution of dislocation formations, the creation of new stable configurations, and the consolidation of nitride phase segregations within them.

The hardening of the UFG surface depends on treatment conditions (annealing before or after ion implantation, number of implantation cycles). It was demonstrated in [27] that a three-time iteration of the cycle (ii + annealing) resulted in the maximum concentration of nitride phases (~0.28%). Subsequent low-temperature annealing (to 400 °C) after ion implantation can lead to the formation of more equilibrium and an ordered grain boundary structure, and, consequently, to the enhancement of both strength and ductile properties [27].

The investigations on various structural states of Ti-6Al-4V alloy have shown that completely new structural state is formed in the substrate surface during sequential PEP and II processes. The new state is associated with an increase in the dislocation density and the creation of a complex dislocation substructure, which indicate successive hardening of the material at all the stages of processing. The UFG state, initially already having a high hardening degree and less equilibrium, is capable of generating a motion of unstable dislocation clusters, leading to relaxation of stresses, their equalization, and some softening with energy pulses from microdischarges during PEP. During subsequent nitrogen ion implantation of the surface of UFG samples, the depth of ion penetration is similar to that of the CG state. The long-range effect as a result of ion bombardment is significantly higher, the surface hardening is retained at a greater depth in comparison with the CG substrate. However, in order to achieve the desired effect of UFG substrate surface hardening, one should pay attention to the processing regimes, which can be varied, for example, by a number of repeated cycles of ion implantation, annealing conditions, etc. This will be the focus of our further investigations.

5. Conclusions

It was found that the initial structural state of the Ti-6Al-4V alloy substrate had a significant effect on the transformation of the dislocation substructure during PEP and subsequent ion implantation with nitrogen:

1. PEP leads to formation of new dislocation configurations and dislocation density enlargement by almost an order of magnitude in the surface of the CG alloy. In thin surface layers of the substrate with UFG structure, the dislocation pile-ups level up, accompanied by a slight decrease in the total density of dislocations;
2. Irrespective of the structural state in the substrate surface after the subsequent high energy nitrogen ion implantation, a "long-range effect" was observed, which manifested itself in strengthening to a depth up to 30–40 μm;
3. Ion implantation with nitrogen leads to nitride and oxide precipitation with a volume fraction of 0.10–0.12% in the near-surface layers, which makes an additional contribution to surface hardening.

Author Contributions: Conceptualization and discussion of the results, M.K.S. and A.M.S.; investigation, V.D.S., I.M.M. and R.R.V.; collection of data and writing the draft, I.P.S. All authors have read and agreed to the published version of the manuscript.

Funding: This research was funded by the Russian Science Foundation, grant number 19-79-10108.

Acknowledgments: Authors are grateful to the personnel of the research and technology Joint Research Center, 'Nanotech', Ufa State Aviation Technical University for their assistance with instrumental analysis.

References

1. Nestler, B.K.; Böttger-Hiller, F.; Adamitzki, W.; Glowa, G.; Zeidler, H.; Schubert, A. Plasma Electrolytic Polishing–an Overview of Applied Technologies and Current Challenges to Extend the Polishable Material Range. *Proc. CIRP* **2016**, *42*, 503–507. [CrossRef]
2. Carter, G.; Katardjiev, I.V.; Nobes, M.I. High-fluence ion irradiation, an overview. In *Materials Modification by High-Fluence Ion Beams*; Kelly, R., da Silva, M.F., Eds.; Springer: Dordrecht, The Netherlands, 1989; Volume 155, pp. 3–27. [CrossRef]
3. Guseva, M.I.; Martynenko, Y.V.; Smyslov, A.M. Deep modification of materials under implantation-plasma treatment Experiment and theory. In Proceedings of the 3rd International Workshop on Plasma-Based Ion Implantation, Rossendorf, Germany, 17–19 September 1996; pp. 101–106.
4. Meletis, E.I.; Nie, X.; Wang, F.L.; Jiang, J.C. Electrolytic Plasma Processing for Cleaning and Metal-coating of Steel Surfaces. *Surf. Coat. Technol.* **2002**, *150*, 246–256. [CrossRef]

Zhecheva, A.; Sha, W.; Malinov, S.; Long, A. Enhancing the microstructure and properties of titanium alloys through nitriding and other surface engineering methods. *Surf. Coat. Technol.* **2005**, *200*, 2192–2207. [CrossRef]

Sharkeev, Y.; Kozlov, E.V. The long-range effect in ion implanted metallic materials: Dislocation structures, properties, stresses, mechanisms. *Surf. Coat. Technol.* **2002**, *158*, 219–224. [CrossRef]

Budzynski, P.; Youssef, A.A.; Sielanko, J. Surface modification of Ti-6Al-4V alloy by nitrogen ion implantation. *Wear* **2006**, *261*, 1271–1276. [CrossRef]

Castagnet, M.; Yogi, L.M.; Silva, M.M.; Ueda, M.; Couto, A.A.; Reis, D.A.P.; Moura Neto, C. Microstructural Analysis of Ti-6Al-4V Alloy after Plasma Immersion Ion Implantation. *Mater. Sci. Forum* **2012**, *727–728*, 50–55. [CrossRef]

Valiev, R.Z.; Zhilyaev, A.P.; Langdon, T.G. *Bulk Nanostructured Materials: Fundamentals and Applications*; Wiley-Blackwell: Hoboken, NJ, USA, 2014; 456p. [CrossRef]

10. Valiev, R.Z.; Estrin, Y.; Horita, Z.; Langdon, T.G.; Zehetbauer, M.J.; Zhu, Y.T. Producing bulk ultrafine-grained materials by severe plastic deformation: Ten years later. *JOM* **2016**, *68*, 1216–1226. [CrossRef]

11. Valiev, R.Z.; Estrin, Y.; Horita, Z.; Langdon, T.G.; Zehetbauer, M.J.; Zhu, Y.T. Fundamentals of superior properties in bulk nanoSPD materials. *Mater. Res. Lett.* **2015**, *4*, 1–21. [CrossRef]

12. Semenova, I.P.; Polyakova, V.V.; Dyakonov, G.S.; Polyakov, A.V. Ultrafine-Grained Titanium-Based Alloys: Structure and Service Properties for Engineering Applications. *Adv. Eng. Mater.* **2020**, *22*, 1900651. [CrossRef]

13. Saitova, L.R.; Semenova, I.P.; Raab, G.I.; Baushev, N.G.; Valiev, R.Z. Method of Thermomechanical Treatment of Two-Phase titanium Alloys. Russia Patent No. 2.285.740 RU C1, 20 October 2006.

14. Tamindarov, D.R.; Smyslova, M.K.; Plotnikov, N.V.; Modina, Y.M.; Semenova, I.P. Surface electrolytic-plasma polishing of Ti-6Al-4V alloy with ultrafine-grained structure produced by severe plastic deformation. *IOP Conf. Ser. Mater. Sci. Eng.* **2019**, *461*, 012079. [CrossRef]

15. Semenova, I.P.; Smyslova, M.K.; Selivanov, K.S.; Valiev, R.R.; Modina, Y.M. Enhanced fatigue properties of Ti-6Al-4V alloy turbine blades via formation of ultra-fine grained structure and ion implantation of surface. *IOP Conf. Ser. Mater. Sci. Eng.* **2017**, *194*, 012035. [CrossRef]

16. Leoni, M.; Confente, T.; Scardi, P. PM2K: A flexible program implementing Whole Powder Pattern Modelling. *Z. Kristallogr Suppl* **2006**, *23*, 249–254. [CrossRef]

17. Scardi, P.; Ortolani, M.; Leoni, M. WPPM: Microstructural analysis beyond the Rietveld method. *Mater. Sci. Forum* **2010**, *651*, 155–171. [CrossRef]

18. Rietveld, H.M. A profile refinement method for nuclear and magnetic structures. *J. Appl. Cryst.* **1969**, *2*, 65–71. [CrossRef]

19. Sonneveld, E.J.; Visser, J.W. Automatic collection of powder data from photographs. *J. Appl. Cryst.* **1975**, *8*, 1–7. [CrossRef]

20. Semenova, I.P.; Saitova, L.R.; Raab, G.I.; Korshunov, A.I.; Zhu, Y.T.; Lowe, T.C.; Valiev, R.Z. Microstructural Features and Mechanical Properties of the Ti-6Al-4V ELI Alloy Processed by Severe Plastic Deformation. *Mater. Sci. Forum* **2006**, *503–504*, 757–762. [CrossRef]

21. Sergueeva, A.V.; Stolyarov, V.V.; Valiev, R.Z.; Mukherjee, A.K. Superplastic behavior of ultrafine-grained a Ti-6Al-4V alloys. *Mater. Sci. Eng.* **2002**, *A323*, 318–325. [CrossRef]

22. Valiev, R.Z.; Alexandrov, I.V.; Enikeev, N.A.; Murashkin, M.Y.; Semenova, I.P. Towards enhancement of properties of UFG metals and alloys by grain boundary engineering using SPD processing. *Rev. Adv. Mater. Sci.* **2010**, *25*, 1–10.

23. Budzynski, P.; Kaminski, M.; Pyszniak, K.; Wiertel, M. Long-Range Effect of Ion-Implanted Materials in Tribological Investigations. *Act. Phys. Pol. A* **2019**, *136*, 290–293. [CrossRef]

24. Balyanov, A.; Kutnyakova, J.; Amirkhanova, N.A.; Stolyarov, V.V.; Valiev, R.Z.; Liao, X.Z.; Zhao, Y.H.; Jiang, Y.B.; Xuc, H.F.; Lowe, T.C.; et al. Corrosion resistance of ultra fine-grained Ti. *Script. Mater.* **2004**, *51*, 225–229. [CrossRef]

25. Sevidova, E.K.; Simonova, A.A. Features of corrosion and electrochemical behavior of Ti with nano and submicrocrystalline structure. *Surf. Eng. Appl. Electrochem.* **2011**, *47*, 70–75. (In Russian) [CrossRef]

26. Guseva, M.I.; Smyslov, A.M. Long-range effect in implantation of N+, B+ and C+ ions in Ti-based alloy. *Surface* **2000**, *6*, 68–71. (In Russian)

27. Semenova, I.P.; Smyslova, M.K.; Selivanov, K.S.; Sitdikov, V.D.; Valiev, R.R. High Energy Surface Implantation by Ion Nitrogen of Ultra-Fine Grained Ti-6Al-4V Alloy for Engineering Application. *Mater. Sci. Forum* **2021**, *1016*, 1305–1311. [CrossRef]

Article

Transformations of the Microstructure and Phase Compositions of Titanium Alloys during Ultrasonic Impact Treatment. Part I. Commercially Pure Titanium

Alexey Panin [1,*], Andrey Dmitriev [1], Anton Nikonov [1], Marina Kazachenok [1], Olga Perevalova [1] and Elena Sklyarova [2]

1   Institute of Strength Physics and Materials Science of Siberian Branch of Russian Academy of Sciences, 634055 Tomsk, Russia; dmitr@ispms.ru (A.D.); anickonoff@ispms.ru (A.N.); kms@ispms.tsc.ru (M.K.); perevalova52@mail.ru (O.P.)
2   National Research Tomsk Polytechnic University, 634050 Tomsk, Russia; skea@tpu.ru
*   Correspondence: pav@ispms.tsc.ru; Tel.: +7-3822-286-979

Abstract: Experimental and theoretical studies helped to reveal patterns of surface roughening and the microstructure refinement in the surface layer of commercial pure titanium during ultrasonic impact treatment. Applying transmission electron microscopy technique, a gradient microstructure in the surface layer of the ultrasonically treated sample, where the grain size is varied from nano- to micrometers was revealed. It was shown that the surface plastic strains of the titanium sample proceeded according to the plastic ploughing mechanism, which was accompanied by dislocation sliding, twinning, and the transformations of the microstructure and phase composition. The molecular dynamics method was applied to demonstrate the mechanism of the phase transformations associated with the formation of stacking faults, as well as the reversible displacement of atoms from their sites in the hcp lattice, causing a change in coordination numbers. The role of the electronic subsystem in the development of the strain-induced phase transformations during ultrasonic impact treatment was discussed.

Keywords: titanium; phase transformation; electronic structure; microstructure; molecular dynamics; ultrasonic impact treatment; transmission electron microscopy

Citation: Panin, A.; Dmitriev, A.; Nikonov, A.; Kazachenok, M.; Perevalova, O.; Sklyarova, E. Transformations of the Microstructure and Phase Compositions of Titanium Alloys during Ultrasonic Impact Treatment. Part I. Commercially Pure Titanium. *Metals* 2021, 11, 562. https://doi.org/10.3390/met11040562

Academic Editor: Francesca Borgioli

Received: 15 March 2021
Accepted: 26 March 2021
Published: 30 March 2021

Publisher's Note: MDPI stays neutral with regard to jurisdictional claims in published maps and institutional affiliations.

1. Introduction

Ultrasonic impact treatment (UIT) is an effective method of surface hardening that significantly improves the functional properties of structural materials, as well as their welded joints [1]. Surface processing with a high-strength striker vibrating at an ultrasonic frequency results in a notable change in the microstructure of metals and alloys due to the refining of existent grains and the formation of new ones. Also, it enhances the density of dislocations and curvature of slip planes, promotes the transformation of the martensitic microstructure, the partial decomposition of supersaturated solid solutions, and the precipitation of second phase particles, as well as raises elastic micro- and macro-stresses, etc. As a result, the surface layer microhardness of the structural materials is increased, and the strength, wear, creep, and corrosion resistance (in addition to other characteristics) are improved [2–5]. Generally, the nature of morphological changes taken place in the materials during the UIT processing is determined not only by the temperature–rate parameters, but also by their microstructures, phase compositions, the presence of carbide-forming elements, possible polymorphic transformations under severe plastic strains, etc.

As a rule, the microstructure evolution in the structural materials during the UIT processing is discussed within the framework of a one-level approach based on the theory of dislocations. In particular, it has been shown that strains of the surface layer of

139

titanium and its alloys (which are widely used in the aerospace, nuclear and other industries) are developed according to the plastic ploughing mechanism during repeated penetrations of a striker into the treated specimen [6–8]. In this case, intense sliding and twinning of dislocations take place, causing the formation of sub-grains with low-angle misorientations, as well as their subsequent transformation into nanocrystalline grains with high-angle misorientations.

According to [9,10], under conditions of strong deviations from equilibrium, self-organized processes with the formation of new space-dissipative structures take place, which provide significant changes in the initial mechanical properties of materials. It has been convincingly shown in the papers of V.E. Panin [11,12] that the nucleation and propagation of all strain defects in a loaded crystal (regardless of their dimension) is based on the development of local transformations of the microstructure and phase composition controlled by the electronic subsystem of the crystal. In the space of interstices of the crystal lattice characterized by a high degree of distortion, new (possible) microstructural states can be formed, which ensure the development of diffusionless shear strains of the crystal under external applied stresses. In doing so, computer simulation is an effective tool for studying atomic mechanisms underlying structural and phase transformations in a loaded solid. Earlier, the authors of the article, using the molecular dynamics method, demonstrated the development of $\beta \rightarrow \alpha \rightarrow \beta$ phase transformations in the Ti-6Al-4V titanium alloy during scratch testing, which made it possible to explain the experimentally observed recovery of a scratch [13].

This paper opens a new series devoted to the study of transformations of the microstructure and phase compositions of titanium alloys during the UIT processing. In particular, the roles of dislocation slips, twinning, and phase transformations in refining the microstructure in the surface layer of commercially pure titanium (CP-Ti) are considered experimentally and using molecular dynamic simulation. The CP-Ti is used as a model material, in which the effect of alloying elements (aluminum, vanadium, etc.) on the instability of the crystal lattice of the α phase solid solution is not taken into account.

2. Materials and Methods

The as-received CP-Ti (the VT1-0 according to the Russian classification) was investigated with the following chemical composition (wt.%): 0.2 Al, 0.4 Zr, 0.3 Mn, 0.01 Cr, 0.06 Si, 0.2 Fe, 0.02 Cu, and 98.8 Ti. The average grain size was 70 μm.

The UIT processing of a plate 50 mm × 50 mm × 1 mm in size was carried out using a spherical striker 10 mm in diameter, fabricated from the hard alloy with composition WC-8 wt.% Co. During the UIT processing, the striker vibrated at a frequency of ~22 kHz and an amplitude of ~40 μm. Its movement speed was 0.2 m/s and load was 200 N.

After the UIT processing, the surface morphology of the sample was examined using a NewView 6200 optical profilometer (Zygo Corp., Middlefield, CT, USA) and a Solver HV atomic force microscope (NT-MDT, Moscow, Russia). A microstructural analysis was performed using a JEM 2100 (JEOL, Akishima, Tokyo, Japan) transmission electron microscope (TEM) with an electron acceleration energy of 200 kV. The cross-sectional TEM samples were manufactured, using a JEOL Ion Slicer EM-091001S (JEOL, Akishima, Tokyo, Japan). During preparation, argon was used as the working gas, the accelerating voltage was 8 kV, and the etching angle was 1.5 to 4°.

Microhardness was measured using a 'PMT-3' hardness tester (LOMO, St. Petersburg, Russia) with a load on the diamond pyramid of 50 g.

The dislocation density p was determined by the rectangular grid method using the following expression:

$$p = (n_1/l_1 + n_2/l_2)\cdot M/t, \tag{1}$$

where n_1 and n_2 were the numbers of intersections; l_1 and l_2 were line lengths; M was magnification; t was the foil thickness.

A computer simulation of the UIT processing was carried out using the LAMMPS software package [14]. On the atomic scale, the simulation of the process was carried out by successive triple indentations of a spherical indenter into an initially defect-free titanium crystallite in the form of a parallelepiped. The indenter was a sphere with a radius of 6.5 nm. The load was directed from the indenter center to atoms within this sphere. The load value was calculated by the $F = -k(R - r)^2$ ratio, where k was the indenter stiffness coefficient; R was the sphere radius; r was distance between the centers of the indenter and atoms [15]. The indenter was moved along the Z axis at a fixed depth of 3.5 nm with a constant indentation rate of 15 m/s. The indentation depth was chosen so that the maximum load on the indenter did not exceed 12 GPa. The crystallographic orientation of grains in the model crystallite was chosen as follows: the X, Y, and Z axes coincided with the $[2\bar{1}\bar{1}0]$, $[01\bar{1}0]$, and $[0001]$ directions, respectively. The equilibrium atomic configuration of the model crystallite was determined by the minimum potential energy calculation. The interaction between atoms was described by the potential [16] assessed within the framework of the embedded atom method. The simulated system was considered as an NVE ensemble, in which the number of atoms, energy and volume were conserved. The movement equations were integrated using the high-speed Verlet algorithm with a time step of 0.001 ps. The total number of atoms exceeded 4.1 million. The Common Neighbor Analysis (CNA) and Dislocation Extraction Algorithm (DXA) of the OVITO software [17] were used to identify the crystal lattice defects produced by multiple indentations of the spherical indenter. The simulation of strains of the model crystallite was carried out considering the nucleation and movement of dislocations along certain crystallographic planes.

3. Experimental Research Results of Changes in the Surface Morphology and the Microstructure of the CP-Ti Sample during the UIT Processing

3.1. Patterns of the Surface Layer Roughening

The penetration of the spherical striker, vibrated at the ultrasonic frequency, into the CP-Ti surface layer was accompanied by extrusion of the material along its perimeter. Since the plate movement was relative to the vibrated striker during the UIT processing, pile-ups of the displaced titanium were strained and re-strained again along the striker perimeter at each subsequent impact [8]. At the ultrasonic frequency of 22 kHz and the movement speed of ~0.2 m/s, the distance between the centers of impact spots was ~10 μm. Since the impact spots were much larger (about 500 μm in diameter), most of the material was plastically displaced at the front of the moved striker at each subsequent impact. Small semicircular pile-ups remained behind the moved striker (Figure 1a,b), the distance between which corresponded to that between the centers of the impact spots, and their height reached 1 μm. In general, dimensions of the pile-ups and the distances between them were determined by the UIT processing parameters and the CP-Ti mechanical properties.

According to detailed studies performed using atomic force microscopy, the micro-roughness on the striker surface also caused local plastic ploughing of the processed material, which resulted in roughening of the semicircular pile-up surfaces due to the formation of numerous protrusions and depressions of various shapes and heights (Figure 1c,d).

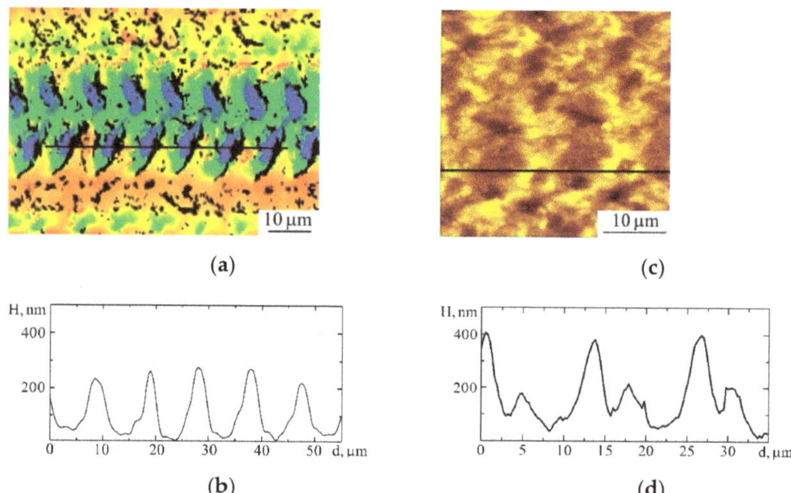

Figure 1. The surface morphology (**a,c**) and corresponding profilograms (**b,d**) of the CP-Ti sample after the ultrasonic impact treatment (UIT) processing. Optical profilometry (**a,b**) and atomic force microscopy (**c,d**).

3.2. TEM Studies of the Gradient Microstructure in the Surface Layer

TEM cross-sectional analysis of CP-Ti samples after the UIT processing revealed the gradient microstructure of their surface layers. Inside the uppermost surface layer, a sub microcrystalline structure of the α phase with grain sizes of less than 200 nm was observed (Figure 2). High-angle misorientations of the α-Ti phase equiaxed grains were confirmed by the quasi-ring structure of an electron diffraction pattern (Figure 2c). The microstructure with a dislocation density of 10^{16} m^{-2} was found inside the α phase grains (Figures 3a and 4a). The bend contours in the TEM images are indicative of high residual stresses developing inside the layer under consideration.

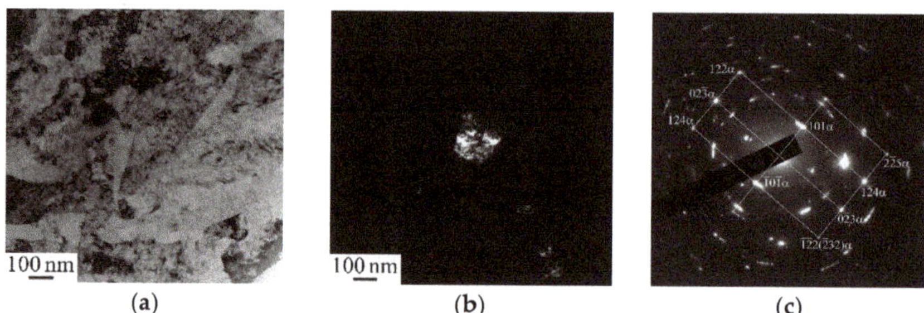

Figure 2. TEM bright- (**a**) and dark-field (**b**) images and associated selected area electron diffraction (SAED) patterns (**c**) of the microstructure of the uppermost surface layer of the CP-Ti sample after the UIT processing. TEM dark-field images were obtained with the $101\,(\bar{2}32)_{\alpha\text{-Ti}}$ reflection.

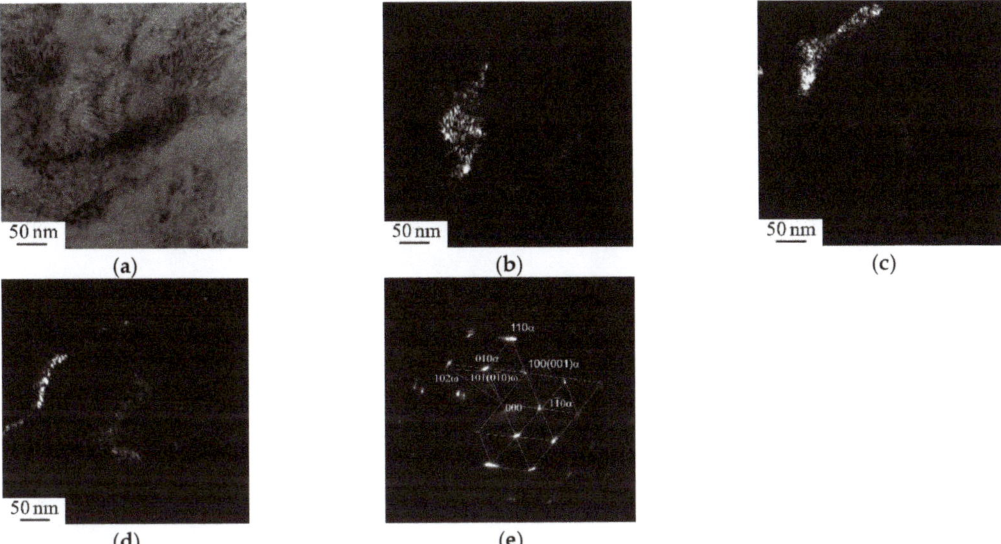

Figure 3. TEM bright- (**a**) and dark-field (**b–d**) images and associated SAED pattern (**e**) of the microstructure of the uppermost surface layer of the CP-Ti sample after the UIT processing. The dark-field images were obtained with the $110(001)_{\omega\text{-Ti}}$ reflection (**b**), the closely spaced $120(001)_{\alpha\text{-Ti}}$ and $102(0\bar{1}0)_{\omega\text{-Ti}}$ ones (**c**), as well as in the $101(0\bar{1}0)_{\omega\text{-Ti}}$ reflex (**d**).

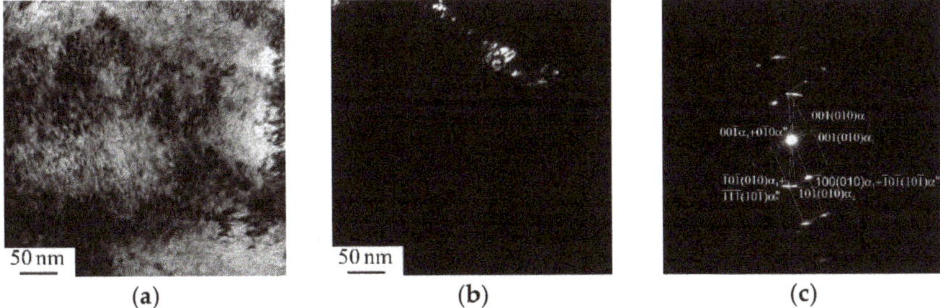

Figure 4. TEM bright- (**a**) and dark-field (**b**) images and associated selected area electron diffraction (SAED) pattern (**c**) of the microstructure of the uppermost surface layer of the CP-Ti sample after the UIT processing. TEM dark-field image was obtained with the $0\bar{1}0(10\bar{1})_{\alpha''\text{-Ti}}$ reflection.

In addition to a refinement of the α phase grains, a significant change in the phase composition was observed in the uppermost surface layer of the CP-Ti sample after the UIT processing. It was due to the high chemical activity of titanium and its polymorphism (the ability to form various crystal microstructures). The ω and α'' phase crystallites with an average transverse size of 10 nm were found in grains and fragments of the α phase bounded by adjacent shear bands (Figures 3 and 4, respectively).

The α phase grains with an average size of 10 μm were observed in the surface layer 10 μm thick, inside which the banded microstructure and microtwins with an average transverse size of 100 nm were revealed (Figure 5a). It should be noted the presence of a large number of extinction contours indicated high internal stresses and distortions of the crystal lattice in the surface layer of the CP-Ti sample after the UIT processing. The dislocation density did not exceed 10^{14} m^{-2}.

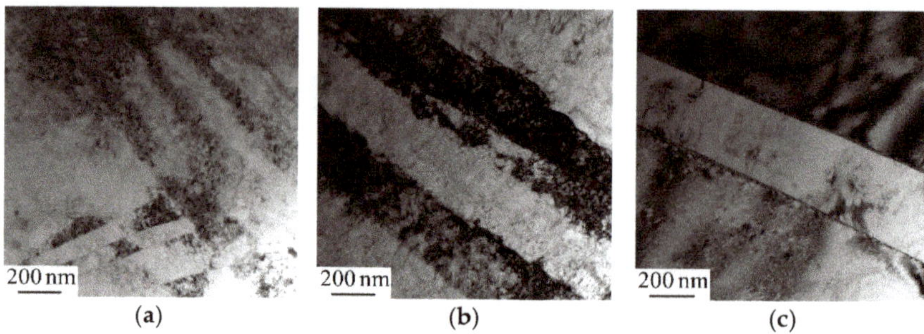

(a) (b) (c)

Figure 5. TEM bright-field images of the microstructure of the CP-Ti sample after the UIT processing, TEM images were obtained at depths of 10 (**a**) and 50 (**b,c**) μm below the surface of the sample.

At a depth of 50 μm, the average size of the α phase grains corresponded to that for the as-received CP-Ti. The transverse dimensions of the fragments separated by adjacent shear bands increased up to 300 nm. In addition, the boundaries of the banded substructure were strongly curved (Figure 5b). A large number of extinction contours were observed inside the fragments. In this case, the dislocation density did not exceed 5×10^{13} m^{-2}. Also, twins with an average transverse size of 300 nm were inside the α phase grains observed at this depth (Figure 5c) [18]. Finally, at a depth of 70–100 μm, the microstructure consisted of equiaxed α phase grains, in which dislocations were identical to the as-received CP-Ti.

The gradient microstructure of the surface layer of CP-Ti samples after the UIT processing makes itself evident in the microhardness variation across the sample. The uppermost surface layer of the samples exhibits the highest microhardness (2.2 GPa). As the depth below the sample surface increases, the microhardness gradually decreases down to a value characteristic of the base metal (1.6 GPa) at a depth of ~100 μm [18].

4. Molecular Dynamics Simulation of Strains of the Surface Layer of the Titanium Crystallite during the UIT Processing

Molecular dynamics simulation enabled a clear demonstration of the patterns of the surface layer roughening on the CP-Ti sample and the development of the microstructural transformations under repeated penetrations of the spherical striker. According to Figure 6, a pile-up of the strained material was crumpled and re-pushed back along its perimeter with each subsequent penetration of the indenter. In this case, small semicircular pile-ups were formed behind the moved indenter. The distance between them, as shown above, corresponded to that between the centers of the impact spots.

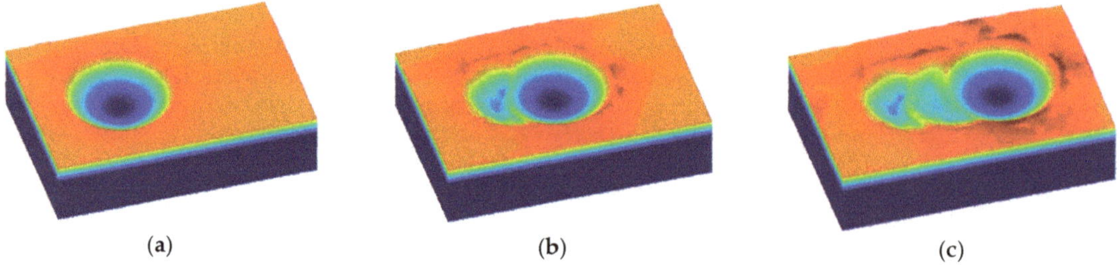

(a) (b) (c)

Figure 6. 3D images of the titanium crystallite after one (**a**), two (**b**) and three (**c**) penetrations of the spherical indenter.

The indenter penetration into the titanium crystallite was accompanied by the nucleation of dislocations in the area of its contact with the sample surface and their translational movement in easy-sliding directions, as well as the formation of twins (Figure 7a). The density of dislocations and twins increased continuously during its subsequent penetration in the sample, causing fragmentation and reorientation of the crystal lattice (Figure 7b,c). In this case, the crystal lattice of the surface layer at the contact spot, as well as at the boundaries between twins and between misoriented fragments, was not identified as a regular atomic structure by the OVITO software due to the fact that some atoms had been displaced from their sites. Inside the fragments, local fcc regions (represented stacking faults) had been formed. However, other areas with the bcc lattice were at the boundaries of the fragments (Figure 8).

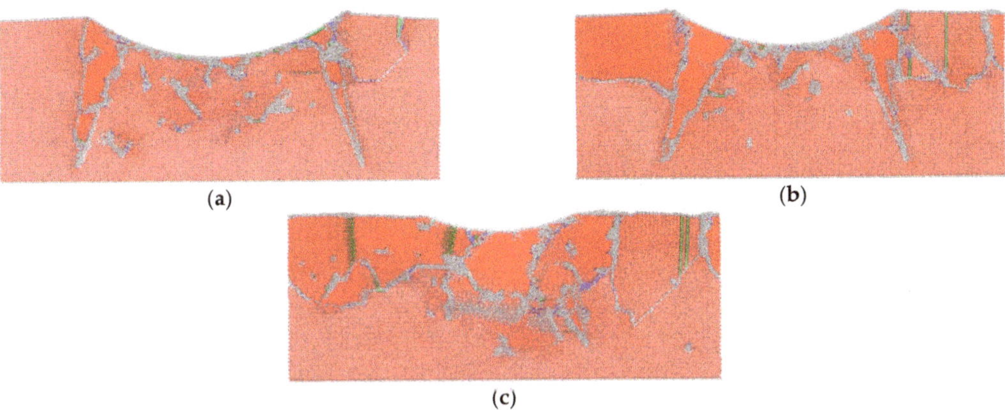

Figure 7. The microstructure of a layer of the titanium crystallite in the YZ plane 2 nm thick with X coordinates of 31 (**a**), 47 (**b**), and 63 nm (**c**) after the first (**a**), second (**b**) and third (**c**) penetrations of the indenter. Atoms that had formed the crystal lattice, in which some of them were displaced from their sites, are marked in light gray.

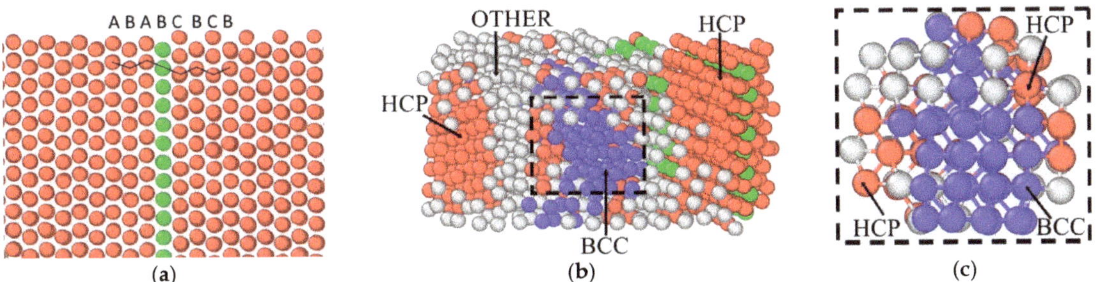

Figure 8. The formation of a stacking fault (**a**) and a local region with the bcc crystal lattice (**b**) in the titanium crystallite after three penetrations of the indenter. The color of atoms corresponds to different local configurations of the atomic lattice: blue is bcc, green is fcc, red is hcp, gray is an unidentified microstructure. Figure (**c**) corresponds to the central region fragment in (**b**) marked by a dotted line, depicting bonds between nearest atoms.

Figure 9 shows a local region with an unidentified crystal lattice (gray). It could be concluded from the analysis that the displacement of atoms from their sites in the hcp lattice was accompanied by a change in their coordination numbers (the number of nearest neighbors), which was an indirect evidence of the microstructure transformation in this defect area.

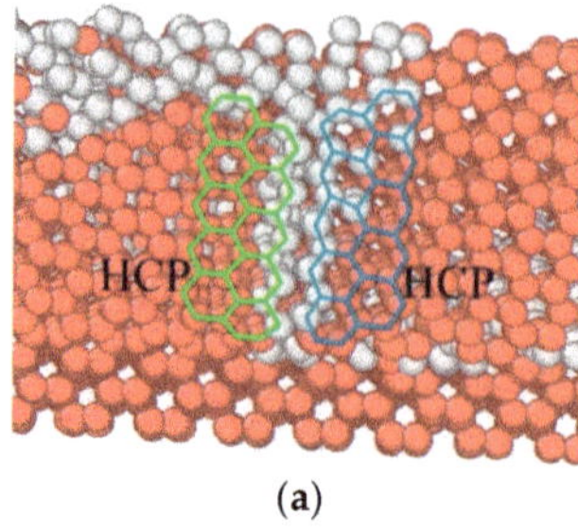

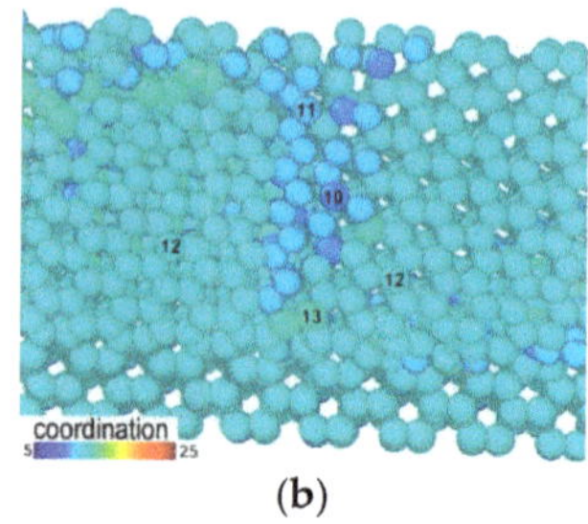

(a)	**(b)**

Figure 9. The spatial arrangement of atoms (**a**) and their coordination numbers (**b**) in the defect region of the titanium crystallite after three penetrations of the indenter.

During repeated penetrations of the indenter into the initially defect-free titanium crystallite, a nonlinear increase in the number of atoms took place, which had a coordination number different from 12. As shown in Figure 10, the number of atoms belonged to the hcp structure gradually decreased that was visualized by computer simulation of the process. However, the number of such atoms slightly increased again at moments when the effect of an impact ceased and the next one began (0.4 and 0.8 ns).

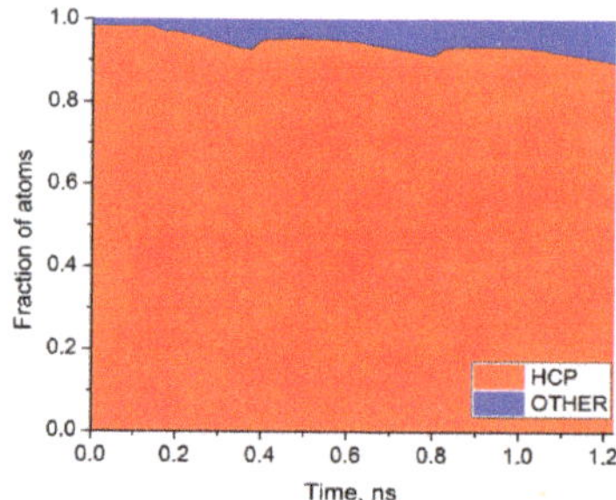

Figure 10. The time dependence of the number of atoms forming the hcp structure (red) and ones whose crystal lattice is different (blue).

5. Discussion

An impact of the spherical WC-Co striker on the CP-Ti sample surface, when only the normal compressive load was applied, could be described in the framework of the typical Hertz problem on the contact of a ball with an elastic half-space. The solution of the Hertz problem enabled the estimation of the contact spot radius a and the striker penetration depth δ using the following expressions [19]:

$$\left(\frac{3F_n R}{4E^*}\right)^{1/3}, \tag{2}$$

$$\delta = a^2/2R' \tag{3}$$

where F_n was the normal compressive load; $E^* = \left(\frac{1-v_1^2}{E_1} + \frac{1-v_2^2}{E_2}\right)^{-1}$ was the reduced Young's modulus of the striker–sample system (E_1, E_2 and v_1, v_2 were the Young's modulus and Poisson's ratio values of the striker and the CP-Ti sample, respectively); R was the curvature radius of the spherical striker.

The reduced Young's modulus of the "WC-Co–CP-Ti" system was 104 GPa. Accordingly, the striker 10 mm in diameter penetrated in the CP-Ti sample to a depth of 3.6 µm with the formation of a pile-up of displaced material along the contact spot perimeter at the initial moment (when the magnetostrictor was turned off and the load was 200 N). In this case, the contact spot radius was 190 μm. During the UIT processing, the magnetostrictor elongation was 20 μm, and, therefore, the striker penetrated the surface layer of the sample to a great depth. Since the contact spot radius observed in the experiment was 500 μm, the penetration depth reached 6.25 μm according to Expression (3). It worth noting that the mathematical model of a cyclic random process [20] can be successfully applied to quantitative estimate the parameters of morphological structures.

As shown by molecular dynamics simulation, the spherical indenter penetration into the titanium crystallite was accompanied by intense dislocation slipping, which caused the crystallite fragmentation into sub-grains with the low-angle misorientation and twinning. During the repeated indenter penetrations, the lengths of the low-angle boundaries gradually increased, and the grain and sub-grain microstructure was formed in the surface layer of the titanium sample. Combined with high compressive residual stresses, this hindered the movement of dislocations into the bulk material along easy-slip planes. As a result, the plastic strain zone gradually narrowed around the indenter. In addition, the material transfer from the contact spot to the free surface was significantly enhanced, causing the formation of pile-ups. According to experimental data, the surface layer thickness, in which dislocation slipping and twinning had been developed during the UIT processing, did not exceed 100 μm. At the same time, the CP-Ti phases had been also transformed in the uppermost surface material due to strains by the plastic ploughing mechanism. This fact was confirmed by the observed ω and α'' nanocrystalline phases.

According to the results of molecular dynamics simulation, the phase transformations developed even after triple indenter penetrations into the titanium crystallite. This was associated with the formation of both stacking faults and local regions with a disordered arrangement of atoms at the boundaries between misoriented fragments of the hcp lattice. In particular, the formation of a stacking fault caused by the movement of a partial dislocation resulted in the appearance of several layers with the three-layer packing (the fcc microstructure) in the hcp titanium crystallite (two-layer packing) [21]. In turn, clusters with the bcc lattice could be formed at the boundaries between the misoriented hcp fragments. It could be assumed that the presence of impurity atoms in the real CP-Ti crystallite, as well as the strong crystal lattice curvature could facilitate the transformation of local bcc regions into the ω and α'' nanocrystalline phases (with the hexagonal and orthorhombic lattices).

During the UIT processing, intense surface strains of the CP-Ti sample was accompanied by the strong distortion of the crystal lattice, mainly at the boundaries of the α phase fragments. In the present studies, the χ curvature-torsion of the α phase lattice was estimated by assessing the bending extinction contours in the TEM images using the following expression:

$$\chi = \frac{\frac{360}{2\pi} \times M}{\langle L \rangle},\tag{4}$$

where M was magnification of the TEM images; $<L>$ was the average contour width. It was found that the χ curvature-torsion of the crystal lattice in the surface layer reached 18 deg/μm under strains by the plastic ploughing mechanism, while χ did not exceed 4 deg/μm in the pronounced shear strain regions (at the depth of 70 μm).

The strong distortion of the α phase crystal lattice, accompanied by a perturbation of the electron density, was the cause of the microstructure and phase transformations in the surface layer of the CP-Ti sample during the UIT processing. The electronic configuration of a Ti atom could be written as [Ar] $3d^2 4s^2$. Accordingly, in the ground state, it had four valence electrons in the $3d$ and $4s$ orbitals. With an increase in the electron gas energy due to the crystal lattice distortion, the $4s$ electrons could be decoupled. As a result, one or both s electrons could occupy the valence orbital of the higher $3d$ sublevel. Thus, it was the transition of electrons from the $4s$ orbital to the $3d$ one that caused a change in the method of overlapping electron shells and, respectively, the strain-induced α phase transformations.

A clear manifestation of the role of the electronic subsystem in the development of the phase transformations was the change of the low-temperature hcp titanium phase to the high-temperature bcc one upon heating up to 882 °C. Fractions of the *d*-orbitals in bond of the hcp and bcc crystals were 0.7 and 0.9, respectively [22]. An increase in the kinetic energy of the electron gas with rising temperature caused the transition of electrons from the *s*-orbital to the *d*-one. It enhanced the proportion of electrons in the *d*-orbital of the bond between titanium ions that was the reason for the change in the degree of the bond orientations and, accordingly, the crystal lattice transformation.

The well-known example of the development of the strain-induced phase transformations in titanium was the $\alpha \rightarrow \omega$ one at high pressures (from 2 up to 12 GPa). In this case, the transformation of the hcp α phase (the unit cell volume was 0.0353 nm^3) into the hexagonal ω one (it was 0.0641 nm^3) under compressive stresses was caused by the convergence of positive titanium ions, which resulted in overlapping of the electron orbitals and a similar increase in the concentration of electrons at the *d*-sublevel. Along with the presence of impurities [23], the environment [24], both loading rate and conditions also affected the pressure level at which the $\alpha \rightarrow \omega$ phase transformation had begun. Thus, the ω phase formation occurred at a pressure of ~9 GPa upon impact loading of titanium samples [23]. However, the $\alpha \rightarrow \omega$ phase transition was observed at 4 GPa in the case of high-pressure torsion [25]. Moreover, the ω phase nanoparticles were found in titanium samples after high-speed channel-angular pressing at 2 GPa [26]. It was shear stresses that had determined the development of the phase transformations at significantly lower pressures, and, therefore, enabled to observe the ω and α'' phases after the UIT processing. Obviously, the presence of translational, and even more so, rotational strain modes had promoted the displacement of atoms in certain directions and had facilitated the formation of various crystal lattices in the α phase.

The presence of clusters with the bcc lattice at the α phase grain boundaries made it possible to suggest that the appearance of the ω and α'' nanocrystalline phases had occurred through the intermediate β phase during the UIT processing. Previously, the predominant precipitation of the ω and α'' phases at the grain boundaries was associated exclusively with an increase in stresses [25].

It should be noted that oxygen, which had penetrated into the surface layer of the CP-Ti sample during the UIT processing [27], suppressed the $\alpha \rightarrow \omega$ transformation [23] but could contribute to a change in the β phase cluster to the ω or α'' one during the UIT processing. This was due to the fact that oxygen could result in raising the total atomic displacements in the β phase clusters because of the static component and, consequently, enhancing the instability of their bcc lattice. In this case, the concentration of oxygen atoms determined the number of atomic displacements and, accordingly, could affect both type ($\beta \rightarrow \omega$ or $\beta \rightarrow \alpha''$) and intensity of the martensitic transformation. It was shown earlier [28] that the $\beta \rightarrow \omega$ or $\beta \rightarrow \alpha''$ transitions developed upon strains of titanium β-alloys by drawing to $\varepsilon = 98\%$. This depended on the size factor of the atomic radii of alloying elements (δR) and, accordingly, the number of atomic displacements. In the case of low δR values, the $\beta \rightarrow \alpha''$ transformation developed, while the $\beta \rightarrow \omega$ occurred at high δR levels. In other words, the β phase was transformed into α'' at low numbers of atomic displacements but changed into ω at large ones. Thus, the $\alpha \rightarrow \beta \rightarrow \omega$ phase transformation, highly likely, was controlled by the process of segregation of oxygen atoms in the studied CP-Ti sample during the UIT processing.

6. Conclusions

Optical profilometry, atomic force microscopy, and molecular dynamics simulation helped elucidate the patterns of the CP-Ti surface roughing upon the UIT processing. The development of the surface plastic strains by the plastic ploughing mechanism caused the formation of small semicircular pile-ups on the processed plate. The distances between them corresponded to those between the centers of the impact spots. The presence of mi croroughnesses on the striker surface facilitated roughing of the semicircular pile-ups due to the formation of numerous protrusions and depressions of various shapes and heights.

It was shown by transmission electron microscopy that plastic strains of the CP-Ti sample surface were localized in the layer less than 100 µm thick. The refinement of the α phase grains, caused by intense dislocation slip and twinning, was characterized by the gradient character. Thus, the α phase submicrocrystalline structure with the grain sizes of less than 200 nm was observed in the uppermost layer. At the depth of 5–10 µm, the α phase grains were found with the average size of 10 µm. Inside them, both the banded microstructure and microtwins with the average transverse size of 100 nm were identified. The mean dimension of the α phase grains was 70 µm at the depth of 50 µm that matched the as-received CP-Ti. In this case, the transverse sizes of the fragments separated by adjacent shear bands increased up to 300 nm. At the depth of 70–100 µm, the microstructure consisted of the equiaxed α phase grains, in which dislocations were identical to the as-received CP-Ti.

In the uppermost surface layer of the CP-Ti sample, the ω and α'' nanocrystalline phases that had been formed under the piled-up materials moved along the striker perimeter. It was found by molecular dynamics simulation that the development of the phase transformations was due to the formation of stacking faults, as well as the bcc phase clusters at the α phase sub-grain boundaries during the UIT processing. It was substantiated that the formation of the ω and α'' phases had occurred through the intermediate β phase under the conditions of the high local curvature of the crystal lattice and had been controlled by the segregation of oxygen atoms.

Author Contributions: Conceptualization, A.P.; software, A.D. and A.N.; investigation, M.K., O.P. and E.S.; writing—original draft preparation, A.P., A.D., M.K.; writing—review and editing, A.P. and A.D. All authors have read and agreed to the published version of the manuscript.

Funding: The work was performed according to the Government research assignment for ISPMS SB RAS, Projects FWRW-2021-0010.

Institutional Review Board Statement: Not applicable.

Informed Consent Statement: Not applicable.

Data Availability Statement: The data presented in this study are available on request from the corresponding author.

Conflicts of Interest: The authors declare no conflict of interest.

References

1. Statnikov, E.S.; Korolkov, O.V.; Vityazev, V.N. Physics and Mechanism of Ultrasonic Impact. *Ultrasonics* **2006**, *44*, e533–e538. [CrossRef]
2. Ye, C.; Telang, A.; Gill, A.S.; Suslov, S.; Idell, Y.; Zweiacker, K.; Wiezorek, J.M.K.; Zhou, Z.; Qian, D.; Mannava, S.R.; et al. Gradient Nanostructure and Residual Stresses Induced by Ultrasonic Nano-Crystal Surface Modification in 304 Austenitic Stainless Steel for High Strength and High Ductility. *Mater. Sci. Eng. A* **2014**, *613*, 274–288. [CrossRef]
3. Wu, B.; Wang, P.; Pyoun, Y.-S.; Zhang, J.; Murakami, R. Effect of Ultrasonic Nanocrystal Surface Modification on the Fatigue Behaviors of Plasma-Nitrided S45C Steel. *Surf. Coat. Technol.* **2012**, *213*, 271–277. [CrossRef]
4. Liao, M.; Chen, W.; Bellinger, N. Effects of Ultrasonic Impact Treatment on Fatigue Behavior of Naturally Exfoliated Aluminum Alloys. *Int. J. Fatigue* **2008**, *30*, 717–726. [CrossRef]
5. Mordyuk, B.N.; Karasevskaya, O.P.; Prokopenko, G.I.; Khripta, N.I. Ultrafine-Grained Textured Surface Layer on Zr–1%Nb Alloy Produced by Ultrasonic Impact Peening for Enhanced Corrosion Resistance. *Surf. Coat. Technol.* **2012**, *210*, 54–61. [CrossRef]
6. Mordyuk, B.N.; Prokopenko, G.I. Ultrasonic Impact Peening for the Surface Properties' Management. *J. Sound Vib.* **2007**, *308*, 855–866. [CrossRef]
7. Dekhtyar, A.I.; Mordyuk, B.N.; Savvakin, D.G.; Bondarchuk, V.I.; Moiseeva, I.V.; Khripta, N.I. Enhanced Fatigue Behavior of Powder Metallurgy Ti–6Al–4V Alloy by Applying Ultrasonic Impact Treatment. *Mater. Sci. Eng. A* **2015**, *641*, 348–359. [CrossRef]
8. Panin, A.V.; Kazachenok, M.S.; Kozelskaya, A.I.; Hairullin, R.R.; Sinyakova, E.A. Mechanisms of Surface Roughening of Commercial Purity Titanium during Ultrasonic Impact Treatment. *Mater. Sci. Eng. A* **2015**, *647*, 43–50. [CrossRef]
9. Prigogine, I. *Non-Equilibrium Statistical Mechanics*; Monographs in Statistical Physics and Thermodynamics; Interscience Publisher: New York, NY, USA, 1962; ISBN 978-0-470-69993-5.

10. Chausov, M.; Maruschak, P.; Pylypenko, A.; Markashova, L. Enhancing Plasticity of High-Strength Titanium Alloys VT 22 under Impact-Oscillatory Loading. *Philos. Mag.* **2017**, *97*, 389–399. [CrossRef]
11. Panin, V.E.; Egorushkin, V.E. Basic Physical Mesomechanics of Plastic Deformation and Fracture of Solids as Hierarchically Organized Nonlinear Systems. *Phys. Mesomech.* **2015**, *18*, 377–390. [CrossRef]
12. Panin, V.E.; Panin, A.V.; Perevalova, O.B.; Shugurov, A.R. Mesoscopic Structural States at the Nanoscale in Surface Layers of Titanium and Its Alloy Ti-6Al-4V in Ultrasonic and Electron Beam Treatment. *Phys. Mesomech.* **2019**, *22*, 345–354. [CrossRef]
13. Shugurov, A.R.; Panin, A.V.; Dmitriev, A.I.; Nikonov, A.Y. Recovery of Scratch Grooves in Ti-6Al-4V Alloy Caused by Reversible Phase Transformations. *Metals* **2020**, *10*, 1332. [CrossRef]
14. Plimpton, S. Fast Parallel Algorithms for Short-Range Molecular Dynamics. *J. Comput. Phys.* **1995**, *117*, 1–19. [CrossRef]
15. Dmitriev, A.I.; Nikonov, A.Y.; Shugurov, A.R.; Panin, A.V. The Role of Grain Boundaries in Rotational Deformation in Polycrystalline Titanium under Scratch Testing. *Phys. Mesomech.* **2019**, *22*, 365–374. [CrossRef]
16. Mendelev, M.I.; Underwood, T.L.; Ackland, G.J. Development of an Interatomic Potential for the Simulation of Defects, Plasticity, and Phase Transformations in Titanium. *J. Chem. Phys.* **2016**, *145*, 154102. [CrossRef]
17. Stukowski, A.; Bulatov, V.V.; Arsenlis, A. Automated Identification and Indexing of Dislocations in Crystal Interfaces. *Model. Simul. Mater. Sci. Eng.* **2012**, *20*, 085007. [CrossRef]
18. Panin, A.V.; Kazachenok, M.S.; Kozelskaya, A.I.; Balokhonov, R.R.; Romanova, V.A.; Perevalova, O.B.; Pochivalov, Y.I. The Effect of Ultrasonic Impact Treatment on the Deformation Behavior of Commercially Pure Titanium under Uniaxial Tension. *Mater. Des.* **2017**, *117*, 371–381. [CrossRef]
19. Johnson, K.L. *Contact Mechanics*, 1st ed.; Cambridge University Press: Cambridge, UK, 1985; ISBN 978-0-521-25576-9.
20. Lytvynenko, I.V.; Maruschak, P.O. Analysis of the State of the Modified Nanotitanium Surface with the Use of the Mathematical Model of a Cyclic Random Process. *Optoelectron. Instrum. Process.* **2015**, *51*, 254–263. [CrossRef]
21. Pei, Z. An Overview of Modeling the Stacking Faults in Lightweight and High-Entropy Alloys: Theory and Application. *Mater. Sci. Eng. A* **2018**, *737*, 132–150. [CrossRef]
22. Altmann, S.L.; Coulson, C.A.; Hume-Rothery, W. On the Relation between Bond Hybrids and the Metallic Structures. *Proc. R. Soc. Lond. Ser. A Math. Phys. Sci.* **1957**, *240*, 145–159. [CrossRef]
23. Hennig, R.G.; Trinkle, D.R.; Bouchet, J.; Srinivasan, S.G.; Albers, R.C.; Wilkins, J.W. Impurities Block the α to ω Martensitic Transformation in Titanium. *Nat. Mater.* **2005**, *4*, 129–133. [CrossRef] [PubMed]
24. Errandonea, D.; Meng, Y.; Somayazulu, M.; Häusermann, D. Pressure-Induced Transition in Titanium Metal: A Systematic Study of the Effects of Uniaxial Stress. *Phys. B Condens. Matter* **2005**, *355*, 116–125. [CrossRef]
25. Zhilyaev, A.P.; Popov, V.A.; Sharafutdinov, A.R.; Danilenko, V.N. Shear Induced ω-Phase in Titanium. *LoM* **2011**, *1*, 203–207. [CrossRef]
26. Zel'dovich, V.I.; Frolova, N.Y.; Patselov, A.M.; Gundyrev, V.M.; Kheifets, A.E.; Pilyugin, V.P. The ω-Phase Formation in Titanium upon Deformation under Pressure. *Phys. Met. Metallogr.* **2010**, *109*, 30–38. [CrossRef]
27. Perevalova, O.B.; Panin, A.V.; Kazachenok, M.S. Concentration-Dependent Transformation Plasticity Effect during Hydrogenation of Technically Pure Titanium Irradiated with an Electron Beam. *Russ. Phys. J.* **2019**, *61*, 1992–2000. [CrossRef]
28. Skiba, I.A.; Karasevska, O.I.; Mordyuk, B.N.; Markovsky, P.E.; Shyvaniuk, V.N. Effect of Strain-Induced B$\rightarrow\omega$ Transformation on Mechanical Behaviour of β-Titanium and β-Zirconium Alloys. *Met. Phys. Adv. Technol.* **2009**, *31*, 1573–1587.

Article

Effect of Surface Modification of a Titanium Alloy by Copper Ions on the Structure and Properties of the Substrate-Coating Composition

Marina Fedorischeva [1,*], **Mark Kalashnikov** [1,2], **Irina Bozhko** [1,2], **Olga Perevalova** [1] and **Victor Sergeev** [1,2]

[1] Institute of Strength Physics and Materials Science SB RAS, 634055 Tomsk, Russia; kmp1980@mail.ru (M.K.); bozhko_irina@mail.ru (I.B.); perevalova52@mail.ru (O.P.); vs@ispms.tsc.ru (V.S.)

[2] Department of Materials Science, National Research Tomsk Polytechnic University, 634050 Tomsk, Russia

[*] Correspondence: fed_mv@mail.ru or fmw@ispms.tsc.ru; Tel.: +79-627-763-406

Received: 28 October 2020; Accepted: 24 November 2020; Published: 27 November 2020

Abstract: To improve the strength properties, adhesion, and the thermal cycling resistance of ceramic coatings, the titanium alloy surface was modified with copper ions under different processing times. It is found that at the maximum processing time, the thickness of the alloyed layer reaches 12 μm. It is shown that the modified layer has a multiphase structure in addition to the main α and β–titanium phases with the intermetallic compounds of the Ti-Cu system. The parameters of the fine structure of the material are investigated by the X-ray diffraction analysis. It has been found that when the surface of the titanium alloy is modified, depletion occurs in the main alloying elements, such as aluminum and vanadium, the crystal lattice parameter increases, the root-mean-square (rms) displacements of the atoms decrease, and the macrostresses of compression arise. A multilevel micro- and nanoporous nanocrystalline structure occurs, which leads to an increase in the adhesion and the thermal cyclic resistance of the ceramic coating based on Si-Al-N.

Keywords: phase composition; structure; surface modification; elemental distribution; ionic treatment; X-ray structural analysis; electron microscopy; multiscale structure

---

1. Introduction

The VT23 alloy is related to the Ti-Al-V-Mo-Cr-Fe system. This is a medium-alloyed (α + β) martensitic-class alloy, which gets the martensitic "α" structure after hardening from the β-region. This alloy has high ductility, which is an important property for technological applications such as drawing, flanging, and other pressure treatment operations [1].

This property of the titanium alloy allows using it as the basis for deposition of heat-shielding coatings such as Zr-Y-O, Si-Al-N [2–6]. However, in order to improve the thermocyclic resistance and the adhesion properties of the coating, it is necessary to prepare the titanium alloy for coating using different ion-beam technology [7–13].

It is known that across the substrate surface adjacent to the coating, there is a sharp jump in the change in the structural phase state and the physical and mechanical properties of the "heat-protective coating—substrate" system. In this interface region, there appears the maximum localization of the elastic stress. In addition, the state of the substrate surface can significantly affect the formation of the structure and the properties of the coating itself [14,15].

To prepare the surface for deposition of a coating, the surface hardening of metals and alloys is widely used by increasing the density of dislocations in a layer up to 10 microns in depth (a long-range effect [16]) or due to the formation of nanocrystalline intermetallic phases [17]. In the latter case, the element pairs of the "ion beam—processed metal" are selected from those systems in which the formation of intermetallides is possible. Thus, in [17], the ionic synthesis of the intermetallic phases based on the Ni-Ti, Ni-Al, Fe-Al, and TiAl systems was discovered. However, the ionic synthesis in these systems is poorly understood.

One of the effective methods of preparing a substrate for the deposition of coatings is the modification of their surface layer with high-energy beams of metal ions. In this case, treatment with ion beams can change the morphology, phase, and the elemental composition of the surface layer [18].

Therefore, an urgent task is to study the effect of the phase composition, the structural state of the surface layer of a titanium substrate on the structure, and the properties of the coating formed on it, as well as on the thermomechanical characteristics of the whole "coating—substrate" system. This article focuses on the study of the surface modification of the VT23 titanium alloy, which was carried out to improve the thermal cycling resistance and the adhesion strength of the Si-Al-N-based coating.

2. Materials and Methods

Samples were treated under a continuous titanium ion beam with accelerating voltage (1200 V ± 20 V) and ion current (about 15 mA) using the vacuum system KVANT-03MI (Techimplant Ltd., Tomsk, Russia), and the vacuum arc ion source with a copper cathode [19]. The samples were placed in the camera on the object table in front of the ion source for ion bombardment. The temperature of the samples during the ion bombardment was 900 ± 100 K. After the ion beam treatment of the substrate, a Si-Al-N coating was deposited using a magnetron sputtering technique. Parameters of treatment of the titanium alloy by copper ions Technological parameters of processing are shown in Table 1.

Table 1. Parameters of treatment of the titanium alloy by copper ions.

Samples	Bias, V	Treatment Time, min	Fluence, Ion/cm^2
Initial Ti alloy	-	-	-
Treated Ti	−900	3	0.9×10^{18}
Treated Ti	−900	6	1.8×10^{18}
Treated Ti	−900	7.5	2.3×10^{18}

The structural phase state of the ion-modified layers of the samples was investigated by the Transmission Electron Microscope TEM method using a JEOL-2100 device (Jeol Ltd., Tokyo, Japan). Foils for the TEM studies were prepared by the cross-section method using an Ion Slicer EM-09100IS (Jeol Ltd., Tokyo, Japan). To classify the structures, the grain size, and the phase composition, the bright-field images together with the corresponding microdiffraction patterns and the dark-field images were used. The phase composition, the crystalline lattice parameters, the root-mean-square atomic displacement (rms), the macrostresses of the surface layer were determined by X-ray using the DRON-7 device St. (UED-Lab, Petersburg, Russia) [20,21]. X-ray investigation of the modified titanium alloy was carried out under continuous 2θ-scanning with the Bragg–Brentano focusing at Co Kα radiation. The data base JCPDS and PDF-2 (International Centre for Diffraction Data, Campus Blvd, PA, USA) was used for interpretation of the diffractograms.

The symmetric and the asymmetric Bragg–Brentano schemes of X-ray investigation were used. The X-ray diffraction (XRD) registration in an asymmetric mode was carried out at grazing angles of X-ray radiation incidence α = 3°. The chemical composition and the element distribution in the titanium surface were determined by the energy dispersive X-ray (EDS) analysis using a microanalyzer INCA-Energy (Oxford Instruments) with the built-in TEM JEOL-2100 and scanning electron microscopy (SEM) LEO EVO-50XVP. The thermal cycling procedure was carried out by heating the samples to 1000 °K in air, followed by holding the sample at this temperature for 1 min and cooling to room temperature.

The VT23 titanium alloy was selected as the material for the study. It contains elements such as: Fe, Cr, Mo, V, Al, as shown in Table 2.

Table 2. The chemical composition of the VT23 alloy (weight %).

Fe	Cr	Mo	V	Ti	Al
0.4–0.8	0.8–1.4	1.5–2.5	4–5	84–89.3	4–6.3

3. Results and Discussion

The VT23 alloy modified with copper ions at different processing times is investigated by X-ray diffraction analysis. Figure 1 shows the diffraction patterns of the titanium alloy modified with copper ions treated for different times (3, 6, and 7.5 min) and the diffraction pattern of the initial titanium alloy VT23. It is seen that the X-ray diffraction patterns of the initial titanium alloy contain the lines of α and β-phases of titanium, which are characteristic of the quenched VT23 alloy. In the composition of the titanium alloy treated with copper ions, in addition to the main phases of α and β-titanium, there are intermetallic phases of the equilibrium state diagram of Ti-Cu [22].

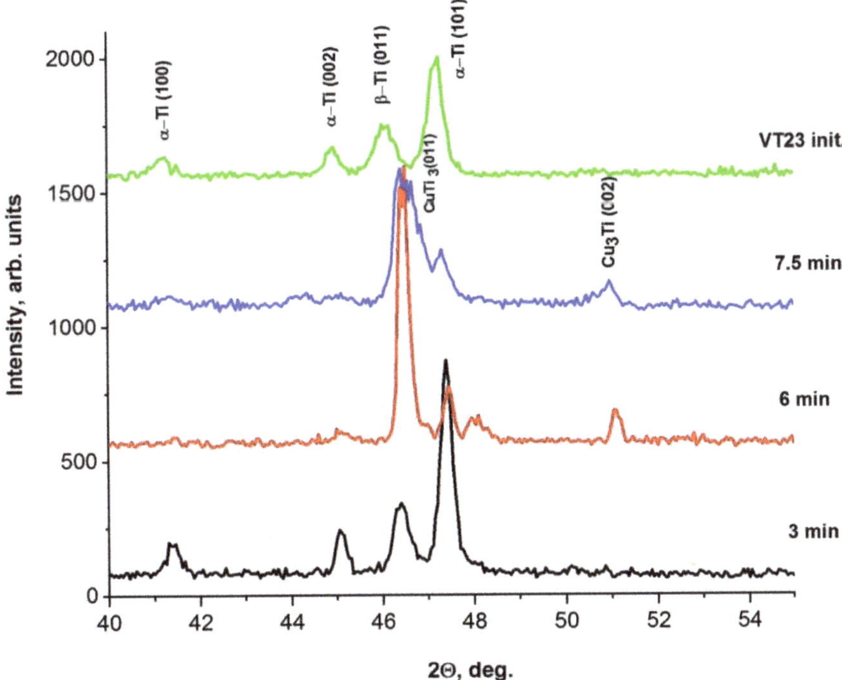

Figure 1. Fragments of the X-ray diffraction patterns of the VT23 alloy in the initial state (1) and upon treatment with copper ions at different processing times (the processing time is indicated on the X-ray diffraction pattern).

It can be seen in the diffraction patterns obtained using the asymmetric X-ray investigation. Figure 1 shows that in the modified layer with a processing time of 7.5 min under the asymmetric X-ray investigation, the intensity of the (111) line of the $CuTi_3$ phase exceeds the intensity of the main peak of the α-phase by several times. This means that the surface layer 1 μm deep mainly consists of the $CuTi_3$ intermetallic compound.

It is interesting to note that at the same processing time, the peak is more extended than all the others. The broadening of the peak can be due to several reasons. First, it is the presence of other phases. At this temperature, in accordance with the phase diagram, the existence of other intermetallic phases, such as Cu_4Ti, Cu_2Ti, is possible, which were subsequently discovered by TEM. Their peaks are close in reflection angles and may overlap. The second reason is a significant grain refinement during more intense bombardment, which we also discovered by the TEM. Concentration inhomogeneity leads to broadening of peaks, since we are dealing with a nonequilibrium process, and the last reason is the creation of higher internal stresses due to defect formation, relaxation and annealing of defects, and accumulation and segregation of impurities. All these reasons can lead to significant peak broadening at maximum processing time.

The crystalline lattice parameter of the α-titanium can be estimated from the average angles of 2θ for the maxima corresponding to this phase (Table 3). The table shows the values of the crystalline lattice parameters of the α-titanium and the c/a ratio for all the investigated alloys.

Table 3. The crystalline lattice parameters, the compressing macrostresses σ //, the rms displacements of atoms $\sqrt{\langle u^2 \rangle}$ for direction <002> of the surface layer at different treatment durations of the titanium alloy with copper ions.

The Structure Parameters	Initial VT23	Treatment Time 3 min	Treatment Time 6 min	Treatment Time 7.5 min
a, nm	0.2893	0.2904	0.2899	0.2918
c, nm	0.4731	0.4705	0.4726	0.4687
c/a	1.63	1.62	1.63	1.61
$\sqrt{<u^2>}$002, nm	0.031	0.023	0.016	0.010
σ//, GPa	-	−1.6	−3.0	-

One can see that the crystalline lattice parameter in the modified titanium is higher than in the VT23 alloy in the initial state. The c/a ratio for this alloy is also higher than for the tabular values of α-titanium (c/a = 1.58). This fact indicates the deformation of the crystal lattice of the titanium alloy during standard heat treatment. The fact that the crystalline lattice parameter of the titanium alloy increases with the ion treatment time suggests that the main dopants, such as aluminum and vanadium, leave the surface layer. At the same time, the rms displacements of the atoms in the modified layer decrease uniformly with increasing ion treatment.

It is shown that the rms displacements of the atoms depend, to a large extent, on the concentration of the alloying elements. As a rule, a decrease in the concentration of the doping atoms in a solid solution leads to a decrease in the rms atomic displacements, which, most likely, is observed in this case with an increase in the duration of the ion treatment. For copper-based solid solutions, it was found that the atomic displacements decrease with a decrease in the concentration of the solid solution [23]. Our assumption has been confirmed by the pattern of the distribution of the elements across the depth of the modified layer.

Figure 2g–i shows that the content of all substitutional elements, such as Al, V, Cr, Mo, first decreases from its initial concentration (table of the composition of the VT23 alloy) in the modified layer, and then gradually reaches its original composition.

The depth of the layer depleted in the alloying elements correlates with the thickness of the modified layer in the first case. With a processing time of 3 min, it is about 150 nm, and then, down to a depth of 540 nm, the content of the alloying elements gradually returns to the previous level (Figure 2g). When the time of the ion treatment is 6 min, the depletion in the alloying elements reaches a depth of about 1 μm and the initial composition, with the thickness of the ion-modified layer being 3.5 μm (Figure 2h). At the maximum processing time, the thickness of the depleted layer is about 1.5 μm and down to 7 μm, with the content of the alloying elements gradually restoring to the initial concentration (Figure 2i). Compressive macrostresses arise in the treated surface, whose magnitudes increase with an increase in the treatment duration of the titanium alloy by an ion beam (Table 3). A similar effect was observed in the work [24], where the authors believe that the stress distribution over the sample depth is such that the alloyed layer is strongly compressed across the surface. The following processes contribute to the formation of a stressed layer (region) upon surface modification: defect formation, relaxation and annealing of defects, accumulation and segregation of impurities, diffusion (thermal and radiation-stimulated), and sputtering.

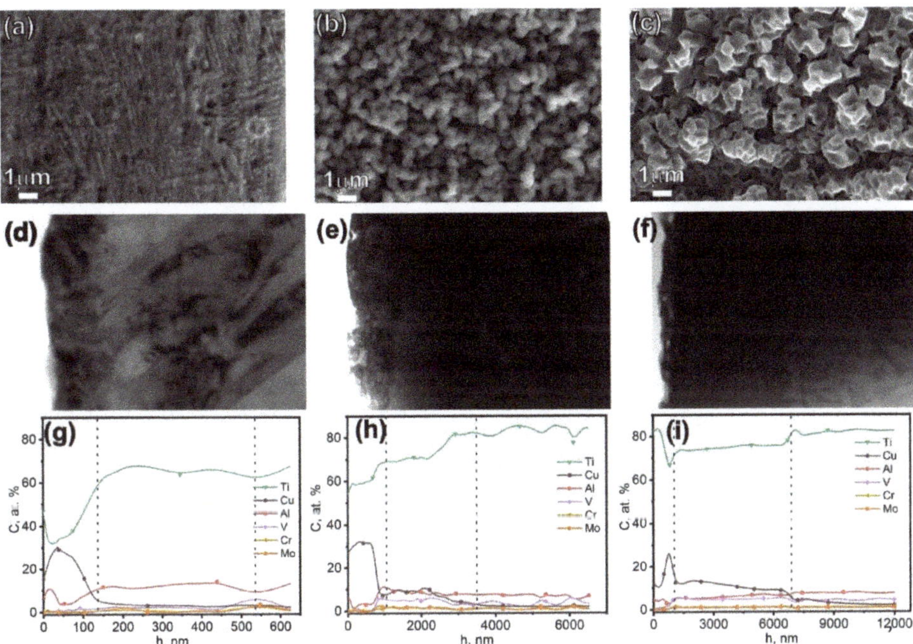

Figure 2. Surface morphology of the treated titanium alloy: 3 min (**a**), 6 min (**b**), 7.5 min (**c**). TEM images of the cross-section of the titanium alloy modified with copper ions with different processing times (**d–f**) and the distribution of the elements in depth: the processing time is 3 min (**g**), 6 min (**h**), 7.5 min (**i**).

Figure 2b,c shows the surface morphology of the titanium alloy treated with copper ions at different processing times. It can be seen that even at a processing time of 3 min, the surface is etched, but the martensite plates are still visible. Upon further processing for 6 min, the structural elements become smaller, the martensitic structure disappears. However, at a processing time of 7.5 min, a structure with a more developed surface takes place; voids are visible between the formations, which should be filled with the material of the deposited coating. In this case, the adhesive properties of the coating are significantly improved, as shown in [25–27].

TEM studies (Figure 3) have confirmed the results of the X-ray phase analysis. It has been established that depending on the processing time, the intermetallic phases of the Cu-Ti equilibrium phase diagram appear in the modified layer. Figure 3a–c shows the cross section of the titanium alloy modified with copper ions for 3 min. One can see that in the upper modified layer, in addition to the main α-Ti phase, there are intermetallic phases of Cu_3Ti, Cu_4Ti_3, and CuTi. The intermetallic phases are identified at a depth of no more than 300 nm (Table 4). The maximum copper content up to 30% is at a depth of 50 nm. Further, in the modified layer, only α-Ti is identified. The depth distribution of the elements shows (Figure 2g) that copper penetrates down to a depth of 600 nm and its amount is not more than 5 at.%. Apparently, it is in solid solution.

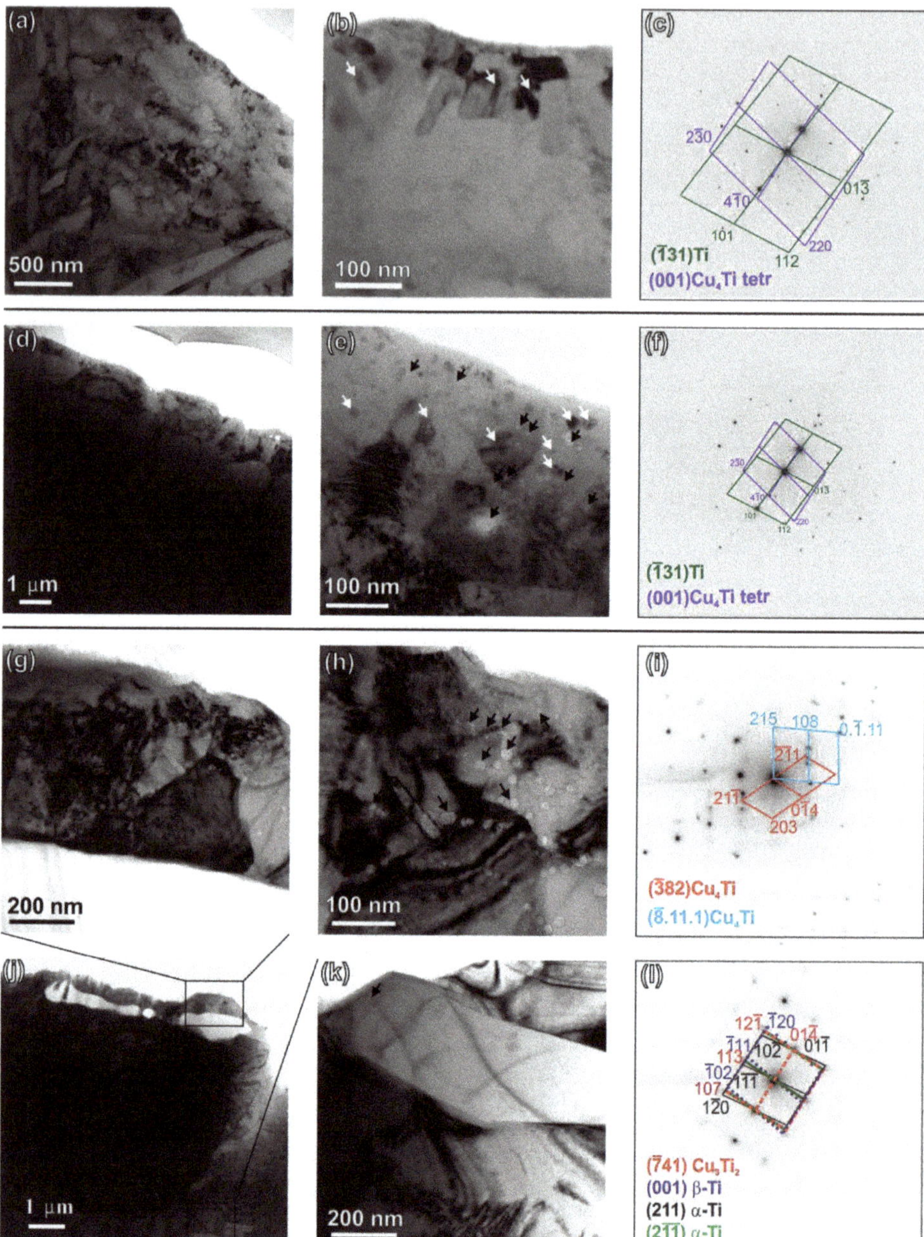

Figure 3. TEM images of the cross-section of the titanium alloy modified with copper ions: during 3 min (**a–c**), bright-field images, microdiffraction, and indication schemes; during 6 min (**d–f**), bright-field images and indication schemes; within 7.5 min (**g–l**). The white arrows show the indicated phases, the black ones—the pores.

Table 4. Phase composition and thickness of Ti modified by Cu ions.

Ion Treatment 3 min		Ion Treatment 6 min		Ion Treatment 7.5 min	
Thickness, nm	Phases	Thickness, nm	Phases	Thickness, nm	Phases
0–150	α-Ti CuTi Cu_4Ti_3 Cu_3Ti	0–250	α-Ti Cu_3Ti Cu_4Ti $CuTi_3$	0–810	α-Ti Cu_2Ti Cu_4Ti
				810–3680	α-Ti CuTi Cu_4Ti
150–500	α-Ti	>250–2000	α-Ti	3680–7500	α-Ti, particles of the Cu_3Ti inside of a-Ti.

For a sample treated with titanium ions for 6 min, the phase formation was found to take place at a depth of 250 nm (Table 4). We can see such phases as Cu_3Ti, Cu_4Ti, $CuTi_3$, and then titanium is observed in the initial state. The maximum amount of copper up to 35 at.% is observed at a distance of 800 nm from the surface, and then it sharply decreases down to 10 at.% and further decreases down to 5 at.% at a distance of 6.5 μm (Figure 2h). In the TEM image, the modified layer is visible up to 3.5 μm from the surface (Figure 3b). In a sample modified during 7.5 min, the pattern is significantly different from the previous ones. The intermetallic phases are observed here almost down to a depth of 4.0 μm. The phases such as Cu_2Ti, Cu_4Ti, CuTi have been identified. Interestingly, the particles of Cu_3Ti within the α-titanium plates are observed in the modified layer at a distance of 6.0 μm from the surface. The maximum amount of copper is observed at a depth of about 1.0 μm, but its concentration at this distance from the surface reaches only 25 at.%, then the copper concentration is 10–12 at.% at a distance of 6.0 μm and down to 12.0 μm the copper concentration remains in the amount of 5 at.% (Figure 2i). It is interesting to note that in this sample, the Cu_3Ti phase is in the α-Ti base phase layer inside the martensite plates at a distance of 7.5 μm from the surface (Table 4).

Figure 3b,e,h shows the pores indicated by black arrows. If in a sample with a short processing time they are only in the upper layer in a small amount, with longer processing times the pores are located everywhere in large quantities. In the sample with a treatment time of 7.5 min, the pore size becomes significantly larger (marked with black arrows). The intermetallic phases present in the modified layer are shown by white arrows.

After a comprehensive study of the titanium alloy modified with copper ions, a Si-Al-N-based coating was applied by magnetron sputtering. For this, the titanium alloy VT 23 is taken in its original state and modified for 7.5 min. Thermal cycling tests were carried out after deposition of the coating. The resistance of the coatings to cracking and peeling under changing temperature was determined from the results of the thermal cycling of the samples according to the following regime: heating the sample up to 1000 °C for 1 min, then cooling for 1 min to 20 °C. For the criterion of the thermal stability of the coatings, the number of the cycles before the destruction of 30% of the coating area of the sample surface was selected [14,28]. After that, the tests were stopped.

The coating, deposited on the initial surface, looks like one containing rectangular hollow cracks. The maximum number of cycles of this coating until catastrophic failure is not more than 500 as shown in Figure 4b. The coating applied to the treated surface contains rounded cracks, which are immediately filled with titanium oxide (in Figure 4d marked with numbers 1 and 2). The amount of oxygen in the cracks (indicated by the number 1) significantly exceeds its value in the coating (number 2). Using the EDS method it was shown that in (1) the oxygen value is about 60 at% and in the coating (2) its amount does not exceed 20 at%. This contributes to the healing of the cracks and prevents their further propagation, which, in turn leads to a high thermal cycling resistance of the coating. It was this coating that withstood 1500 cycles in our investigation (Figure 4d).

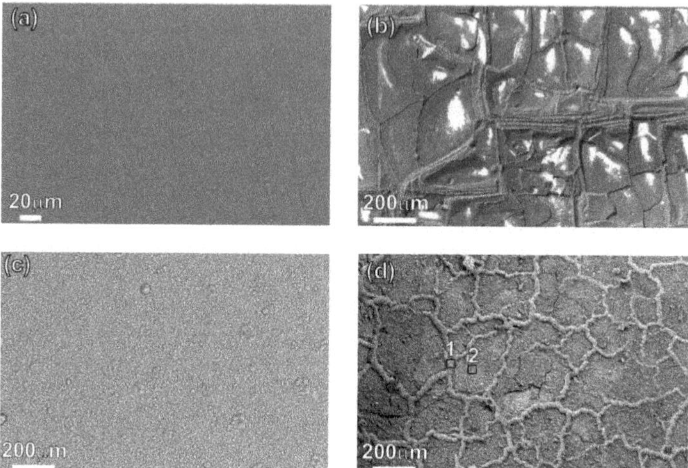

Figure 4. SEM images of coatings on the basis of Si-Al-N: coatings deposited on the initial surface of the titanium alloy: in initial state (**a**), after 500 cycles (**b**). Coating applied to titanium alloy treated with copper ions in initial state (**c**) and after 1500 cycles (**d**).

4. Conclusions

The titanium alloy modified with copper ions at different processing times has been comprehensively studied at various structural levels by the methods of X-ray diffraction analysis and electron and scanning microscopy.

1. It has been established by SEM that the modification of the titanium alloy by copper ions results in the refinement of the structure and the formation of developed micropores.
2. It has been shown by X-ray that the phase composition of the modified surface depends on the processing time. At short times of ionic treatment, the phases rich in titanium ($CuTi_3$) appear. With an increase in the duration of the treatment up to 7.5 min, the reflections of the $CuTi_3$ phase become more intense than those of the titanium substrate, i.e., almost the entire surface consists of an intermetallic compound. The structure parameters of the modified layer change: the crystalline lattice parameters of the titanium alloy increase, the rms displacements of the atom decrease, the compressive macrostresses occur.
3. It has been shown that the depth of the modified layer depends on the processing time. With a minimum processing time of 3 min, the depth of the modified layer is about 150 nm. With a processing time of 6 min, this it is about 3 µm, with a maximum processing time of 7.5 min, it is 7.5 µm, with the copper ions penetrating much deeper (0.6 µm, 7.5 µm, and 12 µm, respectively).
4. It has been found that the intermetallic compounds of the Cu-Ti system are formed not only in the surface layer of the treated surface, but also at a distance of 7.5 µm from the surface located inside the martensite plates, the so-called second-level phases. In the modified surface, the pores are found everywhere for all processing times. As a result, of the above treatment, the multiscale micro- and nanoporous nanocrystalline structure has been produced.
5. The thermocyclic resistance of the Si-Al-N-based coating increases several times if the coating is applied to the surface of the titanium alloy modified with copper ions, which is directly related to an increase in adhesion.

Author Contributions: Conceptualization, V.S. and M.F.; methodology, M.F.; validation, V.S. and O.P.; investigation, M.K., I.B. and M.F.; resources, V.S.; writing—original draft preparation, O.P.; writing—review and editing, M.F., O.P.; funding acquisition, V.S. All authors have read and agreed to the published version of the manuscript.

Funding: This research was supported by Tomsk Polytechnic University Competitiveness Enhancement Program, as well as its part was performed under the government statement of work for ISPMS Project No. III.23.1.1.

Conflicts of Interest: The authors declare no conflict of interest.

References

1. Collings, E.W. The physical metallurgy of titanium alloys. In *American Society for Metals*; University of Michigan: Ann Arbor, MI, USA, 1984; Volume 3, pp. 1–261.

2. Huang, X.; Zakurdaev, A.; Wang, E.D. Microstructure and phase transformation of zirconia-based ternary oxides for thermal barrier coating applications. *J. Mater. Sci.* **2008**, *43*, 2631–2641. [CrossRef]

3. Ji, Z.; Haynes, J.A.; Voelkl, E.; Rigsbee, J.M. Phase Formation and Stability in Reactively Sputter Deposited Yttria-Stabilized Zirconia Coatings. *J. Am. Ceram. Soc.* **2001**, *84*, 929–936. [CrossRef]

4. Bartolome, J.F.; Beltran, J.I.; Gutierrez-Gonzalez, C.F.; Pecharroma, C.; Munoz, M.C.; Moya, J.S. Influence of ceramic–metal interface adhesion on crack growth resistance of ZrO2–Nb ceramic matrix composite. *Acta Mat.* **2008**, *56*, 3358–3366. [CrossRef]

5. Enrico, M.; Chiara, S.; Valter, S. Residual Stresses in Alumina/Zirconia Composites: Effect of Cooling Rate and Grain Size. *J. Am. Ceram. Soc.* **2001**, *84*, 2962–2968.

6. Kalashnikov, M.P.; Fedorischeva, M.V.; Sergeev, V.P.; Neyfeld, V.V.; Popova, N.A. Features of surface layer structure of VT23 titanium alloy under bombardment with copper ions. *AIP Conf. Proc.* **2015**, *1683*, 20076.

7. Schmidt, B.; WetzigIon, K. *Beams in Material Processing and Analysis*; Springer: Wien, Austria; Heidelberg, Germany; New York, NY, USA; Dordrecht, The Netherlands; London, UK, 2013; pp. 1–413.

8. Kucheyev, S.O. *Ion.-Beam Processing*; Material Processing Handbook; Joanna, R., Groza, J.F., Shakelford, E.J., Lavernia, M.T., Eds.; Powers, CRC Press: London, UK; New York, NY, USA, 2007; Volume 2, pp. 25–50.

9. Davis, N.A.; Remnev, G.E.; Stinnett, B.W.; Yatsui, K. Intense ion-beam treatment of materials. *MRS Bull.* **1996**, *21*, 58–62. [CrossRef]

10. Chr, A. Straede Ion implantation as an efficient surface treatment. *Nucl. Instrum. Methods Phys. Res.* **1992**, *B68*, 380–388.

11. Zhang, E.; Wang, X.; Chen, M.; Hou, B. Effect of the existing form of Cu element on the mechanical properties bio-corrosion and antibacterial properties of Ti-Cu alloys for biomedical application. *Mater. Sci. Eng.* **2016**, *C69*, 1210–1221. [CrossRef] [PubMed]

12. Walschus, U.; Hoene, A.; Patrzyk, M.; Lucke, S.; Finke, B.; Polak, M.; Lukowski, G.; Bader, R.; Zietz, C.; Podbielski, A.; et al. A Cell-adhesive plasma polymerized allylamine coating reduces the In Vivo Inflammatory Response Induced by Ti6Al4V modified with plasma immersion ion implantation of copper. *J. Funct. Biomater.* **2017**, *8*, 30. [CrossRef] [PubMed]

13. Pierret, C.; Maunoury, L.; Monnet, I.; Bouffard, S.; Benyagoub, A.; Grygiel, C.; Busardo, D.; Muller, D.; Höche, D. Friction and wear properties modification of Ti-6Al-4V alloy surfaces by implantation of multi-charged carbon ions. *Wear* **2014**, *319*, 19–26. [CrossRef]

14. Sergeev, V.P. *Kinetics and Mechanism of the Formation of Nonequilibrium States of Surface Layers under Conditions of Magnetron Sputtering and Ion. Bombardment*; Nanoengineering surface; Lyachko, N.Z., Psahie, S.G., Eds.; Publishing House of the SB RAS: Novosibirsk, Russia, 2008; pp. 227–276.

15. Panin, V.E.; Sergeev, V.P.; Moiseenko, D.D.; Yu, I. Pochivalov, Scientific basis for the formation of heat-protective and wear-resistant the Si-Al-N / Zr-Y-O multilayer coatings. *Fiz. Mezomekh.* **2011**, *14*, 5–14.

16. Didenko, A.N.; Sharkeev, Y.P.; Kozlov, E.V.; Ryabchikov, A.I. *Long-Range Effects in Ion.-Implantable Metallic Materials*; Kolobov, Y.R., Ed.; Scientific and technical literature: Tomsk, Russia, 2004; pp. 4–326.

17. Fedorischeva, M.V.; Kalashnikov, M.P.; Nikonenko, A.V.; Bozhko, I.A. The structure—Phase state and microhardness of the surface layer of the VT1-0 titanium alloys treated by copper ions. *Vacuum* **2018**, *149*, 150–155. [CrossRef]

18. Kurzina, I.A.; Kozlov, E.V.; Sharkeev, Y.P.; Fortuna, S.V.; Koneva, N.A.; Bozhko, I.A.; Kalashnikov, M.P. *Nanocrystalline Intermetallic and Nitride Structures Formed by Ion.-Plasma Exposure*; Koval, N.N., Ed.; T: Publishing House HTJ1: Tomsk, Russia, 2008; pp. 1–317.

19. Sergeev, V.P.; Yanovsky, V.P.; Paraev, Y.N.; Sergeev, O.V.; Kozlov, D.V.; Zhuravlev, S.A. Installation of ion-magnetron sputtering of nanocrystalline coatings (QUANT). *Phys. Mesomech.* **2004**, *7*, 333–336.

20. Gorelik, O.S.; Rastorguev, L.N.; Skakov, Y.A. *X-Ray and Electron-Optical Analysis*; Metallurgiya: Moscow, Russia, 1970; p. 328.
21. Mirkin, L.I. *Spravochnik Po Rentgenostruktunomu Analizu Polikristallov (The Handbook of X-Ray Analysis of Polycrystals)*; Fizmatlit: Moscow, Russia, 1961; p. 864.
22. Lyakishev, N.P. (Ed.) *Diagrams of Binary Metallic Systems*; Mashinostroenie: Moscow, Russia, 1997.
23. Perevalova, O.B.; Konovalova, E.V.; Koneva, N.A.; Kozlov, E.V. *Influence of Atomic Ordering on the Grain Boundary Ensembles in FCC-Solid Solutions*; NTL Publ.: Tomsk, Russia, 2014; p. 247. (In Russia)
24. Domkus, M.; Pranyavichus, L. *Mechanical Stresses in Implanted Solids*; Mokslas: Vilnius, Lithuania, 1990; p. 158.
25. Xu, W.; Lui, E.W.; Pateras, A.; Qian, M.; Brandt, M. In situ tailoring microstructure in additively manufactured Ti-6Al–4V for superior mechanical performance. *Acta Mater.* **2017**, *125*, 390–400. [CrossRef]
26. Guo, G.; Tang, G.; Ma, X.; Sun, M.; Ozur, G.E. Effect of high current pulsed electron beam irradiation on wear and corrosion resistance of Ti6Al4V. *Surf. Coat. Technol.* **2013**, *229*, 140–145. [CrossRef]
27. Shulov, V.A.; Gromov, A.N.; Teryaev, D.A.; Perlovich, Y.A.; Isaenkova, M.G.; Fasenko, V.A. Texture formation in the surface layer of VT6 alloy targets irradiated by intense pulsed electron beams. *Inorg. Mater. Appl. Res.* **2017**, *8*, 387–391. [CrossRef]
28. Perevalova, O.B. Changes in the elastic-stress state and phase composition of TiAlN coatings under thermal cycling. *Inorg. Mater.* **2018**, *9*, 399–409. [CrossRef]

Publisher's Note: MDPI stays neutral with regard to jurisdictional claims in published maps and institutional affiliations.

Article

Infrasonic Nanocrystal Formation in Amorphous NiTi Film: Physical Mechanism, Reasons and Conditions

Evgeny E. Slyadnikov

ISPMS Institute of Strength Physics and Material Science SB RAS, Akademicheskii pr. 2/4, 634055 Tomsk, Russia; eeslyadnikov@gmail.com; Tel.: +7-923-426-4969

Abstract: The physical mechanism, reasons and conditions of nanocrystal formation in an amorphous NiTi metal film, stimulated by infrasonic action, are formulated. Nanostructural elements of an amorphous medium (relaxation centers) containing disordered nanoregions with two-level systems are considered to be responsible for this process. When exposed to infrasound, a large number of two-level systems are excited, significantly contributing to inelastic deformation and the formation of nanocrystals. The physical mechanism of the nanocrystallization of metallic glass under mechanical action includes both local thermal fluctuations and the additional quantum tunneling of atoms stimulated by shear deformation. A crystalline nanocluster appears as a result of local atomic rearrangement growing increasingly exposed to infrasound. It is possibly unstable in the absence of infrasound. When the radius of the nanocluster reaches a critical value, a potential well appears, in which a conducting electron is localized to form a phason (stable nanocrystal). Estimated values of the phason's radius and the depth of the nanometer potential well is about 0.5 nm and 1 eV, respectively. It forms a condition of stable phason formation. The occurrence of the instability of the amorphous state and following transformation to the nanostructured state is based on the accumulation of the potential energy of inelastic deformation to a critical value equal to the latent heat of the transformation of the amorphous state into the nanostructured state.

Keywords: physical mechanism; nanocrystals; amorphous metal film; infrasound; inelastic deformation; quantum tunneling; localization of the conduction electron

Citation: Slyadnikov, E.E. Infrasonic Nanocrystal Formation in Amorphous NiTi Film: Physical Mechanism, Reasons and Conditions. *Metals* **2021**, *11*, 1390. https://doi.org/10.3390/met11091390

Academic Editor: Giovanni Principi

Received: 8 August 2021
Accepted: 30 August 2021
Published: 1 September 2021

Publisher's Note: MDPI stays neutral with regard to jurisdictional claims in published maps and institutional affiliations.

1. Introduction

Amorphous metal alloys with a nanostructure are possessed by unique mechanical properties occurring due to intensive mechanical processing [1]. For nanostructuring an amorphous NiTi film, low-frequency mechanical vibrations are used at a temperature much lower than the glass transition temperature used for the remaining sample deformation within the elastic region [2–4]. Data obtained by the methods of X-ray diffraction and high-resolution transmission electron microscopy indicate that the initial alloy is homogeneously amorphous, while the structure of the sample subjected to a mechanical vibration amplitude of 1 mkm is close to the structure of the initially untreated sample (the short-range order changes and free volume decreases). In the structure of the sample subjected to mechanical vibrations with an amplitude of 4 mkm, nonspherical clusters with correct atomic positions and sizes of 3–5 nm appear in an amorphous matrix.

It has been established [2] that, under mechanical action, the latent heat of crystallization decreases slightly. It means that the free energy of the state with nanocrystals becomes less in comparison with the free energy of the amorphous state wherein the chemical and phase composition of alloys with and without nanocrystals is the same. Therefore, nanocrystallization caused by infrasonic stimulation is not related to diffusion, but provoked by collective atomic rearrangements during structural relaxation.

Further experimental studies have shown that [3] an increase in the duration of oscillations up to 2 h enlarges the size of clusters and their growth in different directions.

161

Running on mechanical vibrations for 4 h, the film acquires a crystalline structure. consists of irregular shape grains of different crystallographic orientations with amorphous islands embedded.

In [4], the influence of temperature, frequency and amplitude of vibrations (the duration of vibrations is 10 min) on the structure of amorphous alloys based on NiTi was investigated. It was found that, at room temperature, a vibration frequency of 20 Hz, vibration amplitude of 4 μm (10^{-6} m), and crystalline clusters 4–6 nm (10^{-9} m) in size appear in the amorphous matrix. With an increase in the initial temperature of the sample (from 25 to 200 °C), the distribution and size of crystalline clusters practically do not change. The crystallization temperature practically does not change, and the latent heat of crystallization decreases with an increase in the vibration amplitude and the initial temperature of the sample. When the frequency (below 5 Hz) or the vibration amplitude (down to 1 μm) decreases, crystalline clusters do not appear in the amorphous matrix.

Despite the fact that the phenomenon of nanocrystal formation in amorphous NiTi film under the action of mechanical vibrations is described in the literature, the physical mechanisms, reasons and conditions of its occurrence remain unclear.

In accordance with classical concepts [5], the nucleation of crystals in amorphous alloys can occur due to either a homogeneous or heterogeneous mechanism. Approaching the glass transition temperature, the homogeneous nucleation occurs through the fluctuation nucleus formation with a radius larger than the critical one [5]. Heterogeneous nucleation occurs mainly at temperatures below the glass transition temperature placed on defects boundaries, etc., [5,6] and due to "frozen in" crystallization centers [7]. The density of heterogeneous nuclei in metal melts is about $10^{11} - 10^{12}$ m^{-3}. Amorphous alloys, without nanosized crystallites, have a density noticeably exceeding these values (for example, in the $Fe_{80}B_{20}$ alloy, the density of crystallites is 10^{18} m^{-3} [8]). In nanostructured systems (for example, in Cu–Ti alloys [9]), the density of crystallites can reach 10^{25} m^{-3}; therefore, most of the nuclei arise through homogeneous nucleation during annealing.

As analysis shows, the mechanisms of the nanocrystal formation in amorphous NiTi film under stationary annealing and mechanical action differ significantly. During thermal annealing, the sample temperature is much higher than room temperature. The nanocrystallization process is slow and homogeneous. Nanocrystals appear throughout the entire volume by the thermal fluctuation mechanism. Under mechanical action, the medium temperature is room temperature; therefore, the probability of nanocrystal formation by the usual thermal fluctuation mechanism is much less than at the annealing temperature. However, during mechanical vibrations with an amplitude of 4 mkm, significant shear stresses arise in the film. At a critical value of shear stress, the medium loses shear stability and inelastic deformation occurs. Therefore, the mechanism of nanocrystallization of metallic glass under mechanical action should include quantum tunneling of atoms or atomic groups [10–14] stimulated by shear deformation in addition to local thermal fluctuations.

The ability of amorphous metal alloys to cause irreversible deformation and nanocrystal formation is associated with the collectivized character of the metallic interatomic bond [15]. Irreversible processes of local collective atomic rearrangements can occur much more easily in the presence of conducting electrons. Therefore, when constructing the mechanism of the irreversible deformation and nanocrystallization of amorphous alloys, it is necessary to take into account the influence of the electronic subsystem [16–18]. Specifically, for metallic glass NiTi, it is necessary to consider the localization of a conducting electron in a nanometer potential well and the formation of a phason [18].

The study of nanocrystal formation under a different external surrounding is of interest to the modern science of solid state matter. At the same time, as we found in the literature, there is no consideration of the effect of infrasound nanocrystal formation, and its physical mechanisms and possible conditions are not established.

The present work aims to consider the physical mechanisms of nanocrystal formation in amorphous NiTi under the infrasound action. The mechanism of influence of infrasound is not known and has not been previously considered. Herein, we propose a model that takes into account the inhomogeneity of the amorphous medium, the so-called relaxation centers, capable of rearranging according to the mechanism of athermal shear deformations stimulated by infrasound. As a result of the atomic rearrangement of the relaxation center, a more ordered structure is formed. As a consequence, a potential well arises, in which a conductivity electron is localized.

As we supposed, the analytical formulation of the problem is of two objectives: (1) in the framework of the model to formulate the equations of the density operator, taking into account the relaxation term and the field of static shear deformation; (2) to determine the probabilities of transitions of an atom in a double-well potential by thermal and athermal, i.e., quantum, mechanisms.

At the same time, numerical estimations might be provided in order to make the process of crystallization predictable. (3) It is likely to know the value of the shear strain, epsilon, at which quantum tunneling (disordered nanocluster transforms to a nanocrystal) occurs according to the athermal type; (4) It is necessary to determine conditions for the appearance of a localized state of conductivity electron at the potential well of the nanocrystal. There is a need in quantitative estimations of the critical values of the potential well depth and the radius of a stable nanocrystal.

2. Physical Mechanism of Inelastic Deformation of Metal Glasses

Inelastic deformation occurring as a response to the applied shear stress [19–24] is explained by local static displacements of a group of atoms (or an atom) from the initial equilibrium positions to the new ones at a distance less than interatomic spacing.

The first microscopic model of elementary cooperative inelastic rearrangements in metallic glasses is based on the concept of a uniformly distributed free volume. It was proposed by Argon in 1979 [19]. In the model [20], it is assumed that relaxation centers appear due to the inhomogeneity of the glass structure, particularly in the regions with an excess free volume relative to the "ideal structure". At present, the model of the activation energies spectrum [21], as well as the model of directed structural relaxation oriented by an external force [22], is often used.

The model [23] proposes a method of description of the local structure of amorphous alloys on the basis of the concept of n, p, and τ defects. In [24], a polycluster model of the structure of an amorphous state was proposed. The elementary rearrangements of atomic configurations occur at intercluster boundaries under the action of external forces. At high temperatures, the rearrangements of configurations are, possibly, a mechanism of thermal fluctuations, whereas at low temperatures quantum tunneling is more physically suitable.

The structural elements of an amorphous medium containing free volume (relaxation centers [20]) rearranging not only as a result of thermal fluctuations but due to quantum tunneling are of particular interest. Atom moving in a double-well potential (Figure 1) or a group of atoms with two configurations possessing different energies considered as two-level systems [10–14,24]. The atom transition from one well to another (or the changing of one configuration of atoms to another), which means a transition from one energy level to another, can occur due to quantum tunneling. These effects are more probable when a two-level system is close to degenerating. Specifically, the levels of zero-point vibrations of an atom, in each of the neighboring wells, should differ slightly without taking into account tunneling, (i.e., there is a resonance detuning Δ_0). A two-level system is characterized by an initial resonance detuning $\Delta_0 << 1$ eV and an initial frequency of tunnel junctions $\omega_K^0 = \omega_0 \exp\left(-\frac{a}{a_{dB}}\right)$. Here, a_{dB} is the de Broglie wavelength and a is the width of the potential barrier in the double-well potential, $\omega_0 \sim 10^{13}$ Hz. The frequency, ω_K^0, on the one hand, must be much lower Δ_0 in order for the initial splitting of the levels $E_0 = \sqrt{\Delta_0^2 + \hbar^2(\omega_K^0)^2}$ to be small $\sim\Delta_0$. On the other hand, ω_K^0 should not be too small for the tunneling to manifest itself during experiment.

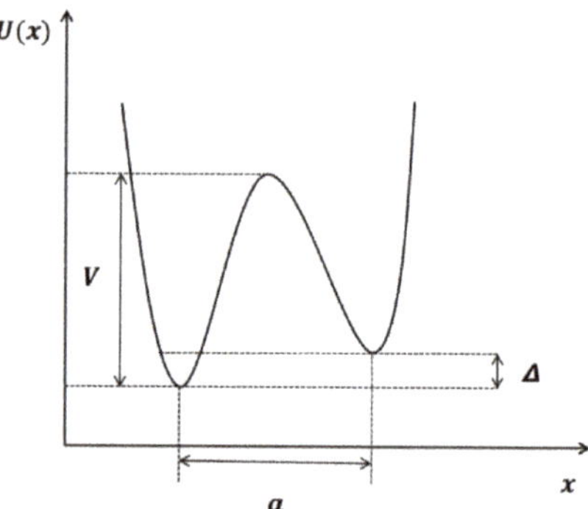

Figure 1. Dependence of the double-well potential $U(x)$ on the spatial coordinate x. Δ—resonanc detuning; V—interwell barrier; a—width of the potential barrier.

In the framework of the one-particle model of the interaction of phonons with two level systems [10–14,24], the initial resonance detuning Δ_0 and the initial frequency of tunneling transitions ω_K^0 are renormalized. During the passage of a sound wave, the atoms that form the potential curve of the two-level system are displaced. As a consequence individual wells of the curve are distorted, which leads to a change in the resonance detuning Δ. The shape of the potential barrier also changes and shifts to the frequency of tunnel junctions ω_K. The distance between the two levels of the two-level system become to be equal to $E = \sqrt{\Delta^2 + \hbar^2 \omega_K^2}$. For the linear approximation of deformation, we have

$$\delta\Delta = \sum_{i,j} B_{ij}\varepsilon_{ji}, \ \delta\omega_K = \sum_{i,j} D_{ij}\varepsilon_{ji}, \ \varepsilon_{ij} = \frac{1}{2}\left(\frac{\partial u_i}{\partial x_j} - \frac{\partial u_j}{\partial x_i}\right) \tag{1}$$

Here, B_{ij}, D_{ij} are tensors of strain potentials, which depend on the local structure of the glass in the place where the two-level system is located. ε_{ij}—strain tensor, which is expressed through the components of the displacement vector u_i The energy per atom per one deformation cycle is $e_a \cong 10^{-6}$ eV; therefore, $\delta\Delta \cong e_a \cong 10^{-6}$ eV. Since at the initial moment of time $\Delta_0 < 0$, it is possible to estimate $|\Delta_0| \cong \delta\Delta \cong 10^{-6}$ eV and the maximum value of $\omega_K \cong 10^9$ s^{-1}.

To estimate the characteristic parameters of the atomic tunneling, we will perform a modeling calculation. Let us suppose an atom in the relaxation center is in a two-well potential or there is a group of atoms in which there are two configurations with slightly different energies [12–14,24]. The potential energy of such a subsystem can be represented as the sum of two single-well potentials U_L and U_R. At the initial moment of time (in the absence of mechanical load), the left and right potential holes have different depths—the left hole is deeper than the right one. The action of an external mechanical force on the system leads to a change in the distances between the atoms of the medium (deformation which in turn changes the shape of the double-well potential.

Let the wave functions Ψ_L, Ψ_R and energy eigenvalues E_L^0, E_R^0 be known for the Schrödinger equation with single-well potentials U_L, U_R. For simplicity, when superimpos ing the field of the static shear component of deformation ε, only the energy eigenvalues E_L^0, E_R^0 are renormalized, according to the rule $E_L = E_L^0 + B\varepsilon$, $E_R = E_R^0 - B\varepsilon$. The complete Hamiltonian of a one-dimensional subsystem with a two-well potential $U_L(\varepsilon) + U_R(\varepsilon)$ $U_L(\varepsilon) = U_L + B\varepsilon$, $U_R(\varepsilon) = U_R - B\varepsilon$ in the field of a static shear strain component ε has the form.

$$H = \frac{-\hbar^2}{2m}\frac{\partial^2}{\partial x^2} + U_L(\varepsilon) + U_R(\varepsilon), \ \Psi = a_L\Psi_L + a_R\Psi_R, \ \Delta = E_L - E_R = E_L^0 - E_R^0 + 2B\varepsilon \tag{2}$$

Due to the assumed weak overlap of wave functions Ψ_L, Ψ_R, $(S = \int dx \Psi_L \Psi_R \ll 1)$, solving the Schrödinger equation with Hamiltonian (2), we find the stationary levels $E_\pm$ (Figure 2) and stationary states $\Psi_\pm$, in which an atom is delocalized between two wells:

$$E_\pm = \frac{1}{2}[E_L + E_R \pm \sqrt{D}], \quad D = \Delta^2 + 4U_{LR}^L U_{LR}^R, \quad U_{LR}^L = \int dx \Psi_L U_L(\varepsilon) \Psi_R \quad (3)$$

$$\Psi_+ = [\Delta + \sqrt{D}]\Psi_L + 2U_{LR}^R \Psi_R, \quad \Psi_- = 2U_{LR}^L \Psi_L - [\Delta + \sqrt{D}]\Psi_R \quad (4)$$

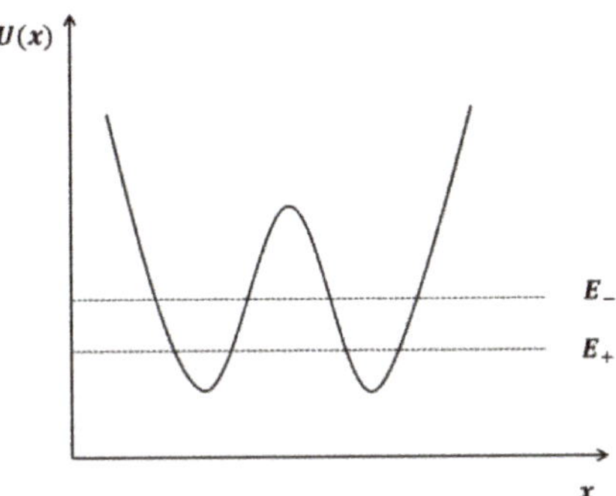

Figure 2. Ground state for a symmetric ($\Delta = 0$) double-well potential $U(x)$. Quantum tunneling removes the initial degeneracy of the levels. E_+—energy eigenvalue for a symmetric wave function Ψ_+; E_-—energy eigenvalue for an antisymmetric wave function Ψ_-.

Let us investigate the tunneling transition of an atom from one potential well to another in the field of the static shear component of deformation ε at a finite temperature using the formalism of the density operator ρ [25]. In this case, the atom making transitions between the wells interacts with the thermostat. Using eigenvectors Ψ_L, Ψ_R as basis vectors, leaving only the first-order terms in the parameter S, we obtain differential equations for the matrix elements of the density operator ρ:

$$i\hbar \frac{\partial \rho_{LL}}{\partial t} = U(\rho_{RL} - \rho_{LR}) - \frac{i\hbar}{\tau_{LL}}(\rho_{LL} - \rho_{LL}^0), \quad U \equiv U_{LR}^L = U_{LR}^R, \quad \tau_{ij} = \tau_{ji}, \quad i,j = L,R, \quad (5)$$

$$i\hbar \frac{\partial \rho_{RR}}{\partial t} = U(\rho_{LR} - \rho_{RL}) - \frac{i\hbar}{\tau_{LL}}(\rho_{RR} - \rho_{RR}^0), \quad \rho_{LL}^0 = \frac{1}{[1 + \exp\left(\frac{E_L^0 - E_R^0 + 2B\varepsilon}{kT}\right)]}, \quad \rho_{i \neq j}^0 = 0, \quad (6)$$

$$i\hbar \frac{\partial \rho_{LR}}{\partial t} = \left(\Delta - \frac{i\hbar}{\tau_{LR}}\right)\rho_{LR} - U(\rho_{LL} - \rho_{RR}), \quad \rho_{RR}^0 = \frac{\exp\left(\frac{E_L^0 - E_R^0 + 2B\varepsilon}{kT}\right)}{[1 + \exp\left(\frac{E_L^0 - E_R^0 + 2B\varepsilon}{kT}\right)]}, \quad \rho_{LR} = \rho_{RL}^* \quad (7)$$

The diagonal matrix element ρ_{LL} has the meaning of the probability of finding an atom in the left well. For simplicity, the system under study is characterized by two relaxation times: the longitudinal relaxation time of the system $\tau_{LL} = \tau_{RR}$ with an atom in the left (right) potential well and $\tau_{LR} = \tau_{RL}$ the transverse relaxation time of off-diagonal matrix elements.

When the time of the tunneling transition of an atom significantly exceeds the time of longitudinal or transverse relaxation, quantum coherence is violated. The probability of detecting a particle in a certain state can be found from the solution of the classical kinetic equations obtained by reducing the system Equations (5)–(7). The relaxation time has an exponential dependence on temperature $\omega_0^{-1} \exp\left(\frac{V}{kT}\right)$, where V is the height of the interwell barrier. When the temperature is high (the relaxation time is short) and the she[...] deformation is less than the critical one $\varepsilon < \varepsilon_*$, $\varepsilon_* = -\frac{E_L^0 - E_R^0}{2B}$, the two-level system relax[...] into a thermodynamically equilibrium state with $\rho_{LL}^0 > \frac{1}{2}$, $\rho_{RR}^0 < \frac{1}{2}$, in which the atom in the left well with a higher probability than in the right well. When the deformation equal to the critical $\varepsilon = \varepsilon_*$, the two-level system relaxes into the equilibrium state wi[...] $\rho_{LL}^0 = \frac{1}{2}$, $\rho_{RR}^0 = \frac{1}{2}$, in which the atom is with equal probability in the left and right wel[...] When the deformation is greater than the critical one $\varepsilon > \varepsilon_*$, the two-level system relax[...] to the equilibrium state with $\rho_{LL}^0 < \frac{1}{2}$, $\rho_{RR}^0 > \frac{1}{2}$, in which the atom is in the right well wi[...] a greater probability than in the left well.

If the period of quantum oscillations is much shorter than the time of longitudin[...] and transverse relaxation, then the process of tunneling transition is coherent. The tunn[...] transition probability is an oscillating function of time, i.e., described by quantum kinetic[...] A linear combination of stationary wave functions $\Psi(x,t)$, satisfying the time-dependen[...] Schrödinger equation with Hamiltonian (2) and the initial condition $\Psi(t=0) = \Psi_L$, make[...] it possible to find the probability of finding an atom in the left well:

$$W(t) = a^2(t) = 1 - \frac{16[\Delta + \sqrt{D}]^2 U^2}{([\Delta + \sqrt{D}]^2 + 4U^2)^2} \sin^2\left(\frac{\sqrt{D}}{2\hbar}t\right), \; a(t) = \int dx\Psi(x,t)\Psi_L(x) \quad (8)$$

The function $W(t)$ oscillates from unity to the minimum value, which is determined by th[...] difference in energies in two states Δ and the magnitude of the product U^2. If $\Delta = 0$, the[...] $W(t) = 1 - \sin^2\left(\frac{U}{\hbar}t\right)$. The probability of finding an atom in the left well varies from one t[...] zero with a period $\tau = \frac{\pi\hbar}{U}$, and the atom spends the same time in both wells. If $\Delta^2 \gg 4U[...]$ then $W(t) = 1 - 4\frac{U^2}{\Delta^2}\sin^2\left(\frac{U}{\hbar}t\right)$. The atom is in the left potential well almost all of the tim[...]

When the shear strain ε is zero, the case $\Delta = \Delta_0 = E_L^0 - E_R^0 < 0$, $\Delta^2 \gg 4U^2$ is realize[...] and the atom is only in the left well. At a critical value of deformation $= \varepsilon_*$, the case $\Delta = [...]$ is reached and the atom tunnels into the right well. Further increase in deformation lead[...] to the case $\Delta > 0$ $\Delta^2 \gg 4U^2$, and the atom is only in the right well. The static displacemen[...] of an atom is an elementary act of structural relaxation of a medium under load and i[...] accompanied by stress relaxation. This leads to a change in the positions of atoms fror[...] the nearest environment and "freezing" of the changed shape of the double-well potentia[...] This process is reversible; after a slow removal of the load, the atom returns to the left we[...] again. This is how the elementary inelastic reversible deformation proceeds in metalli[...] glass at $T = 0$.

Metallic glass contains a large number of relaxation centers containing two-level system[...] which are described by a double-well potential, resonance detuning Δ, and tunnel transitio[...] frequency ω_K. Two groups of such double-well potentials can be distinguished—soft (with a[...] interwell barrier $V \sim 10^{-4} - 10^{-2}$ eV) and hard (with $V \sim 10^{-1} - 1$ eV) [12–14,24]. In th[...] temperature range $T \cong 1 - 300$ K, thermal jumps between the states of a two-level system with a barrier $V \sim 10^{-4} - 2.4 \cdot 10^{-2}$ eV are activated. Therefore, the quantum tunnelin[...] mechanism is more probable for a two-level system with a hard potential.

Let us estimate the temperature T_* at which the frequency of the quantum sub-barrie[...] tunneling of an atom ω_K becomes equal to the frequency of the above-barrier fluctuatio[...] hopping ω_T. Using the Gamow formula and Arrhenius's law, equating the exponent[...] we get

$$\omega_K = \omega_0 \exp\left(-\frac{a}{a_{dB}}\right), \; \omega_T = \omega_0 \exp\left(-\frac{V}{kT}\right), \; a_{dB} = \frac{\hbar}{\sqrt{2mV}}, \; T_* = \sqrt{\frac{V}{2m}}\frac{\hbar}{ak} \quad (9[)]$$

From (9), it can be seen that T_* increases $\sqrt{V}$ proportionally and decreases $\sqrt{\frac{1}{m}}$, $\frac{1}{a}$ proportional[...]

For a two-level system with a hard potential, the characteristic values ar[...] $V \cong 1$ eV, $m \cong 10^{-25}$ kg, and $a_{dB} \cong 0.56 \cdot 10^{-12}$ m. With a barrier width $a = 3 \cdot 10^{-1[...]}$ $1.5 \cdot 10^{-11}$, $0.9 \cdot 10^{-11}$ m (0.1, 0.05, 0.03 interatomic distance), we obtain, respectivel[...]

$T_* \cong 65, 216, 721$ K and the quantum tunneling frequency $\omega_K \cong 10^{-10}$ 20, 100, 000 Hz. For local structural rearrangements with small displacements of atoms (much less than the interatomic distance), the tunneling effect is significant for room temperatures, and becomes the main at lower temperatures. Therefore, a possible physical mechanism of the inelastic deformation of metallic glasses (at room temperatures and below) is the quantum tunneling of some atoms in a double-well potential, or within an atomic group.

Under mechanical action on an amorphous material, a large number of two-level systems are excited, making a significant contribution to inelastic deformation and the formation of nanocrystals. The ability of amorphous metal alloys to deform irreversibly and to form nanocrystals is associated with the collectivized metallic character of the interatomic bond. In the presence of conduction electrons, irreversible processes of local collective atomic rearrangements can be carried out much more easily.

3. Electron Localized States in Potential Nanometer Well and Phason Formation

Let us assume that the electron moves in a three-dimensional spherically symmetric potential well having experienced the action of central force. In the simplest case, the potential energy $U(r)$ depends only on the distance between the electron and the force center r. Let the three-dimensional potential well have the form $U(r) = -U_0$ at $r \leq a$ and be equal to zero at $> a$. In a spherical coordinate system, the Schrödinger equation for stationary states has the form [26]

$$\frac{1}{r^2}\frac{d}{dr}\left(r^2\frac{d}{dr}\Psi\right) + \frac{2m}{\hbar^2}(E - U(r))\Psi = 0 \tag{10}$$

As solutions to Equation (10), which are finite at $r = 0$ and vanish at $r \rightarrow \infty$, we can take:

$$\Phi = B\sin(kr), \text{ at } r \leq a, \Phi = C\exp(-\alpha r), \text{ at } r > a, k = \sqrt{\frac{2m(E + U_0)}{\hbar^2}}, \alpha = \sqrt{\frac{-2mE}{\hbar^2}} \tag{11}$$

Since we consider the particle inside a potential well, it is necessary to count < 0. Thus, the problem of Equations (10) and (11) has been reduced to the problem of the motion of an electron in a one-dimensional potential well. Therefore, the energy levels are determined in the same way. The only difference now is in the necessity to discard states with even wave functions and leave only states with odd wave functions.

The fundamental difference between a one-dimensional potential well and a three-dimensional one is that for a one-dimensional well there is at least one eigenvalue of energy with even wave function. In the case of a spherically symmetric rectangular well, this may not be the case. It can be seen from the relation $k^2 + \alpha^2 = \frac{2mU_0}{\hbar^2}$ if $\frac{2mU_0}{\hbar^2}a^2 < \left(\frac{\pi}{2}\right)^2$, i.e.,

$$U_0 < \frac{\pi^2\hbar^2}{8ma^2} \tag{12}$$

that the curve given by the equation $tg\left(\frac{ka}{2}\right) = -\frac{k}{\alpha}$ will never intersect the circle given by the equation $k^2 + \alpha^2 = \frac{2mU_0}{\hbar^2}$. This means that when condition (12) is satisfied, not a single level of the discrete energy spectrum will appear in the potential well (due to the fact/the depth of the potential is not enough/the power of the well is too low).

Substituting the values of the fundamental constants $\hbar \cong 1 \cdot 10^{-34}$ J·s, the electron mass $m \cong 9 \cdot 10^{-31}$ kg, for the radii of the first, second, and third coordination sphere $a_1 = 0.35 \cdot 10^{-9}$ m, $a_2 = 0.7 \cdot 10^{-9}$ m, and $a_3 = 1.05 \cdot 10^{-9}$ m, we obtain $U_0(a_1) = 0.7$ eV, $U_0(a_2) = 0.18$ eV, and $U_0(a_3) = 0.08$ eV, respectively. From physical considerations, it follows that the perfect local order can propagate starting from the first coordination sphere. The minimum radius of the potential well is equal to the interatomic distance $a_1 = 0.35 \cdot 10^{-9}$ m. For a localized state of an electron to appear in a potential well with a minimum radius, the depth of the potential well of the resulting nanocrystal should be approximately 1 eV.

At a high interfacial surface energy inherent in solid-phase transformation, the fluctu ation formation of a region of a new crystalline phase leads to a significant change in th thermodynamic potential and may turn out to be unstable (nonequilibrium). However, i the presence of the attraction of a conduction electron in the formed region of a new cry talline phase, the electron can be localized in there. In this case, during the formation of new region, if the decrease in the electron energy exceeds the growth of the thermodynami potential energy of a new crystalline phase, then the fluctuation is stabilized. The arisin thermodynamically stable formation of the new phase region with a localized electron in (nonequilibrium in the absence of an electron) is called a phason [18].

In Section 1 of this work, it is shown that the formation of the nanoregion of th crystalline phase in an amorphous metal system can also occur by quantum tunneling stim ulated by shear deformation. Therefore, the process of the collective atomic rearrangeme of the relaxation center into a more ordered configuration (a transition between states in two-level system, stimulated by shear deformation) can be accompanied by the localizatio of a conduction electron in the resulting potential well.

Let us consider a center of relaxation of radius R, which has experienced a shea deformation, to which the potential well with depth $-U$ corresponds to (with respe to the average potential energy of the amorphous state). During the transition betwee states in a two-level system, the change in the thermodynamic potential $\Delta\Phi$ is expresse by the formula

$$\Delta\Phi(R) = \frac{4}{3}\pi R^3 \varphi + 4\pi R^2 \sigma_s + E_e(R), \quad E_e(R) = -U + \frac{\hbar^2 k^2}{2m_*} \tag{13}$$

where $E_e(R)$ is the energy of an electron in a potential well with a depth $-U$ and m_* i the effective mass of an electron. Here, $\varphi < 0$ is the difference between the densities c thermodynamic potentials of the new ordered state of atoms in the relaxation center an the initial disordered amorphous state, and σ_s is the interphase surface energy. Using th condition $2R = \lambda = \frac{2\pi}{k}$, where λ is the wavelength of the electron, and substituting k int the expression for the kinetic energy of the electron (13), we find $\frac{\hbar^2 k^2}{2m_*} = \frac{\hbar^2 \pi^2}{2m_* R^2}$.

$\Delta\Phi$ becomes negative (the deformed relaxation center is stable) at U, greater tha a certain value U_0. The minimum value $\Delta\Phi$ corresponds to the critical value R_0, whic determines the radius of the stably existing deformed relaxation center. Let us determin the value of R_0 and U_0. Equating $\frac{\partial \Delta\Phi(R)}{\partial R}$ to zero, for a large interfacial surface energ $|\varphi|R_0 \ll 2\sigma_s$, we obtain

$$R_0 = \left(\frac{\hbar^2 \pi}{8\sigma_s m_*}\right)^{\frac{1}{4}}, \quad U_0 = 8\pi\sigma_s R_0^2 = (2\pi)^{\frac{2}{3}}\sqrt{\frac{\hbar^2 \sigma_s}{m_*}} \tag{14}$$

For the value of $\sigma_s = 20$ erg/cm^2 (1 erg $= 10^{-7}$ J; 1 cm $= 10^{-2}$ m), which is characte istic of metals [18], and the condition $|\varphi|R_0 \ll 2\sigma_s$, the critical radius of the deforme relaxation center is $R_0 \cong 0.5$ nm, and the value U_0 is approximately 1 eV. It is clear fro physical considerations that the minimum radius of a stable deformed relaxation center i approximately equal to the radius of the second coordination sphere. Since the theoreticall estimated radius of a phason coincides in order of magnitude with the experimentall founded value of the radius of the nanocrystal [2–4,15], it is reasonable to assume that th possible mechanism for the nanocrystal formation in an amorphous NiTi metal film is th phason formation.

4. Discussion: Physical Representation, Reasons and Conditions for Nanocrystal Formation, and Model Parameters Estimation

In the framework of the model, the equations of the density operator are formulate taking into account the relaxation term and the field of static shear deformation. Th probabilities of the transitions of an atom in a double-well potential by thermal an athermal, i.e., quantum, mechanisms are determined. The obtained results allow us to formulate a physical idea of the nanocrystal formation and make conclusions about proper conditions for the crystallization.

At the initial moment of time, an amorphous metallic medium contains a large number of nanoregions (relaxation centers) as well as free volume and has a disordered atomic structure. The single action of infrasound (shear deformation) on the relaxation center, an atom (located in a double-well potential) or a group of atoms (which have two configurations with slightly different energies) provokes both thermal fluctuation (thermal mechanism) and quantum tunneling to new equilibrium positions (athermal mechanism). At a critical value of deformation ε_* (corresponding to the material proportionality limit), the resonance detuning becomes zero and the atom tunnels. At room temperature, local structural rearrangement with small static displacements of atoms (much less than the interatomic distance) occurs with a tunneling frequency higher than the thermal hopping frequency. Thus, at room temperature and lower temperatures, the athermal (quantum) mechanism of inelastic deformation prevails. This conclusion is consistent with the experimental result that with an increase in the initial temperature of the sample (from 25 to 200 °C), the distribution and size of crystalline clusters practically does not change [4]. As a result of inelastic deformation, a nanocluster with a crystalline atomic order appears at the relaxation center, but it may turn out to be unstable in the absence of infrasound.

Since the relaxation time and time of tunnel transition depend on temperature differently, both mechanisms (thermal and athermal) are realized upon the deformation of metallic glasses. With the decreasing of temperature, the thermal motion of atoms is getting to be frozen; as a consequence, the relaxation rate decreases exponentially accordingly to the Arrhenius law. Hence, at low temperatures, the rate of the tunneling transition exceeds the relaxation rate and coherent tunneling will occur. As the temperature rises, the relaxation rate increases faster than the tunneling rate. These values are initially compared, and then the relaxation rate can significantly exceed the tunneling rate. The transition process becomes thermofluctuational, and an ordered nanocluster arises due to thermal fluctuations. Thus, by changing the temperature, one can change the kinetics of the process of inelastic deformation of metallic glasses from quantum to classical.

The conditions for the appearance of a localized state of a conduction electron at the potential well of a nanocrystal are determined. The critical values of the potential well depth and the radius of a stable nanocrystal are calculated. Analysis of the obtained results allows us to formulate the following conclusions.

Under repeated exposure to infrasound, a nanocluster with a crystalline atomic order grows and reaches a critical radius. The nanocluster achieved the value of the critical radius and creates a local potential well in which the conduction electron appears in a bounded state. If the electron is attracted to the region of the unstable crystalline phase and is localized in there, then a decrease in the electron energy can be compensated for the surface energy of the interface. Thus, stable formation arises, also known as a phason (stable nanocrystal). The physical reason for the appearance of the stable nanocrystal is in the conduction electron localization by the potential well created by the nanocrystal.

The radius of the phason is determined by the value of the interfacial surface energy σ_s, and an increase σ_s leads to a decrease in the radius of the phason. Physically, the radius of the phason cannot be less than the radius of the first coordination sphere. For the value $\sigma_s = 20$ erg/cm^2, which is characteristic of metals [18], the critical radius of the phason is $R_0 \cong 0.5$ nm. The critical value of the depth of the potential well U_0, i.e., the condition under which a phason with $R_0 \cong 0.5$ nm is formed, is approximately 1 eV.

Since the theoretical estimation of the critical radius of a phason coincides in order of magnitude with the experimentally founded value of the radius of a nanocrystal [2–4,15], it is reasonable to assume that the mechanism of the formation of a nanocrystal in a metallic glass is the mechanism of the phason formation. When conduction electrons are localized on nanocrystals formed as a result of inelastic deformation, they lower the energy of the

system [2,4]. As a result, the state with nanocrystals becomes stable and a locally stab
nanostructure appears (with respect to the globally stable crystal structure [3]).

It has been experimentally established [2–4,15] that the morphology of nanocrystals d
pends on the method of nanocrystallization. The difference in the morphology of nanocry
tals, isotropic upon annealing [15] and anisotropic upon exposure to infrasound [2–4],
associated with different mechanisms of nucleation and growth of nanocrystals. Durir
nanocrystallization by high-temperature annealing, the nucleation and growth of nanocry
tals are controlled by thermal fluctuation processes. However, at room temperature, th
contribution of thermal fluctuation processes is small, inelastic deformation and nanocry
tallization occur due to quantum transitions in nanoscale two-level systems, stimulate
by infrasonic vibrations. The anisotropy of the nanocrystal shape is determined by th
strain potential tensor, which depends on the local glass structure in the place where th
two-level system is located.

Let us estimate the density of the stored potential energy of static displacements
atoms (inelastic deformation) under infrasonic action on the NiTi amorphous film. Th
dimensions of the film are known from the experiment [2–4]: length $l = 1 \cdot 10^{-2}$ m, widt
$r = 1.6 \cdot 10^{-3}$ m, and thickness $d = 4 \cdot 10^{-5}$ m. The time of the cyclic mechanical action
10 min, and the frequency of the action is 20 Hz. The amplitude of the tensile displacemen
of the infrasound is $A = 4 \cdot 10^{-6}$ m, and the relative longitudinal elongation of the fil
is $= \frac{A}{l} = 4 \cdot 10^{-4}$. It is known that the potential energy density of homogeneous long
tudinal elastic deformation is equal to $= \frac{E}{2} \cdot \varepsilon^2$, where E is the modulus of longitudina
elasticity (Young's modulus). The NiTi amorphous film is an alloy in which the mai
elements are nickel (45 percent) and titanium (41 percent). It is known that $E_{Ni} \approx 220$ GP
$E_{Ti} \approx 110$ GPa, and their average value is equal to $E \approx 165$ GPa. Taking into account th
average value $E \approx 165$ GPa and the relative elongation $\varepsilon = 4 \cdot 10^{-4}$, the potential energ
density is equal to $e \approx 1.3 \cdot 10^4$ J/m^3.

Further, let us estimate the numerical value of the potential energy of uniform lor
gitudinal elastic deformation per atom of the medium. Since the concentration of atom
in the medium is approximately equal to $0.9 \cdot 10^{29}$ atom/m^3, the energy per atom pe
one deformation cycle is equal to $e_a \approx 1 \cdot 10^{-6}$ eV/atom. During the time of exposur
to infrasound, $n = 12,000$ cycles of mechanical deformation occur; therefore, the energ
supplied to the system during the entire exposure time is $n \cdot e_a \approx 1.2 \cdot 10^{-2}$ eV/atom. It i
more probable that the small part (estimations are about 10 percent) of this energy can b
converted into potential energy of inelastic deformation. Therefore, the value of the store
potential energy of inelastic deformation is $0.1 \cdot n \cdot e_a \approx 1.2 \cdot 10^{-3}$ eV/atom, which is muc
less than the kinetic energy of an atom at room temperature $kT_0 \approx 0.024$ eV/atom.

The latent heat of amorphous–crystalline transformation, determined through th
temperature difference between amorphous and crystalline states $\Delta T \approx 150$ K in the TiC
alloy [27], is $L_a = k\Delta T \approx 0.012$ eV/atom, which we can consider as an example. Fc
the NiTi alloy, the latent heat of transformation is $L_a \approx 0.01$ eV/atom [4]. It has bee
experimentally established [2–4] that the latent heat of amorphous–crystalline transfo
mation slightly decreases after 12,000 cycles of infrasonic exposure. Therefore, the fre
energy of the amorphous state with nanocrystals slightly decreases in comparison wit
the free energy of the amorphous state. It is reasonable to assume that the latent heat c
transformation of an amorphous state into a nanostructured state (nanocrystals in an amo
phous matrix) is much less than $L_a \approx 0.01$ eV/atom, for example, $0.1 \cdot L_a \approx 0.001$ eV/aton
Thus, the latent heat of transformation of an amorphous state into a nanostructural stat
$0.1 \cdot L_a \approx 0.001$ eV/atom coincides in order of magnitude with the value of the store
potential energy of inelastic deformation $0.1 \cdot n \cdot e_a \approx 1.2 \cdot 10^{-3}$ eV/atom.

As was experimentally established in [4], when the frequency is below 5 Hz or th
vibration amplitude is down to 1 μm, crystalline clusters do not appear in the amorphou
matrix. Therefore, the following conclusion can be formulated: the occurrence of th
instability of the amorphous state and following transformation to the nanostructure
state is based on the accumulation of the potential energy of inelastic deformation to

critical value equal to the latent heat of the transformation of the amorphous state into the nanostructured state.

5. Conclusions

The relaxation centers including disordered nanoregions with two-level systems are the causes giving rise to the nanocrystal formation. When exposed to infrasound, the two-level systems, of which there are many, are excited, significantly contributing to inelastic deformation and the formation of nanocrystals. The physical mechanism of the nanocrystallization of metallic glass under mechanical action includes both local thermal fluctuations and the additional quantum tunneling of atoms stimulated by shear deformation. A crystalline nanocluster appears as a result of local atomic rearrangement growing increasingly exposed to infrasound. It is possibly unstable in the absence of infrasound. When the radius of the nanocluster reaches a critical value, the potential well appears. Bounding the conducting electron, it forms the phason. The estimated values (in case of the NiTi film) of the phason's radius and the depth of the nanometer potential well are 0.5 nm and 1 eV, respectively.

We consider the fact that the proposed microscopic mechanism makes it possible to formulate physical reasons and conditions for the appearance of nanocrystals, and we describe results and regularities obtained experimentally both quantitatively and qualitatively.

Funding: The work was performed according to the government research assignment for ISPMS SB RAS, project topic number FWRW-2019-0031.

Institutional Review Board Statement: Human or animal studies have not been conducted

Informed Consent Statement: Human studies have not been conducted.

Data Availability Statement: All the necessary data confirming the obtained results are contained in the manuscript.

Conflicts of Interest: There is no conflict of interest.

References

1. Gunderov, D.; Astanin, V. Influence of HPT deformation on the structure and properties of amorphous alloys. *Metals* **2020**, *10*, 415. [CrossRef]
2. Belyaev, S.; Resnina, N.; Rubanik, V.; Shelyakov, A.; Niapomniashchay, V.; Ubyivovk, E.; Kasatkin, I. Influence of low-frequency vibrations on the structure of amorphous Ti40.7Hf9.5Ni44.8Cu5 alloy. *Mater. Lett.* **2017**, *209*, 231–234. [CrossRef]
3. Belyaev, S.; Rubanik, V.; Resnina, N.; Rubanik, V.; Ubyivovk, E.; Demidova, E.; Uzhekina, A.; Kasatkin, I.; Shelyakov, A. Crystallization of amorphous Ti40.7Hf9.5Ni41.8Cu8 alloy during the low-frequency mechanical vibrations at room temperature. *Mater. Lett.* **2020**, *275*, 128084–128087. [CrossRef]
4. Belyaev, S.; Rubanik, V.; Resnina, N.; Rubanik, V.; Ubyivovk, E.; Demidova, E.; Uzhekina, A.; Kasatkin, I.; Shelyakov, A. Variation in the structure of the amorphous NiTi-based alloys during mechanical vibrations. *J. Non-Cryst. Solids* **2020**, *542*, 120101. [CrossRef]
5. Christian, J.W. *The Theory of Transformations in Metals and Alloys*; Pergamon Press Ltd.: Oxford, UK, 1975.
6. Burton, W.K.; Cabrera, N.; Frank, F.C. The growth of crystals and the equilibrium structure of their surfaces. *Philos. Trans. R. Soc.* **1951**, *243*, 299–358. [CrossRef]
7. Köster, U.; Herold, U. Crystallization of Metallic Glasses. In *Topics Applied Physics*; Glassy Metals, I., Guntherrodt, H.-J., Beck, H., Eds.; Springer-Verlad: Berlin, Germany, 1981; pp. 225–259. [CrossRef]
8. Greer, A.L. Crystallisation kinetics of Fe80B20 glass. *Acta Metall.* **1982**, *30*, 171–192. [CrossRef]
9. Köster, U.; Meibhardt, J.; Birol, Y.; Aronin, A. Crystallization of Cu50Ti50 glasses and undercooled melts. *Z. Metallkd.* **1995**, *86*, 171–175. [CrossRef]
10. Anderson, P.W.; Halperin, B.I.; Varma, C.M. Anomalous low-temperature thermal properties of glasses and spin glasses. *Philos. Mag.* **1972**, *25*, 1–9. [CrossRef]
11. Phillips, W.A. Tunneling states in amorphous solids. *J. Low Temp. Phys.* **1972**, *7*, 351–360. [CrossRef]
12. Sussmann, J.A. Phonon Induced tunneling of ions in solids. *Phys. kondens. Mater.* **1964**, *2*, 146–160. [CrossRef]
13. Klinger, M.I. Low-temperature properties and localized electronic states of glasses. *Sov. Phys. Usp.* **1987**, *30*, 699–715. [CrossRef]
14. Slyadnikov, E.E. Pretransition state and structural transition in a deformed crystal. *Phys. Solid State* **2004**, *46*, 1095–1100. [CrossRef]
15. Luborsky, F.E. *Amorphous Metallic Alloys*; Butterworths: London, UK, 1983.
16. Peierls, R. *Quantum Theory of Solids*, 1st ed.; Clarendon Oxford Press: Oxford, UK, 1955.

17. Mott, N.F.; Davis, E.A. *Electronic Processes in Non-Crystalline Materials*, 2nd ed.; Oxford at Clarendon Press: Oxford, UK, 1979.
18. Krivoglaz, M.A. Fluctuon states of electrons. *Sov. Phys. Usp.* **1974**, *16*, 856–877. [CrossRef]
19. Argon, A.S. Plastic-deformation in metallic glasses. *Acta Metall.* **1979**, *27*, 47–58. [CrossRef]
20. Taub, A.I.; Spaepen, F. The kinetics of structural relaxation of a metallic glass. *Acta Metall.* **1980**, *28*, 1781–1788. [CrossRef]
21. Gibbs, M.R.J.; Evetts, J.E.; Leake, J.A. Activation energy spectra and relaxation in amorphous materials. *J. Mater. Sci.* **1983**, *18*, 278–288. [CrossRef]
22. Kosilov, A.T.; Khonik, V.A. Directional structural relaxation and homogeneous flow of freshly quenched metallic glasses. In Proceedings of the Relaxation Phenomena in Solids, Seminar 'Relaksatsionnye Yavleniya v Tverdykh Telakh', Voronezh, Russia, 11 February 1993. Reference Number: AIX-25-051385; EDB-94-109615.
23. Egami, T.; Vitek, V. Local structural fluctuations and defects in metallic glasses. *J. Non-Cryst. Solids* **1984**, *62*, 499–510. [CrossRef]
24. Bakai, A.S. The polycluster concept of amorphous solids. In *Glassy Metals III. Topics in Applied Physics*; Springer: Berlin/Heidelberg, Germany, 1994; Volume 72, pp. 209–255. [CrossRef]
25. Pantel, R.; Puthof, G. *Fundamentals of Quantum Electronics*; Wiley: New York, NY, USA, 1969.
26. Landau, L.D.; Lifshitz, E.M. *Quantum Mechanics. Non-Relativistic Theory*, 3rd ed.; Pergamon Press Ltd.: London, UK, 1991.
27. Rogachev, A.S.; Vadchenko, S.G.; Shchukin, A.S.; Kovalev, I.D.; Aronin, A.S. Self-propagating crystallization waves in the TiC amorphous alloy. *JETP Lett.* **2016**, *104*, 726–729. [CrossRef]

Article

Performance Analysis of Electrochemical Micro Machining of Titanium (Ti-6Al-4V) Alloy under Different Electrolytes Concentrations

Geethapriyan Thangamani [1], Muthuramalingam Thangaraj [2,*], Khaja Moiduddin [3,*], Syed Hammad Mian [3], Hisham Alkhalefah [3] and Usama Umer [3]

1   Department of Mechanical Engineering, SRM Institute of Science and Technology, SRM Nagar, Kattankulathur 603203, Kanchipuram, Chennai, TN, India; geethapt@srmist.edu.in
2   Department of Mechatronics Engineering, SRM Institute of Science and Technology, SRM Nagar, Kattankulathur 603203, Kanchipuram, Chennai, TN, India
3   Advanced Manufacturing Institute, King Saud University, Riyadh 11421, Saudi Arabia; smien@ksu.edu.sa (S.H.M.); halkhalefah@ksu.edu.sa (H.A.); uumer@ksu.edu.sa (U.U.)
*   Correspondence: muthurat@srmist.edu.in (M.T.); khussain1@ksu.edu.sa (K.M.); Tel.: +91-9994872013 (M.T.); +966-540542861 (K.M.)

Abstract: Titanium alloy is widely used in modern automobile industries due to its higher strength with corrosion resistance. Such higher strength materials can be effectively machined using unconventional machining processes, especially the electro-chemical micro machining (ECMM) process. It is important to enhance the machining process by investigating the effects of electrolytes and process parameters in ECMM. The presented work describes the influence of three different combinations of Sodium Chloride-based electrolytes on machining Titanium (Ti-6Al-4V) alloy. Based on the ECMM process parameters such as applied voltage, electrolytic concentration, frequency and duty cycle on response, characteristics are determined by the Taguchi design of experiments. The highest material removal rate (MRR) was achieved by the Sodium Chloride and Sodium Nitrate electrolyte. The combination of Sodium Chloride and Citric Acid achieve highest Overcut and Circularity. The optimal overcut was observed from the Sodium Chloride and Glycerol electrolyte due to the presence of glycerol. The better conicity was obtained from Sodium Chloride and Citric Acid in comparison with other electrolytes. A Sodium Chloride and Glycerol combination could generate better machined surface owing to the chelating effect of Glycerol.

Keywords: titanium alloy; electrolyte; ECMM; sodium chloride; glycerol

Citation: Thangamani, G.; Thangaraj, M.; Moiduddin, K.; Mian, S.H.; Alkhalefah, H.; Umer, U. Performance Analysis of Electrochemical Micro Machining of Titanium (Ti-6Al-4V) Alloy under Different Electrolytes Concentrations. *Metals* **2021**, *11*, 247. https://doi.org/10.3390/met11020247

Academic Editor: Artur Shugurov

Received: 30 December 2020
Accepted: 29 January 2021
Published: 2 February 2021

Publisher's Note: MDPI stays neutral with regard to jurisdictional claims in published maps and institutional affiliations.

1. Introduction

Titanium (Ti-6Al-4V) alloy is used widely in automobile, aerospace, power sectors and the pharmaceutical industry due to its property of resistance towards corrosion in the wide range of applications [1]. Nowadays, as the demand of micro components has grown in the industry, electro-chemical micro machining (ECMM) is appearing as a promising technique to machine these miniature components with exceptional precision. In the process, the dissolution of the workpiece takes place on the application of voltage between the anode and cathode according to the electrolysis laws given by Michael Faraday. The reaction products precipitated during the electrolysis process are separated from the electrode gap by a fast-flowing electrolyte [2]. ECMM and electrical discharge micro machining (EDMM) have many advantages such as accuracy of the machined components, higher rate of machining, featureless finish and absence of wear on the tools [3]. The processes also aid in machining three-dimensional microstructures. The ECM process includes corrosion products despite not being defined exclusively as a corrosion process owing to anodic dissolution although mostly driven by redox [4,5].

173

The process of ECMM is immensely affected by the choice of electrolyte, which is used to perform the corrosion process [6]. Sen and Shan (2005) presented the electrochemical machining process for drilling of macro to micro holes with accurate surface and for the application in the computer, electronic, micro-mechanics and aerospace industries. The paper represents the new developments and recent trends for machining micro-level quality holes in difficult-to-machine materials [7]. Rajurkar et al. (1998) discussed the improvement of accuracy by the reduction of sludge using passive electrolytes. The waste generation from the machining surface has minimum machining allowance to increased localization effect for improving performance of the ECMM process. The passive electrolytes like NaNO have been used to achieve better machining characteristics such as higher unit removal and machining exactness [8]. Sodium chlorate ($NaClO_3$), sodium chloride (NaCl) and sodium nitrate ($NaNO_3$) are frequently used electrolytes in the process owing to their ability to form a significant quantity of sludge as the reactional by-product, which creates issues like clogging [9]. A significant amount of research has already been performed focusing on the use of different types of electrolytes in the ECMM process. It was suggested that high accuracy for machined component and low quantity of sludge can be achieved by the use of electrolytes of lower concentration [10]. Ayappan et al. found that oxygenated aqueous NaCl provides better surface finish than aqueous NaCl in the machining of alloy steel specimen [11]. Tang et al. used various compositions of $NaNO_3$ and $NaClO_3$ to machine special stainless-steel workpiece to study the various response characteristics [12]. Kirchner et al. used HF and HCl to generate 3D microstructures on the stainless-steel specimen [13]. Huaiqian et al. investigated the utilization of pure water as an electrolyte to generate hole and cavities on a stainless steel plate [14]. Naoki Shibuya et al. has fabricated micro pin using $NaNO_3$ as an electrolyte [15]. It was evident from the aforementioned research that response characteristics in the ECMM process are immensely affected by the type and concentration of electrolytes. Spieser and Ivanov (2013) discussed the recent development and issues in the ECM process to maintain a narrow gap between the tool electrode and workpiece during the machining process to obtain accuracy. The smaller value of IEG is used to control the micro-spark generation in the machining surface. It was stated that the overall machining efficiency has been mainly influenced by input process parameter. The stress-free micro-hole has been produced with complex shapes for automobile and aerospace applications [16]. Rajurkar et al. (1999) described that an endeavor has been made to develop an ECMM process for carrying out innovative research to control the ECMM process parameters to get better machining accuracy. The developments of the ECMM process are in the field of design of the tool, pulse period for current, micro-size shaping, surface finishing, numerically controlled ECM, ECM environmental concern, hybrid process and industrial application for various industries [17]. Kozak and Rajurkar (2004) included the study of pulse electrochemical micro machining process for stress, burr- and crack-free micro components on the machining surface. The influence of process parameters such as applied voltage and feed rate on performance measures has been studied. The reduction of inter-electrode gap is used to increase localized dissolution for improved accuracy in the ECMM process [18,19]. The surface quality of the metal implants such as titanium alloy specimens in the medical field should be better as much as possible [20,21]. It can be made possible by enhancement of the electrochemical micro machining process.

From the referred literature, it has been understood that very little attention has been given to analyze the influence of electrolytes on quality measures using the ECMM process. Many research works are available that analyze the effect of electrolyte concentration. Nevertheless, only few works are available that investigate the influence of electrolytes in different proportions along with the process parameters on machining titanium alloy specimens in the ECMM process [22]. Hence, the present investigation was carried out. The objective of the study is to examine the influence of a combination of different electrolytes like Sodium Chloride (NaCl) + Sodium Nitrate ($NaNO_3$), Sodium Chloride (NaCl) + Glycerol ($C_3H_8O_3$) and Sodium Chloride (NaCl) + Citric Acid ($C_6H_8O_7$) in the

ECMM process on machining Titanium (Ti-6Al-4V) alloy using copper tool electrode. Various output parameters such as MRR, overcut, conicity and circularity are measured by varying the input parameters such as applied voltage, electrolyte concentration, frequency and duty cycle. The micro structural analysis was carried out to understand the variation of grains around the machined area of each specimen.

2. Materials and Methods

The ECMM setup (Sinergy nano system, Navi Mumbai, India) consists of a power input, control unit to set the process parameters, electrolyte supply system and a machining chamber which contains the tool and the workpiece along with the electrolyte, as shown in Figure 1 [23]. Titanium (Ti-6Al-4V) alloy with the size of (50 mm × 50 mm × 3 mm) was chosen as the workpiece due to its importance in manufacturing industries.

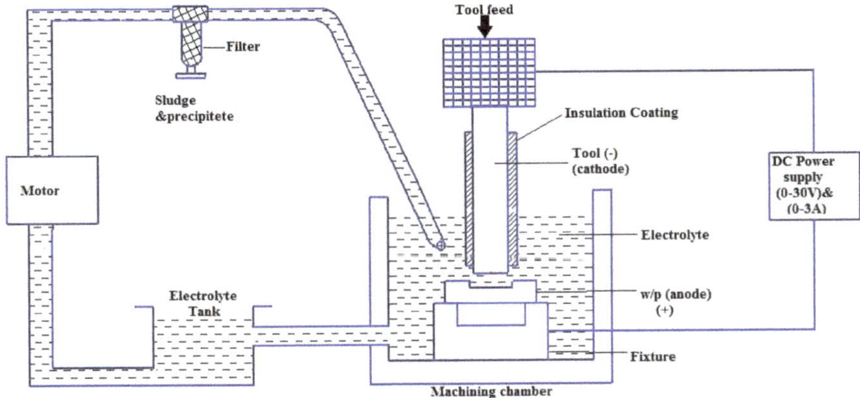

Figure 1. Schematic representation of the electro-chemical micro machining (ECMM) process.

The workpiece is fixed using the plastic clamps to avoid the imbalanced surface calibration and the acrylic tank is sealed to avoid the leakage of electrolytes. The cathode is brought to the closest non-contact distance possible near the anode manually by using the screw mechanism. The ECMM process is started after introducing the electrolyte in the machining chamber such that the space between the tool and workpiece is well immersed. The ECMM process is performed by the sinking of the tool, forming its replica. In the present study, the machining experiments were performed with different combinations of Sodium Chloride (NaCl) + Sodium Nitrate (NaNO$_3$), Sodium Chloride (NaCl) + Glycerol (C$_3$H$_8$O$_3$) and Sodium Chloride (NaCl) + Citric Acid (C$_6$H$_8$O$_7$) (Anmol Chemicals, Mumbai, India) under Taguchi-based L$_{18}$ orthogonal array (OA) as shown in Table 1 [22]. The Copper tool electrode was used as a tool electrode with diameter of 0.1 mm. The performance measures such as metal removal rate (MRR), overcut (OC), conicity (CN) and circularity (CY) were considered as response parameters in the present study. MRR was calculated as weight loss of workpiece specimens that happened during machining time. The values of OC, CN and CY were computed with the help of an Optic Light Vision Measurement System (SIPCON, Chennai, India) and digital micrometer (Mitutoyo, Osaka, Japan) [23]. The cut region image was acquired using a more accurate Vision measuring system. The overcut was directly computed from the image using a calibrated scale measurement technique. The scanning electron microscope-based images (Hitachi, Tokyo, Japan) were utilized to analyze the topography of the machined workpiece surface. The process parameter combinations based on Design of Experiments are shown in Table 2. As per the Taguchi Design of Experiments approach, L$_{18}$ orthogonal array (OA) has been chosen since the present study does contain five process factors with three levels [24].

Table 1. Selection of process variables with their levels.

Combination	Symbol	NaCl + NaNO$_3$			NaCl + C$_3$H$_8$O$_3$			NaCl + C$_6$H$_8$O$_7$		
Process Parameter		I	II	III	I	II	III	I	II	III
Applied voltage(V)	V	10	12	14	10	12	14	10	12	14
Electrolyte Concentration (g/L)	EC	10 + 7	10 + 9	10 + 3	10 + 4	10 + 7	10 + 8	10 + 16	10 + 5	10 + 3
Micro-tool feed rate (µm/s)	MF	0.1	0.5	1	0.1	0.5	1	0.1	0.5	1
Frequency (Hz)	F	50	60	70	50	60	70	50	60	70
Duty Cycle (%)	DC	33	50	66	33	50	66	33	50	66

Table 2. Experimental methodology of the present study.

S. No	V	EC			MF	F	DC
		NaCl + NaNO$_3$	NaCl + C$_3$H$_8$O$_3$	NaCl + C$_6$H$_8$O$_7$			
1	10	17	14	26	0.1	50	33
2	10	19	17	15	0.5	60	50
3	10	13	18	13	1	70	66
4	12	17	14	26	0.1	60	50
5	12	19	17	15	0.5	70	66
6	12	13	18	13	1	50	33
7	14	17	14	26	0.5	50	66
8	14	19	17	15	1	60	33
9	14	13	18	13	0.1	70	50
10	10	17	14	26	1	70	50
11	10	19	17	15	0.1	50	66
12	10	13	18	13	0.5	60	33
13	12	17	14	26	0.5	70	33
14	12	19	17	15	1	50	50
15	12	13	18	13	0.1	60	66
16	14	17	14	26	1	60	66
17	14	19	17	15	0.1	70	33
18	14	13	18	13	0.5	50	50

3. Results and Discussion

The Titanium (Ti-6Al-4V) alloy specimens have been machined using ECMM proces In this section, the effect of process parameters and influence of various electrolytes o response parameters such as MRR, overcut, conicity and circularity are discussed. Figure demonstrates the impact of different combinations of electrolytes on different qualit measures in the ECMM process.

3.1. Effect of Electrolytes Andprocess Parameters on MRR

It can be inferred from the experimental data obtained that electrolyte concentratio was the strongest influencing factor on MRR for the combination of NaCl + NaNO$_3$ an NaCl + C$_3$H$_8$O$_3$. The higher electrolyte concentration increases conductivity, which i turn increases the stray current, which results in a decrease of localization, consequentl decreasing the MRR [25]. When machining is carried out with the second combinatio of electrolytes, glycerol acts as a chelating agent, which increases bonding of suspende ions resulting in an increase of current density and subsequently the volume of materia removed. Applied voltage appeared as the most contributing parameter on MRR fo the combination of NaCl + C$_6$H$_8$O$_7$. On application of a high voltage, the stray curren reduces, which results in the increases of localization, thus increasing the MRR [26]. Th main material removal mechanism in the electro-chemical machining process is the anodi dissolution. Sodium ions are considered as the aggressive anions in electro-chemica micro machining process, which induce more liberation of the anodic particles from th machining surface. This has resulted in higher material removal rate in the ECM process. was evident from the Figure 3 that NaCl + NaNO$_3$ has generated better MRR than the othe two combinations of electrolytes due to the presence of more Sodium (Na+) ions, which are easily displaced by elements of titanium alloy according to the reactivity series. From the experimental data obtained, it is seen that the first electrolyte combination has generated 56% and 24% higher MRR than NaCl + $C_3H_8O_3$ and NaCl + $C_6H_8O_7$, respectively [27].

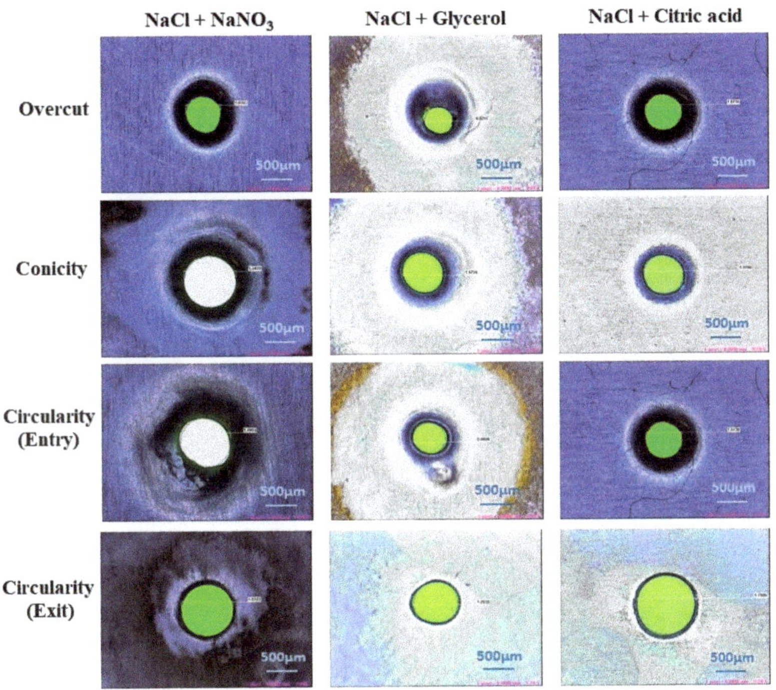

Figure 2. Effects of different electrolyte combinations on quality measures.

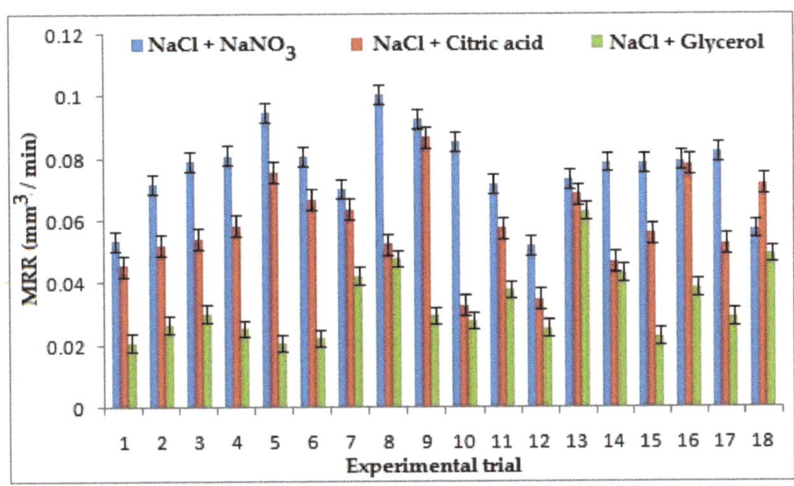

Figure 3. Effects of Electrolytes and process parameters on metal removal rate (MRR).

3.2. Effect of Electrolytes Andprocess Parameters on Overcut

Applied voltage appeared as the strongest influencing parameter on overcut as show[n] in Figure 4 for the combination of NaCl+ NaNO$_3$, because on the application of hig[h] voltage, stray current decreases and side current decreases which results in the reductic[n] of overcut. For the combination of NaCl + C$_3$H$_8$O$_3$, the frequency was found as th[e] strongest contributing factor as it depends upon the duration of pulse ON and puls[e] OFF time. With an increase of frequency, the pulse duration decreases and duty cyc[le] ratio increases, which shows that high pulse ON time will result in increased curre[nt] supply, which in turn decreases the localization, increasing the stray current, resultir[g] in high overcut. Electrolyte concentration appeared as the strongest influencing fact[or] for the combination of NaCl + C$_6$H$_8$O$_7$ because higher electrolyte concentration leads [to] a generation of a larger number of ions during the ECMM process, which leads to larg[e] ionization and decreases the stray current causing a reduction in overcut [28,29]. The lo[w] value of overcut is considered better in machining. It is apparent from Figure 4 that th[e] combination of NaCl + C$_3$H$_8$O$_3$ yields better overcut than the other two combinations [of] the electrolyte. It is evident from the experimental data that it produces 40% and 22[%] improvement in overcut than the combination of NaCl + NaNO$_3$ and NaCl + C$_6$H$_8$[O] because of the chelating effect of Glycerol, which influences greater bonding of ions ar[d] thereby reduces the stray current during machining, resulting in lesser overcut [30]. Durir[g] the machining, chromium chloride and chromium hydroxide get removed as sludge.

$$\text{At Anode}: \ Cr \rightarrow Cr^{6+} + 6e^-$$

$$\text{Sludge Reaction}: \ Cr^{6+} + 6Cl \rightarrow Cr(Cl)_6$$

$$Cr^{6+} + 6(OH) \rightarrow 6Cr(OH)_6$$

$$CrCl_2 + 2OH \rightarrow Cr(OH)_2 + 2Cl$$

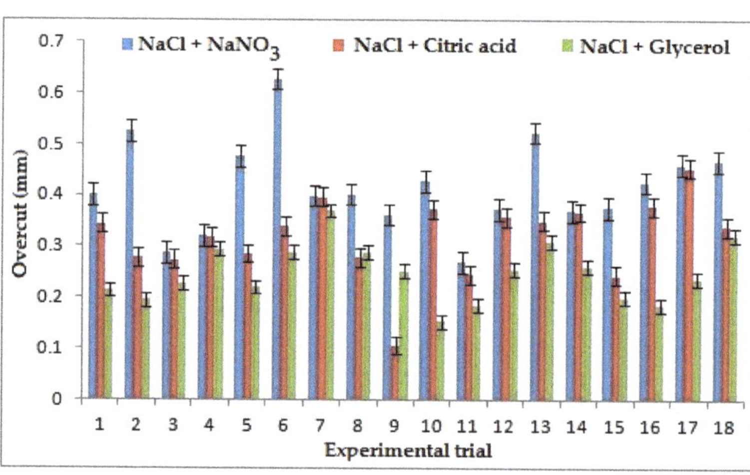

Figure 4. Effects of electrolytes and process parameters on overcut.

3.3. Effect of Electrolytes Andprocess Parameters on Conicity

The strongest influencing factor on conicity for the combination of NaCl + NaNO$_3$ i[s] duty cycle, because with the increment of pulse ON time (T$_{on}$) the flow of current betwee[n] the electrode and workpiece increases. When T$_{on}$ is lower than T$_{off}$ the machining tim[e] is less, which increases flush around the machining area and results in better conicit[y] However, for the combination of NaCl + C$_3$H$_8$O$_3$ frequency appears as the stronges[t] contributing parameter. It was seen that conicity rises with increases in frequency, whic[h] results in the reduction of pulse duration, which resulted in higher T_{on}. This could result in higher conicity of the ECMM process. Electrolyte concentration was observed as the strongest influencing parameter for the combination of $NaCl + C_6H_8O_7$ because ionization rate reduces with a decrease in concentration of electrolytes, which results in the increases of stray current, thus generating high conicity [29]. The lower value of conicity was considered better in machining. It was evident from Figure 5 that the combination of $NaCl + C_6H_8O_7$ produced better conicity than the other two combinations. It was observed that the conicity could be improved by 70% and 39% in comparison to the combination of $NaCl + NaNO_3$ and $NaCl + C_3H_8O_3$. The presence of NaCl causes ionization in solution and citric acid reduces smutting around the machined hole with less sludge formation which results in a well-defined profile with less conicity [28].

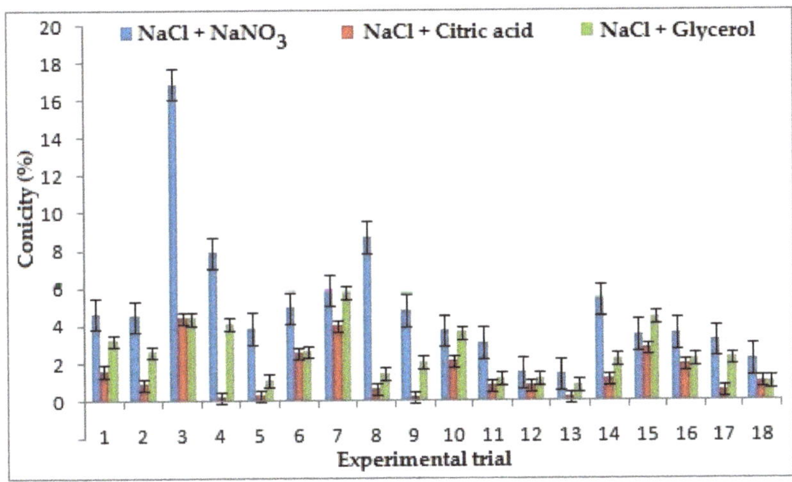

Figure 5. Effects of electrolytes and process parameters on conicity.

3.4. Effect of Electrolytes and Process Parameters on Circularity at Entry and Exit

During the tool entry, frequency was considered as the most influential parameter on circularity for the combination of $NaCl + NaNO_3$ as shown in Figure 6, because as frequency increases it results in an increase of pulse ON time, which in turn increases the duration of current passed, which consequently reduces localization and increases circularity. However, for the combination of $NaCl + C_3H_8O_3$ duty cycle is observed as the most influential parameter because as the pulse ON time increases the current flow between the electrode and work piece increases. It leads to a higher dissolution of ions which results in a decrease of flush time that helps remove the sludge formed around the machining area. Frequency is found as the most influential parameter for $NaCl + C_6H_8O_7$ as low frequency decreases the pulse ON time, which reduces the duration of machining between each pulse cycle and increases flushing and formation of sludge on the surface, thereby giving better machining and reducing circularity. However, for the tool exit, applied Voltage is considered as the most influential parameter for all three combinations of electrolytes because high voltage helps in dissolution of ions which increases the stray current and reduces the localization, resulting in increase of circularity.

A low value of Circularity is considered better in machining processes. It was inferred that $NaCl + NaNO_3$ has improved the circularity at entry by 62% when compared to other combinations of electrolytes. The dissolution rate of $NaCl + NaNO_3$ in the solvent is lesser, which reduces the stray current on the surface, thereby machining a proper through hole resulting in less circularity. However, the best circularity at exit is presented by $NaCl + NaNO_3$ as shown in Figure 7. It reduces the circularity by 76% and 79% comparatively to the combination of $NaCl + C_6H_8O_7$ and $NaCl + C_3H_8O_3$, respectively. This can be

attributed to the presence of sodium nitrate, which reduces the sludge formation on th
sides of machined hole thereby giving a better profile of the hole and reducing circulari

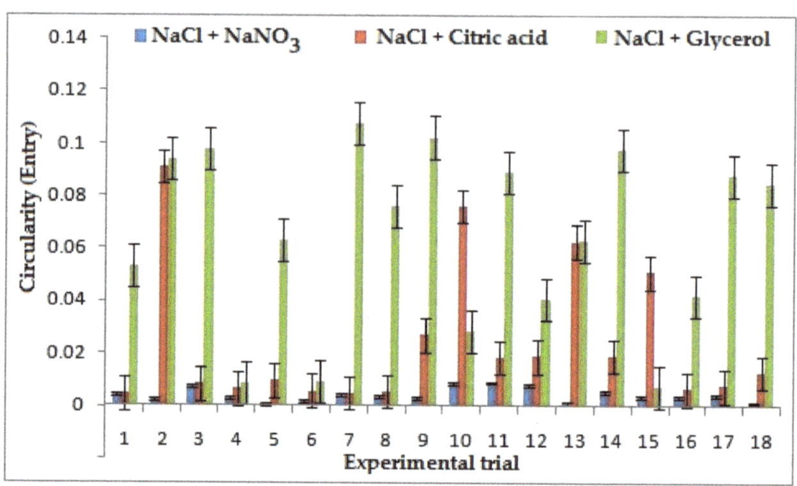

Figure 6. Effects of electrolytes and process parameters on circularity at entry.

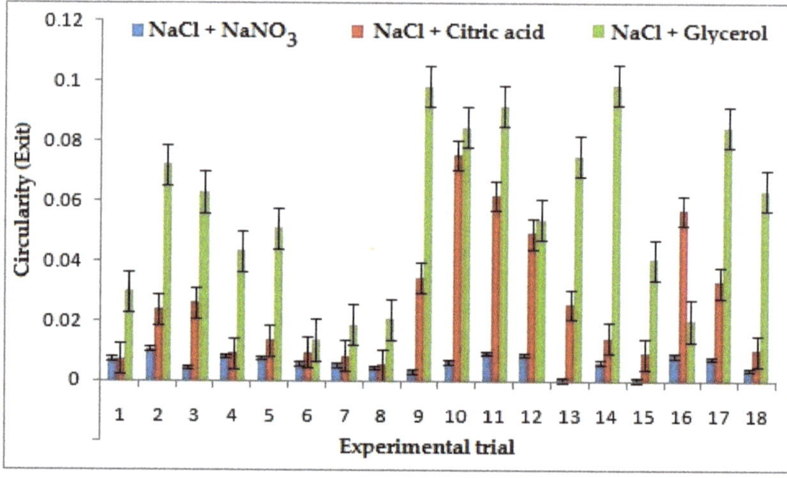

Figure 7. Effects of electrolytes and process parameters on circularity at exit.

3.5. SEM Analysis of Machined Surface

The surface profiles of machined workpieces were obtained and analyzed with scan
ning electron microscope (SEM) images. Figure 8 depicts the surface topography imag
of machined specimens with $NaCl + NaNO_3$, $NaCl + C_3H_8O_3$ and $NaCl + C_6H_8O_7$ elec
trolytes using SEM analysis. The circularity value is mostly due to the size of the crate
developed during the electro-chemical micro machining process. The size of the crate
is established by the current density and the time duration (pulse-on time + pulse-of
time ratio) of applied electrical energy [31]. The $NaCl + NaNO_3$ has higher electrica
conductivity nature to produce a larger peak and valley surface and it has achieved highe
circularity [32,33]. It has been understood that the higher pulse duration and higher curren
density can generate a better machined workpiece surface. It has been observed that th

NaCl + C3H$_8$O$_3$ electrolyte has produced a better micro hole over the workpiece surface due to its decrease of electrical conductivity and reduced quantity of current passing through the electrochemical cell. It has been attained that $NaCl + C_6H_8O_7$ with higher duty cycle (66%) has obtained moderate overcut with tolerable circularity of drilled micro hole over the machined workpiece surface. It has been observed that $NaCl + C_3H_8O_3$ with lower duty cycle (33%) has generated higher overcut due to its thermal and electrical conductivity to create better circularity. The higher pulse-off time has generated lower circularity error because it has higher flushing time to remove the sludge particle from the machining surface to reduce the crater size, which leads to better surface quality.

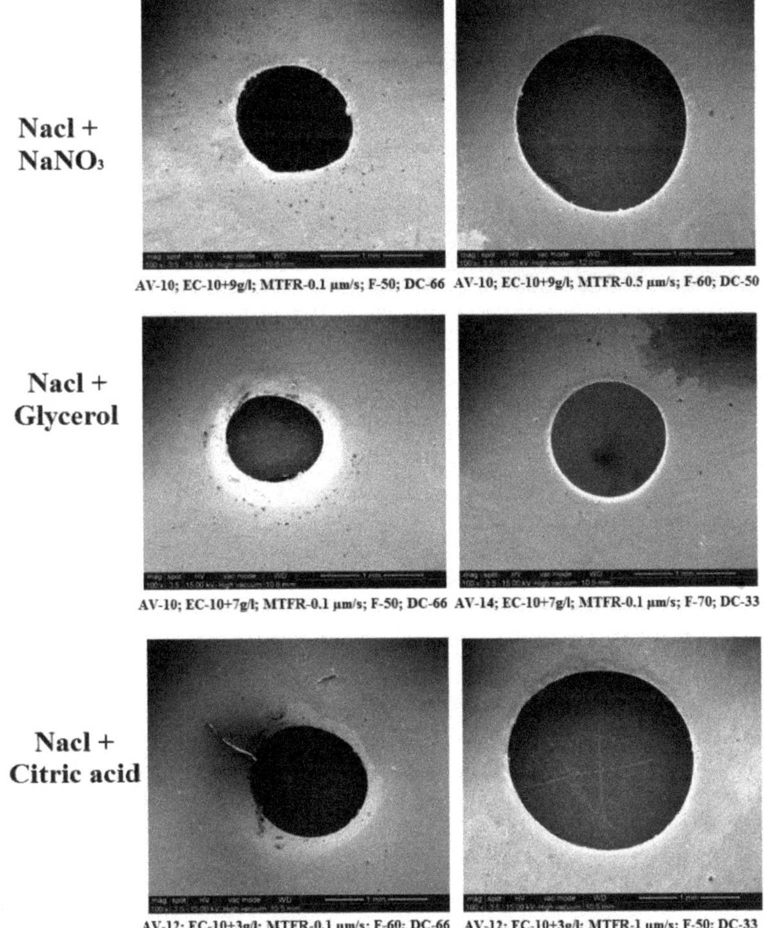

Figure 8. Effects of different electrolyte combinations on circularity using the ECMM process.

4. Conclusions

In the present study, an endeavor has been made to find the effects of electrolytes and process parameters involved in the ECMM process on machining Titanium (Ti-6Al-4V) alloy for obtaining better performance measures. From the experimental results, the following conclusions have been drawn.

- NaCl + NaNO$_3$ combined electrolyte can provide 56% and 24% higher MRR tha
  NaCl + Glycerol and NaCl + Citric acid, respectively, due to the dominance of th
  Sodium ions present in the electrolyte since it possesses higher reactive nature;
- NaCl + Glycerol mixed electrolyte creates 40% and 22% lower Overcut than electroly
  combinations NaCl + NaNO$_3$ and NaCl + Citric acid, respectively, due to the presen
  of glycerol, which acts as a good chelating agent by increasing the bonding of ions;
- NaCl + Citric Acid can improve reduce the conicity by 70% and 39% compared
  that of NaCl + NaNO$_3$ and NaCl + Glycerol due to the presence of citric acid, whic
  reduces the smutting around the machining surface;
- From the SEM analysis, it has been inferred that the NaCl + C$_3$H$_8$O$_3$ electroly
  produces better micro hole over the workpiece surface compared to other electrolyt

Author Contributions: Conceptualization, M.T. and G.T.; methodology, K.M. and M.T.; softwa
M.T.; validation, M.T. and G.T.; formal analysis, H.A. and M.T.; investigation, G.T. and S.H.M
resources, S.H.M.; writing—original draft preparation, M.T. and K.M.; project administration, M
and U.U.; funding acquisition, K.M. All authors have read and agreed to the published version
the manuscript.

Funding: This research was financially supported by the Deanship of Scientific Research, King Sau
University: Research group No: RG-1440-034.

Acknowledgments: The authors extend their appreciation to the Deanship of Scientific Research
King Saud University for funding this work through research group no. (RG-1440-034).

Conflicts of Interest: The authors declare no conflict of interest.

References

1. Suresh, H.S.; Sudhir, G.B.; Dayanand, S.B. Analysis of electrochemical machining process parameters affecting material remov
   rate of hastelloy C-276. *Int. J. Adv. Res. Sci. Eng. Technol.* **2014**, *5*, 18–23.
2. Haisch, T.; Mittemeijer, E.; Schultze, J.W. Electrochemical machining of the steel 100Cr6 in aqueous NaCl and NaNO$_3$ solution
   Microstructure of surface films formedby carbides. *Electrochim. Acta* **2001**, *47*, 235–241. [CrossRef]
3. Huo, J.; Liu, S.; Wang, Y.; Muthuramalingam, T.; Pi, V.N. A Influence of process factors on surface measures on electrical discharg
   machined stainless steel using TOPSIS. *Mater. Res. Express* **2019**, *6*, 086507. [CrossRef]
4. Rajurkar, K.P.; Sundaram, M.M.; Malshe, A.P. Review of electrochemical and electro discharge machining. *Procedia CIRP* **2013**, *6*, 13–2
   [CrossRef]
5. Chiou, Y.C.; Lee, R.T.; Chen, T.J.; Chiou, J.M. Fabrication of high aspect ratio micro-rod using a novel electrochemical microm
   chining method. *Precis. Eng.* **2012**, *36*, 193–202. [CrossRef]
6. Mileham, A.R.; Jones, R.M.; Harvey, S.J. Changes of valency state during Electrochemical Machining Production Points. *Precis. En*
   1982, *4*, 168–170. [CrossRef]
7. Sen, M.; Shan, H.S. A review of electrochemical macro-to micro-hole drilling processes. *Int. J. Mach. Tool. Manuf.* **2005**, *45*, 137–15
   [CrossRef]
8. Rajurkar, K.P.; Zhu, D.; Wei, B. Minimization of machining allowance in electrochemical machining. *Ann. CIRP* **1998**, *47*, 165–16
9. Thanigaivelan, R.; Arunachalam, R.M.; Karthikeyan, B.; Loganathan, P. Electrochemical micromachining of stainless steel wit
   acidified sodium nitrate electrolyte. *Procedia CIRP* **2013**, *6*, 351–355. [CrossRef]
10. De-Silva, A.K.M.; Altena, H.S.J.; McGeough, J.A. Influence of electrolyte concentration on copying accuracy of precision-ECM
    CIRP Ann. Manuf. Technol. **2013**, *52*, 165–168. [CrossRef]
11. Ayyappan, S.; Sivakumar, K. Experimental investigation on theperformance improvement of electrochemical machining proces
    using oxygen-enriched electrolyte. *Int. J. Adv. Manuf. Technol.* **2014**, *75*, 479–487. [CrossRef]
12. Tang, L.; Yang, S. Experimental investigation on the electrochemicalmachining of 00Cr12Ni9Mo4Cu2 material and multi-objectiv
    parameters optimization. *Int. J. Adv. Manuf. Technol.* **2013**, *67*, 2909–2916. [CrossRef]
13. Kirchner, V.; Cagnon, L.; Schuster, R.; Etrl, G. Electrochemical machining of stainless steel microelements with ultrashort voltag
    pulses. *Appl. Phys. Lett.* **2008**, *79*, 1721–1723. [CrossRef]
14. Huaiqian, B.; Jiawen, X.; Ying, L. Aviation-oriented Micromachining Technology Micro-ECM in pure Water. *Chin. J. Aeronau*
    2008, *218*, 455–461. [CrossRef]
15. Shibuya, N.; Ito, Y.; Natsu, W. Electrochemical Machining of Tungsten Carbide Alloy Micro-pin with NaNO$_3$ Solution. *Int.
    Precis. Eng. Manuf.* **2012**, *13*, 2075–2078. [CrossRef]
16. Spieser, A.; Ivanov, A. Recent developments and research challenges in electrochemical micromachining (µECM). *Int. J. Ad*
    Manuf. Technol. **2013**, *69*, 563–581. [CrossRef]

Rajurkar, K.P.; Zhu, D.; McGeough, J.A.; Kozak, J.; Silva, A.D. New developments in electro-chemical machining. *Ann. CIRP* **1999**, *48*, 567–579. [CrossRef]

Kozak, J.; Rajurkar, K.P.; Makkar, Y. Study of pulse electrochemical micromachining. *J. Manuf. Proc.* **2004**, *6*, 7–14. [CrossRef]

Geethapriyan, T.; Kalaichelvan, K.; Muthuramalingam, T. Influence of coated tool electrode on drilling Inconel alloy 718 in Electrochemical micro machining. *Procedia CIRP* **2016**, *46*, 127–130. [CrossRef]

Ginestra, P.; Ferraro, R.M.; Hauber, K.Z.; Abeni, A.; Giliani, S.; Ceretti, E. Selective Laser Melting and Electron Beam Melting of Ti6Al4V for Orthopedic Applications: A Comparative Study on the Applied Building Direction. *Materials* **2020**, *13*, 5584. [CrossRef]

Lowther, M.; Louth, S.; Davey, A.; Hussain, A.; Ginestra, P.; Carter, L.; Eisenstein, N.; Grover, L.; Cox, S. Clinical, industrial, and research perspectives on powder bed fusion additively manufactured metal implants. *Addit. Manuf.* **2019**, *28*, 565–584. [CrossRef]

Davydov, A.D.; Kabanova, T.B.; Volgin, V.M. Electrochemical Machining of Titanium. Review. *Russ. J. Electrochem.* **2017**, *53*, 941–965. [CrossRef]

Geethapriyan, T.; Muthuramalingam, T.; Kalaichelvan, K. Influence of Process Parameters on Machinability of Inconel 718 by Electrochemical Micromachining Process using TOPSIS Technique. *Arab. J. Sci. Eng.* **2019**, *44*, 7945–7955. [CrossRef]

Muthuramalingam, T.; Ramamurthy, A.; Moiduddin, K.; Alkindi, M.; Ramalingam, S.; Alghamdi, O. Enhancing the Surface Quality of Micro Titanium Alloy Specimen in WEDM Process by Adopting TGRA-Based Optimization. *Materials* **2020**, *13*, 1440.

Senthilkumar, C.; Ganesan, G.; Karthikeyan, R. Study of electrochemical machining characteristics of Al/SiCp composites. *Int. J. Adv. Manuf. Technol.* **2009**, *43*, 256–263. [CrossRef]

Geethapriyan, T.; Muthuramalingam, T.; Vasanth, S.; Thavamani, J.; Srinivasan, V.H. Influence of Nanoparticles-Suspended Electrolyte on Machinability of Stainless Steel 430 Using Electrochemical Micro-machining Process. *Lect. Notes Mech. Eng. Adv. Manuf. Process.* **2018**, 433–440. [CrossRef]

Mingcheng, G.; Yongbin, Z.; Lingchao, M. Electrochemical Micromachining of Square Holes in StainlessSteel in H_2SO_4. *Int. J. Electrochem. Sci.* **2019**, *14*, 414–426.

Saxena, K.K.; Qian, J.; Reynaerts, D. A review on process capabilities of electrochemical micromachining and itshybrid variants. *Int. J. Mach. Tool. Manuf.* **2018**, *127*, 28–56. [CrossRef]

Thakur, A.; Tak, M.; Mote, R.G. Electrochemical micromachining behavior on 17-4 PH stainless steel using different electrolytes. *Procedia Manuf.* **2019**, *34*, 355–361. [CrossRef]

Leese, R.J.; Ivanov, A. Electrochemical micromachining: Anintroduction. *Adv. Mech. Eng.* **2016**, *8*, 1–13. [CrossRef]

Geethapriyan, T.; Kalaichelvan, K.; Muthuramalingam, T.; Rajadurai, A. Performance Analysis of Process Parameters on Machining α-β Titanium Alloy in Electrochemical Micromachining Process. *Proc. Inst. Mech. Eng. Part B J. Eng. Manuf.* **2018**, *232*, 1577–1589. [CrossRef]

Tang, L.; Li, B.; Yang, S.; Duan, Q.; Kang, B. The effect of electrolyte current density on the electrochemical machining S-03 material. *Int. J. Adv. Manuf. Technol.* **2014**, *71*, 1825–1833. [CrossRef]

Chun, K.H.; Kim, S.H.; Lee, E.S. Analysis of the relationship between electrolyte characteristics and electrochemical machinability in PECM on invar (Fe-Ni) fine sheet. *Int. J. Adv. Manuf. Technol.* **2016**, *87*, 3009–3017. [CrossRef]

Article

Improvement of the Crack Propagation Resistance in an α + β Titanium Alloy with a Trimodal Microstructure

Changsheng Tan [1,*], Yiduo Fan [1], Qiaoyan Sun [2] and Guojun Zhang [1,*]

[1] School of Materials Science and Engineering, Xi'an University of Technology, Xi'an 710048, China;
2190120044@stu.xaut.edu.cn

[2] State Key Laboratory for Mechanical Behavior of Materials, Xi'an Jiaotong University, Xi'an 710049, China;
qysun@xjtu.edu.cn

* Correspondence: cstan@xaut.edu.cn (C.T.); zhangguojun@xaut.edu.cn (G.Z.); Tel.: +86-29-82668614 (C.T.);
Fax: +86-29-82663453 (C.T.)

Received: 9 July 2020; Accepted: 31 July 2020; Published: 6 August 2020

Abstract: The roles of microstructure in plastic deformation and crack growth mechanisms of a titanium alloy with a trimodal microstructure have been systematically investigated. The results show that thick intragranular α lath and a small number of equiaxed α phases avoid the nucleation of cracks at the grain boundary, resulting in branching and fluctuation of cracks. Based on electron back-scattered diffraction, the strain partition and plastic deformation ahead of the crack tip were observed and analyzed in detail. Due to the toughening effect of the softer equiaxed α phase at the grain boundary, crack arresting and blunting are prevalent, improving the crack growth resistance and generating a relatively superior fracture toughness performance. These results indicate that a small amount of large globular α phases is beneficial to increase the crack propagation resistance and, thus, a good combination of mechanical property is obtained in the trimodal microstructure.

Keywords: titanium alloy; trimodal microstructures; strain partition; crack propagation

---

1. Introduction

Due to their high strength, good corrosion and fatigue resistance, titanium alloys have been extensively used for aerospace engineering [1,2]. During applications in the aircraft industry, two typical microstructures of bimodal microstructure and lamellar microstructure are widely used for titanium alloys [2,3]. Generally, the crack growth resistance is significantly influenced by the volume fraction and size of the equiaxed α phase as well as the grain boundary [4–6] in bimodal microstructures and equiaxed microstructures. The propagation of microvoids can be restricted by softer coarse α particles [7]. For titanium alloys, this is significantly strengthened by the fine secondary α phase [3,8]; crack growth is mainly affected by the thickness of the lamellar α phase, grain boundary α (GB α) and the α colony size of the lamellar microstructure [4,9]. High fracture toughness could be achieved in a lamellar microstructure with large α plates and the finest lamellar spacing [10–12]. It has been reported [11,13,14] that α plates are an effective microstructure unit for controlling fracture toughness as they can effectively deflect the crack propagation path. In conclusion, the fracture toughness of titanium alloy is extremely sensitive to microstructural parameters, such as the prior β grain size, α morphology, the width of the grain boundary of α phase and α laths and so on. It has been found that the lamellar shape of α phase promotes high toughness, while an equiaxed α phase results in low toughness; however, the ductility is degraded with the lamellar α phase and improved with the equiaxed α phase [15]. Due to these contradictions, the microstructure that has high fracture toughness may lead to an unsatisfactory decrease in other properties.

Recently, a new type of microstructure, named "trimodal microstructure", has been reported. This microstructure contains a primary globular α phase, a lamellar α phase and a transformed β matrix (β_{trans}: secondary α phase and β phase) [16,17]. Hosseini et al. [18] found that the excellent comprehensive mechanical properties of the trimodal microstructure are achieved when compared with the widmanstätten microstructure and bimodal microstructure. However, there is little investigation of the deformation behavior ahead of the crack tip, crack formation and crack growth in the trimodal microstructure during loading. Especially in the two-phase titanium alloy with high strength and toughness.

Therefore, in this work, research on the plastic deformation ahead of the crack tip as well as on the detailed essential relationship between microstructure and crack propagation behavior of the Ti-6Al-2Sn-2Zr-3Mo-1Cr-2Nb-0.1Si titanium alloy with the trimodal microstructure was carried out systematically. The present results can be applied to predict the microstructural features that are required to obtain the desired mechanical properties.

2. Experimental Program

The initial titanium alloy was supplied by the Northwest Institute for Non-ferrous Metals Research of China. The ingot with diameter of 450–500 mm was obtained after 3 vacuum-consumable electric arc smelting processes. Subsequently, forging was performed more than a dozen times on the ingot and then the bar with a diameter of about 350 mm was obtained after forging. The titanium alloy was finally forged in a two-phase region of 30–50 °C (lower than the phase transition point), and it was then strengthened with a solution and aging treatment. Vacuum smelting can remove defects and obtain component uniformity. The chemical composition (wt.%) of H and O was 0.001% and 0.075%. The content of C was low than 0.005%. In present work, the phase transition temperature (Tβ) was determined by the metallographic method with continuous heating. The α to β transformation temperature was about 945 ± 5 °C. Firstly, these samples were heated at 930 °C for 2 h and then air cooled. Several primary equiaxed α grains (α_p) were retained and some α laths with 0.5 to 1 μm in width were obtained during this process. Then, the low temperature aging at 600 °C for 4 h was performed and cooled by air to room temperature. This process was performed to obtain the secondary α precipitation from the residual β matrix. Consequently, a tri-modal microstructure (TM) was developed consisting of an intermixture of primary equiaxed α grains (α_p), α lath (α_l) and a transformed β matrix (secondary α phase (α_s) in the β matrix). A plate-shaped tensile specimen (width: 6 mm; thickness: 2 mm; gage length: 15 mm) was determined according to the national standard of the People's Republic of China (GB/T 228-2010) and performed on INSTRON 1195 Testing Machine (INSTRON, Boston, MA, USA). At least three individual tests were carried out to increase the accuracy of the tensile property. The compact tensile specimens with a size of 25 mm × 60 mm × 62.5 mm were used in the present work. After the corresponding heat treatment, the specimens with a notch of a "V" shape were machined into the compact tensile specimens. Firstly, a prefabricated crack of 2 mm in length was carried out on MTS 810 machines with sinusoidal waveforms at room temperature (a stress ratio of R = 0.1). The tensile test was carried out on the Instron 5895 testing machine (INSTRON, Boston, MA, USA). Three specimens were carried out to increase the accuracy of K_{IC}. Finally, scanning electron microscopy (SEM, HITACHI SU6600, Tokyo, Japan) were applied to observe the fractographic surface and analyze the crack growth behavior. The microstructures were etched by a corrosion solution of HF:HNO$_3$:H$_2$O = 1:2:5. The microstructures ahead of the crack tip were investigated in detail using a field emission gun SEM (Carl Zeiss Microscopy GmbH 73447, Carl Zeiss AG, Jena, Germany) equipped with an electron backscattered diffraction (EBSD) system.

3. Results

3.1. Microstructures before Deformation

Figure 1 displays the SEM microstructural features before deformation of the TM. The TM contains three different α morphologies, namely, the equiaxed primary α phase (α_p), the lamellar α phase (α_l) and the acicular secondary α phase (α_s), as shown in Figure 1a. A few α_p are located at the prior β grain boundary. Figure 1b displays the magnified image of the morphology of α_l and α_s in the β_{trans} matrix, as indicated by the blue box. The grain size, volume fraction and aspect for α_p are about 5.9 μm, 5.0% and 1.5, respectively. The thickness for α_l and α_s is about 554 nm and 133 nm, respectively. The volume fraction for α_l and the β_{trans} matrix is about 32.0% and 62.0%, respectively.

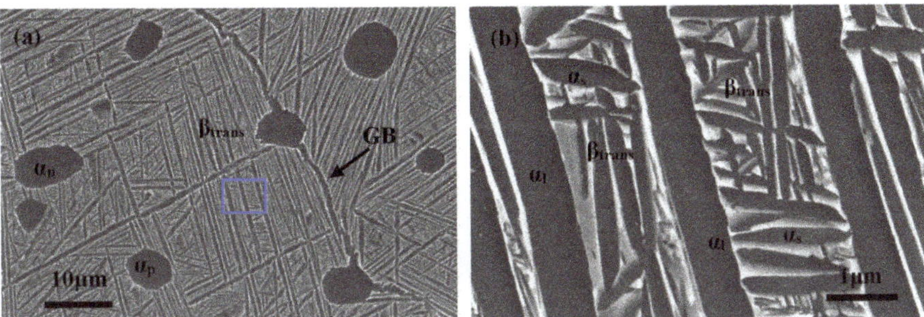

Figure 1. The SEM features of the trimodal microstructure (TM) before deformation: (**a**) three different α morphologies, namely, the primary equiaxed α phase (α_p), the lamellar α phase (α_l) and the secondary α phase(α_s); (**b**) the morphology of α_l and α_s in the β_{trans} matrix.

3.2. Mechanical Properties of the Alloy with a TM

The tensile engineering stress–strain curves at room temperature of the TM are shown in Figure 2a, which indicates that the average value of yield strength, tensile strength and elongation of the TM are about 1067 MPa, 1186 MPa and 12.1%, respectively. Figure 2b displays the force–displacement curve. The fracture toughness of the TM is approximately 62 MPa·m$^{1/2}$ and is significantly higher than the bimodal microstructure and bi-lamellar microstructure [19], which indicates that a small number of the equiaxed primary α phases are beneficial to the improvement of fracture toughness. The mechanical properties and the error of these measurements are listed in Table 1. It can be concluded that a good comprehensive mechanical property is achieved for the TM in the present work, which has been reported in other studies [16,18].

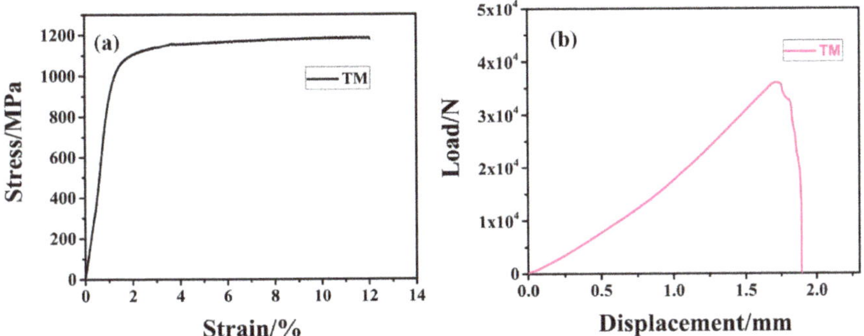

Figure 2. The comparison of mechanical properties of the TM and BLM: (**a**) the engineering stress–strain curves; (**b**) the force–disposition curves.

Table 1. The mechanical properties of the bimodal microstructure and bi-lamellar microstructure.

Microstructures	Yield Strength/MPa	Tensile Strength/MPa	Elongation/%	Fracture Toughness/MPa·m$^{1/2}$
TM	1067 ± 24	1186 ± 4	12.1 ± 1	62 ± 1

3.3. Fractographic Analyses and the Crack Propagation Behavior

The characteristics of fracture surfaces will provide important information to clarify the fracture mechanism during the failure process. To reveal the influence of the microstructure on the fracture mechanisms, the samples with different combinations of α phase were opted for the fracture analysis. Fracture morphology of the TM is shown in Figure 3. As can be seen, the fracture surface could be clearly separated into two apparent zones the: crack source zone and crack growth zone (Figure 3a). To further observe the detailed information regarding the fracture surface, some local fractographs of the TM were obtained, as shown in Figure 3.

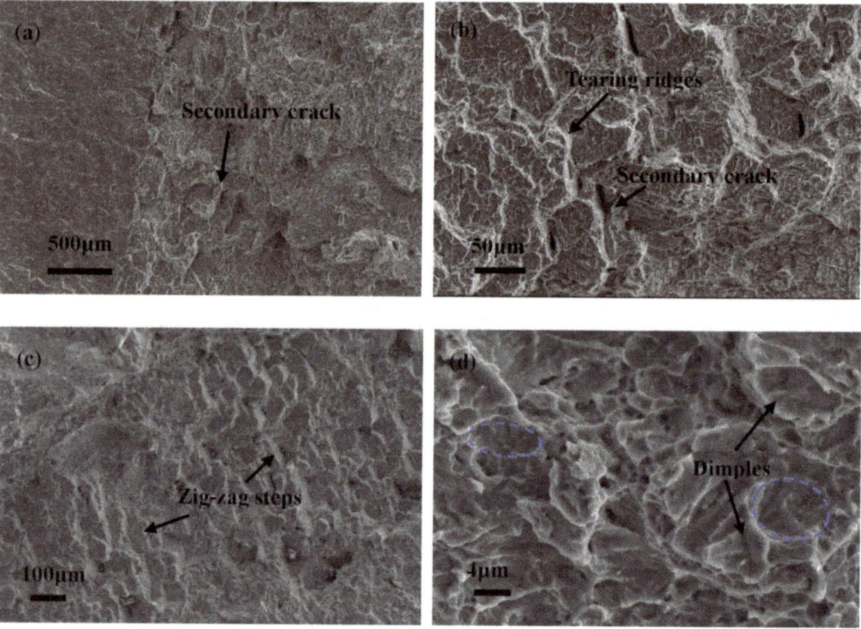

Figure 3. The SEM morphological characteristics of the fracture toughness of the TM fractures: (**a**) the transition region from fatigue crack to tensile crack propagation, (**b**) tear ridges and secondary cracks, (**c**) large fracture steps with a zig-zag fracture pattern, (**d**) big dimples surrounded by ridges.

The fracture surface of the TM is characterized by a high amount of ductile tearing ridges and secondary cracks resulting in transgranular crack propagation of the whole crack propagation region, which indicates a dimple-type fracture (ductile fracture), as shown in Figure 3a,b. The large number of tearing ridges that appear on the fracture surface and the fibrous zones imply a transgranular fracture (Figure 3b), which has previously been reported in titanium alloy [3,20]. The fracture toughness can be improved by the tearing ridges, which indicates that the higher the number of ridges, the higher the achievable K_{IC} of the titanium alloy [21]. Furthermore, a large amplitude of fracture steps with zig-zag fracture patterns are present in the crack propagation region (Figure 3c), which indicates that the crack alters direction and causes crack bifurcation, zigzagging and formation of secondary cracks, that is, much more fracture energy is consumed [22,23]. Thus, it can be concluded that the considerable steps were induced by large crack deviation. Figure 3d exhibits a significant number of inhomogeneous and deep ductile dimples and microvoids, which reveals that the fracture is caused by the typical ductile mechanism of microvoid nucleation, growth, and coalescence [24].

It has been reported that the fracture toughness was mainly affected by the tortuosity of the crack path and plastic deformation ahead of the crack tip [25]. In general, the improvement in tortuosity and plasticity is beneficial to the process of enhancing the fracture toughness of material [26]. The crack front profiles of the fractured specimen of TM are shown in Figure 4. Figure 4a is the main crack propagation path of the TM. The letters b, c, d and e denote the regions that are selected for detailed observation and analysis. The larger deflection and bifurcation of the crack with the local regions are achieved, as shown in Figure 4b, and the fluctuation of the crack can be up to 60 μm. Due to the existence of the equiaxed α phase at the grain boundary, the main crack does not further propagate along the β grain boundary, but changes the direction of propagation away from the GB, even though some microcracks initiate at the equiaxed α interface in front of the crack tip (Figure 4c). This result indicates that the equiaxed α phase at the GB could be the obstacle to the crack growth and lead to branching of the crack of about 90 μm in length. Figure 4d displays the characteristics of the secondary crack initiation near the main crack. It is suggested that although the crack is created at the grain boundary, it connects with the interface crack of intragranular α laths, thus avoiding the propagation along the β grain boundary. It has been reported that cracks tend to grow by passing through rather than cutting the thick α lath [15]. This can be further illustrated by Figure 4e, in which the crack mainly propagates along the long axis of the α laths. Additionally, it deviates from the grain boundary with an angle of 58° when it encounters the equiaxed α phase located at the GB (Figure 4e), which improves the crack growth resistance and is beneficial to improving the fracture toughness of the microstructure. These results indicate that a small amount of equiaxed primary α phases located at the grain boundary is instrumental in the deflection and bifurcation of the main crack propagation.

Figure 5a shows that the interface between the primary α lath and the β_{trans} matrix is the preferable location for crack creation. Cracks mainly form at the α lath interface and propagate along the long axis of the primary α lath, forming cracks related to the orientation of the primary α lath. The width of the primary α lath in the TM is almost equal to the thickness of the grain boundary α, and there are a small number of equiaxed α phases with large sizes at the β grain boundary, as shown in Figure 5b. Figure 5c,d (c and d regions in Figure 5b) represent the plastic deformation behavior of the primary equiaxed α phase at the GB and trimodal β grain boundaries, respectively. The equiaxed α phase at the β grain boundary is relatively soft, and reaches yielding first, producing abundant of slip bands. It has been reported that the multi-slip bands with different orientation are preferably created in the equiaxed primary α phase [27,28]. Cracks easily grew along the slip bands in the equiaxed α phase, because the slip bands provided a low energy channel for crack propagation [29]. As displayed in Figure 5c, there is a certain intersection angle of about 70 degrees between the slip bands in the equiaxed α phase and the direction of the β grain boundary. It changes the propagation direction away from the GB, as the main crack propagation encounters the equiaxed α phase because the slip band is not parallel to the β grain boundary. This result can be validated by Figure 4e. Excellent plastic deformation of the equiaxed α phase results in the blunting effect at the GB, which reduces the crack growth rate [30]. Figure 5d shows that the dislocation slip band occurs both in the GB α phase and the primary α lath, that is, the plastic deformation is not only confined to the GB α phase. At the trimodal β grain boundary, the plastic deformation takes place both at the GB α and the primary α phase near the grain boundary simultaneously, which reduces the local plastic deformation at the grain boundary, making the strain distribution near the grain boundary more uniform, and reducing the strain concentration at β grain boundary to some extent.

Figure 6 shows the plastic deformation and strain distribution in front of the main crack. Three β grains named grain 1, grain 2, and grain 3 are observed in Figure 6a. Several primary α_p particles are located at the grain boundary, such as α_{p1}, α_{p2}, α_{p3}, α_{p5}, and α_{p7} (Figure 6a). Two α_l colonies are observed (named colony1 and colony2). From the inverse pole figure (IPF) map of the α phase, the crystal orientations of α_{p1} to α_{p7} are (−13 −21), (−4.11 −70), (17 −8 −6), (−12 −10), (17 −80), (−13 −20) and (−12 −1 −1), respectively (the direction is perpendicular to the surface). This indicates that the anisotropy displays between these α_p particles. It displays the IPF map of the β grains in Figure 6c, which shows that the crystal orientation for grain 1, grain 2 and grain 3 is (315), (435) and (546), respectively. The Schmid factors of basal slip for α_{p1} to α_{p7} are 0.3, 0.11, 0.25, 0.14, 0.36 and 0.5, respectively, while value of colony1 and colony2 is 0.43 and 0.44, respectively (Figure 6d). However, relatively higher Schmid factor values of prismatic slip of the α_p particles are observed in Figure 6e, except for α_{p2} with 0.01 and α_{p7} with 0.25. Additionally, the Schmid factor values for colony1 and colony2 are 0.34 and 0.31, respectively. Figure 6f shows the strain distribution ahead of the crack tip, which indicates that the plastic strain is relatively inhomogeneous and forms a partition with the β grain interior instead of concentrating at the grain boundary. Even if slightly higher strain concentration is observed at the boundaries of colony1 and colony2 in Figure 6f, this strain partition effect effectively avoids the crack initiation and propagation at the grain boundary. Thus, the crack initiation and propagation mainly happens at the boundaries of α_l colonies within the β grains, as can be seen in Figure 5a.

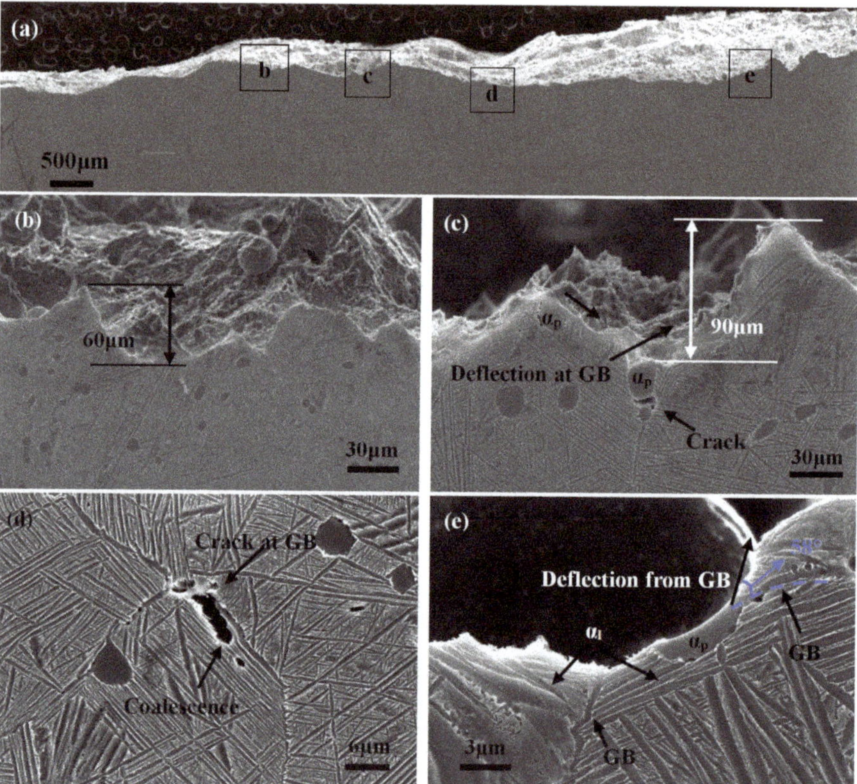

Figure 4. The crack front profiles of the fractured specimen of TM: (**a**) the propagation path of the main crack; (**b**) high fluctuation of about 60 μm of the propagation path is observed in microregion, (c) the crack initiates from the α_p interface but not connects with the main crack, which indicates that α_p inhibits the crack growth along the GB; (**d**) although the crack forms at the GB, it propagates into the grain interior through connecting with the microcrack in α_l, (**e**) the main cracks propagate along the grain boundaries and deviate from grain boundaries with about 58° when encounter α_l and α_p.

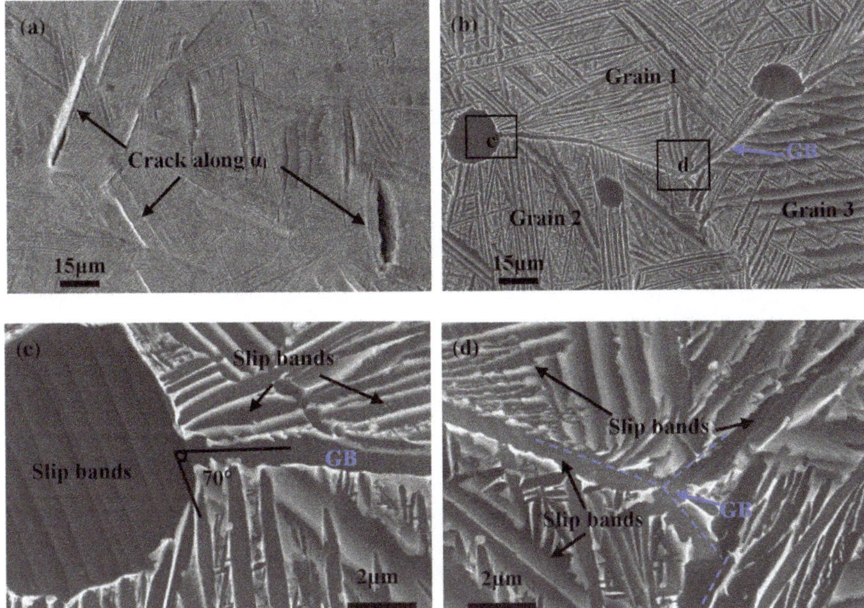

Figure 5. The plastic deformation and crack initiation behavior of the TM: (**a**) the crack initiates in α_l; (**b**) the equiaxed alpha phase (α_p) in the β GB; (**c**) the multi-slip bands in the globular α phase, (**d**) the dislocation slip bands occur both in the GB α phase and α_l; the plastic deformation is not confined to the GB α phase.

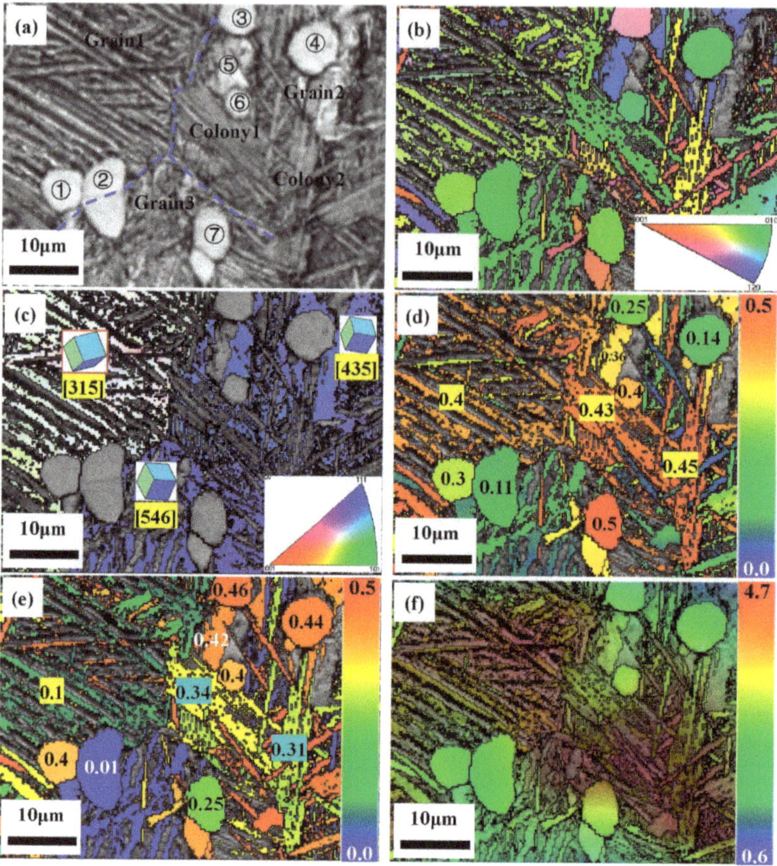

Figure 6. The plastic deformation and strain distribution in front of the main crack: (**a**) the α_p particles located at the grain boundary; (**b**) the inverse pole figure (IPF) map of the α phase; (**c**) the IPF map of the β phase; (**d**) the Schmid factor of basal slip of the α phase; (**e**) the Schmid factor of prismatic slip of the α phase; (**f**) the strain distribution near the grain boundary.

4. Discussion

The main difference of trimodal microstructure is that it contains a small amount of primary equiaxed α phases (about 5% in volume fraction) besides the primary α lath and the β_{trans} matrix. As Chan [30] reported, the softer equiaxed α phase can preferentially coordinate plastic incompatibility and cause blunting of cracks to achieve toughening. Consequently, a certain strain can be produced in the adjoin matrix, and the strain distribution near the grain boundary is more uniform and does not cause the formation of microcracks on the adjacent grain boundary or phase boundary, thus improving the fracture toughness. This can be seen in Figure 5, which shows that the plastic deformation is not only confined to the GB α phase, but occurs both in the α_p phase and the neighbor primary α lath. Based on EBSD, the strain partition within the grain interior that could reduce the stress concentration at the grain boundary to some extent was observed (Figure 6f).

Due to the thick α_l phase (32.0% in volume fraction), cracks nucleated mainly at the α_l phase within the grain interior in the trimodal microstructure instead of at the β grain boundary, which was different with the lamellar microstructure or widmannstatten microstructure [31], as shown in Figure 5a. Furthermore, it has been reported that cracks are both deviated and arrested when they reached an

α phase unfavorably oriented for prismatic slip in a two-phase titanium alloy [32]. As can be seen in Figure 6e, a very low Schmid factor value of approximately 0.01 of prismatic slip was obtained in the primary α_{p2} phases. It seems that crack arresting and crack path deviation will happen when the crack tip encounters these unfavorably oriented phases. Retardation of the crack growth occurs due to the crack arresting and deviation, as it requires more energy to expand the crack to a lower stress position, subsequently improving the crack propagation resistance of the trimodal microstructure. As the crack continues to grow, the crack tip tends to stop propagation, blunting or deviate from the grain boundary when it penetrates the equiaxed α phase, as shown in Figure 4c,e. Crack propagation will deflect along the long axis direction of the primary α lath or the direction of slip bands within the equiaxed primary α phase at the grain boundary, avoiding propagation along the β grain boundary and promoting the transgranular fracture. It can be seen that a high fluctuation (up to 90 µm) of the crack path is observed in the microregion of the TM, which increases the flexibility of crack growth and enhances the resistance of crack growth (Figure 4c).

Moreover, although the width of the GB α in the TM is about 640 nm, it displays a distinct characteristic of discontinuous and zig-zag features as marked by the blue dotted lines in Figure 7a. The width and continuity of the GB α can significantly influence the fracture toughness [33]. Researchers have reported that the thicker and continuous GB α would lead to a preferable crack propagation path and induce a detrimental influence on the fracture toughness [4,22]. However, the crack propagation is much more difficult to pass through the discontinuous grain boundary α, which is beneficial to the heightening of fracture toughness [14,33]. In contrast, in the widmannstatten microstructure, because of the lack of toughening effect of the primary equiaxed α phase, cracks are easy to initiate at grain boundaries and propagate along the β grain boundary, which results in low plasticity [31]. According to the present experiments and theoretical analysis, the schematic diagram is carried out to illustrate the effect of α morphology on the crack nucleation and growth behavior of titanium alloy, as shown in Figure 7b. It indicates that the crack mainly initiates at the primary α lath (α_l), and it avoids the initiation of the crack at the β grain boundary. Cracks will change the direction of propagation when they encounter the equiaxed α_p phase at the β grain boundary, which leads to a tortuous crack growth path. This study can provide theoretical support to tailor the microstructure and the mechanical properties of titanium alloys that contain both the α and β phase. For instance, a small number of equiaxed primary α phases are needed if high ductility and fracture toughness are required. It also indicates that a large primary equiaxed α phase is not always detrimental to fracture toughness. If the appropriate amount of the equiaxed α_p phase is obtained, the fracture toughness of the trimodal or bimodal microstructure may be higher than that of the lamellar microstructure.

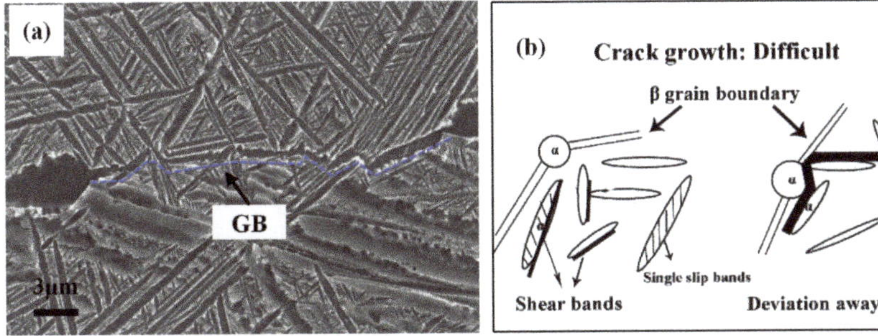

Figure 7. The features of the grain boundary α phase and the schematic diagram of crack growth behavior: (**a**) the discontinuous and "zig-zag" GB α phase, (**b**) the crack mainly initiates at α_l, and it will change the direction of propagation when it encounters the equiaxed α phase at the β grain boundary, which leads to an intergranular fracture.

5. Conclusions

The plastic deformation and crack propagation behavior of the titanium alloy with a trimodal microstructure were systematically investigated during fracture toughness tests. According to the present work, the following conclusions are drawn: A higher fracture toughness of 62 MPa·m$^{1/2}$ is obtained for the trimodal microstructure, which offers a preferable combination of strength (1186 MPa) and ductility (12.1%). In addition to dimples, a large number of tearing edges and secondary cracks are produced in the trimodal microstructure, showing transgranular fracture characteristics. The coarser and longer intragranular α lath and a small number of equiaxed α phases, as well as the discontinuous GB α, lead to a high number of branches and fluctuation of the cracks. Because of the toughening effect of the softer phase at the GB, the equiaxed α phase can preferentially coordinate plastic incompatibility and cause arresting and blunting of cracks, which improves the crack growth resistance in the trimodal microstructure. These results indicate that a small amount of the primary globular α phases located at the grain boundary will be good for improving the resistance to crack propagation. The present work can provide a theoretical support to tailor the microstructure and mechanical properties of titanium alloys in the future.

Author Contributions: Conceptualization, C.T. and G.Z.; data curation, C.T. and Q.S.; investigation, C.T. and Y.F.; project administration, G.Z.; supervision, C.T.; visualization, Q.S.; writing–original draft, C.T.; writing–review and editing, Q.S. and G.Z. All authors have read and agreed to the published version of the manuscript.

Funding: This project was financially funded by China Postdoctoral Science Foundation (2020M673614XB), the Natural Science Basic Research Program of Shaanxi (Program No. 2020JQ-618), the State Key Laboratory for Mechanical Behavior of Materials (20202211) and the National Natural Science Foundation of China (Grants No. 51671158 and 51674196).

References

1. Banerjee, D.; Williams, J.C. Perspectives on titanium science and technology. *Acta Mater.* **2013**, *61*, 844–879. [CrossRef]
2. Leyens, C.; Peters, M. *Titanium and Titanium Alloys: Fundamentals and Applications*; John Wiley & Sons: Hoboken, NJ, USA, 2003.
3. Liu, Y.W.; Chen, F.W.; Xu, G.L.; Cui, Y.W.; Chang, H. Correlation between Microstructure and Mechanical Properties of Heat-Treated Ti-6Al-4V with Fe Alloying. *Metals* **2020**, *10*, 854. [CrossRef]
4. Renon, V.; Henaff, G.; Larignon, C.; Perusin, S.; Villechaise, P. Identification of Relationships between Heat Treatment and Fatigue Crack Growth of $\alpha\beta$ Titanium Alloys. *Metals* **2019**, *9*, 512. [CrossRef]
5. He, S.T.; Zeng, W.D.; Xu, J.W.; Chen, W. The effects of microstructure evolution on the fracture toughness of BT-25 titanium alloy during isothermal forging and subsequent heat treatment. *Mater. Sci. Eng. A* **2019**, *745*, 203–211. [CrossRef]
6. Niinomi, M.; Kobayashi, T.; Inagaki, I.; Thompson, A.W. The effect of deformation-induced transformation on the fracture toughness of commercial titanium alloys. *Metall. Trans. A* **1990**, *21*, 1733–1744. [CrossRef]
7. Li, X.H.; He, J.C.; Ji, Y.J.; Zhang, T.C.; Zhang, Y.H. Study of the Microstructure and Fracture Toughness of TC17 Titanium Alloy Linear Friction Welding Joint. *Metals* **2019**, *9*, 430. [CrossRef]
8. Xu, T.W.; Zhang, S.S.; Cui, N.; Cao, L.; Wan, Y. Precipitation Behavior of Ti15Mo Alloy and Effects on Microstructure and Mechanical Performance. *J. Mater. Eng. Perform.* **2019**, *28*, 7188–7197. [CrossRef]
9. Wen, X.; Wan, M.P.; Huang, C.W.; Tan, Y.B.; Lei, M.; Liang, Y.L.; Cai, X. Effect of microstructure on tensile properties, impact toughness and fracture toughness of TC21 alloy. *Mater. Des.* **2019**, *180*, 107898. [CrossRef]
10. Shi, X.H.; Zeng, W.D.; Zhao, Q.Y. The effects of lamellar features on the fracture toughness of Ti-17 titanium alloy. *Mater. Sci. Eng. A* **2015**, *636*, 543–550. [CrossRef]
11. Wen, X.; Wan, M.P.; Huang, C.W.; Lei, M. Strength and fracture toughness of TC21 alloy with multi-level lamellar microstructure. *Mater. Sci. Eng. A* **2019**, *740*, 121–129. [CrossRef]
12. Richards, N.L. Quantitative evaluation of fracture toughness-microstructural relationships in α-β titanium alloys. *J. Mater. Eng. Perform.* **2004**, *13*, 218–225. [CrossRef]

13. Cvijović-Alagić, I.; Gubeljak, N.; Rakin, M.; Cvijović, Z.; Gerić, K. Microstructural morphology effects on fracture resistance and crack tip strain distribution in Ti-6Al-4V alloy for orthopedic implants. *Mater. Des.* **2014**, *53*, 870–880. [CrossRef]

14. Fan, J.K.; Li, J.S.; Kou, H.C.; Hua, K.; Tang, B. The interrelationship of fracture toughness and microstructure in a new near β titanium alloy Ti-7Mo-3Nb-3Cr-3Al. *Mater. Charact.* **2014**, *96*, 93–99. [CrossRef]

15. Hirth, J.P.; Froes, F.H. Interrelations between fracture toughness and other mechanical properties in titanium alloy. *Metall. Trans. A* **1977**, *8A*, 1165–1176. [CrossRef]

16. Gao, P.F.; Cai, Y.; Zhan, M.; Fan, X.G.; Lei, Z.N. Crystallographic orientation evolution during the development of tri-modal microstructure in the hot working of TA15 titanium alloy. *J. Alloys Compd.* **2018**, *741*, 734–745. [CrossRef]

17. Gao, P.F.; Fan, X.G.; Yang, H. Role of processing parameters in the development of tri-modal microstructure during isothermal local loading forming of TA15 titanium alloy. *J. Mater. Process. Technol.* **2017**, *239*, 160–171. [CrossRef]

18. Hosseini, R.; Morakabati, M.; Abbasi, S.M.; Hajari, A. Development of a trimodal microstructure with superior combined strength, ductility and creep-rupture properties in a near α titanium alloy. *Mater. Sci. Eng. A* **2017**, *696*, 155–165. [CrossRef]

19. Tan, C.S.; Sun, Q.Y.; Zhang, G.J. Role of microstructure in plastic deformation and crack propagation behaviour of an α/β titanium alloy. *Vacuum* **2020**, major revise.

20. Zhang, S.; Liang, Y.L.; Xia, Q.F.; Ou, M.G. Study on Tensile Deformation Behavior of TC21 Titanium Alloy. *J. Mater. Eng. Perform.* **2019**, *28*, 1581–1590. [CrossRef]

21. Jia, R.C.; Zeng, W.D.; He, S.T.; Gao, X.X.; Xu, J.W. The analysis of fracture toughness and fracture mechanism of Ti60 alloy under different temperatures. *J. Alloys Compd.* **2019**, *810*, 151899. [CrossRef]

22. Terlinde, G.; Rathjen, H.J.; Schwalbe, K.H. Microstructure and fracture toughness of the aged, β-Ti Alloy Ti-10V-2Fe-M. *Metall. Trans. A* **1988**, *19*, 1037–1049. [CrossRef]

23. Niinomi, M.; Kobayashi, T. Fracture characteristics analysis related to the microstructures in titanium alloys. *Mater. Sci. Eng. A* **1996**, *213*, 16–24. [CrossRef]

24. Ren, L.; Xiao, W.L.; Chang, H.; Zhao, Y.Q.; Ma, C.L.; Zhou, L. Microstructural tailoring and mechanical properties of a multi-alloyed near β titanium alloy Ti-5321 with various heat treatment. *Mater. Sci. Eng. A* **2018**, *711*, 553–561. [CrossRef]

25. Suresh, S. Fatigue crack deflection and fracture surface contact: Micromechanical models. *Metall. Trans. A.* **1985**, *16*, 249–260. [CrossRef]

26. Xu, J.W.; Zeng, W.D.; Zhou, D.D.; Ma, H.Y.; Chen, W.; He, S.T. Influence of α/β processing on fracture toughness for a two-phase titanium alloy. *Mater. Sci. Eng. A* **2018**, *731*, 85–92. [CrossRef]

27. Tan, C.S.; Sun, Q.Y.; Xiao, L.; Zhao, Y.Q.; Sun, J. Cyclic deformation and microcrack initiation during stress controlled high cycle fatigue of a titanium alloy. *Mater. Sci. Eng. A* **2018**, *711C*, 212–222. [CrossRef]

28. Huang, J.; Wang, Z.R.; Xue, K.M. Cyclic deformation response and micromechanisms of Ti alloy Ti-5Al-5V-5Mo-3Cr-0.5Fe. *Mater. Sci. Eng. A* **2011**, *528*, 8723–8732. [CrossRef]

29. Sangid, M.D. The physics of fatigue crack initiation. *Int. J. Fatigue* **2013**, *57*, 58–72. [CrossRef]

30. Chan, K.S. Toughening mechanisms in titanium aluminides. *Metall. Trans. A* **1993**, *24A*, 569–583. [CrossRef]

31. Ankem, S.; Greene, C.A. Recent developments in microstructure: Property relationships of beta titanium alloys. *Mater. Sci. Eng. A* **1999**, *263*, 127–131. [CrossRef]

32. Bantounas, I.; Lindley, T.C.; Rugg, D.; Dye, D. Effect of microtexture on fatigue cracking in Ti–6Al–4V. *Acta Mater.* **2007**, *55*, 5655–5665. [CrossRef]

33. Ghosh, A.; Sivaprasad, S.; Bhattacharjee, A.; Kar, S.K. Microstructure–fracture toughness correlation in an aircraft structural component alloy Ti-5Al-5V-5Mo-3Cr. *Mater. Sci. Eng. A* **2013**, *568*, 61–67. [CrossRef]

Article

First Principles Study of Bonding Mechanisms at the TiAl/TiO$_2$ Interface

Alexander V. Bakulin [1,2,*], **Sergey S. Kulkov** [2], **Svetlana E. Kulkova** [1,2], **Stephen Hocker** [3] and **Siegfried Schmauder** [3]

[1]   Institute of Strength Physics and Materials Science, Siberian Branch of Russian Academy of Sciences, pr. Akademichesky 2/4, Tomsk 634055, Russia; kulkova@ms.tsc.ru

[2]   National Research Tomsk State University, pr. Lenina 36, Tomsk 634050, Russia; sskulkov@ispms.tsc.ru

[3]   Institute for Materials Testing, Materials Science and Strength of Materials, University of Stuttgart, Pfaffenwaldring 32, 70569 Stuttgart, Germany; stephen.hocker@imwf.uni-stuttgart.de (S.H.); siegfried.schmauder@imwf.uni-stuttgart.de (S.S.)

*   Correspondence: bakulin@ispms.tsc.ru; Tel.: +7-3822-286-952

Received: 16 September 2020; Accepted: 27 September 2020; Published: 29 September 2020

Abstract: The adhesion properties of the TiAl/TiO$_2$ interface are estimated in dependence on interfacial layer composition and contact configuration using the projector augmented wave method. It is shown that a higher value of the work of separation is obtained at the interface between the Ti-terminated TiAl(110) surface and the TiO$_2$(110)$_O$ one than at that with the Al-terminated alloy. An analysis of structural and electronic factors dominating the chemical bonding at the interfaces is carried out. It is shown that low bond densities are responsible for low adhesion at both considered interfaces, which may affect the spallation of oxide scale from the TiAl matrix.

Keywords: interface; adhesion; titanium aluminides; titanium dioxide; electronic structure; first principles calculations

---

1. Introduction

Among the big family of intermetallic compounds, Ti-Al alloys attract much attention from both experimental and theoretical researchers. In most studies, γ-TiAl is investigated due to its excellent combination of mechanical properties such as low density, high specific strength, and stiffness, as well as good high-temperature creep resistance [1,2]. However, potential application of the alloy in aeronautical and space technologies is restricted by the poor high-temperature oxidation resistance, which is connected with the growth of non-protective Al$_2$O$_3$ + TiO$_2$ mixed oxide layers on the internal scale [3,4]. To improve the oxidation resistance of the γ-TiAl alloy, numerous treatment techniques, including the addition of alloying elements and surface modification, have been applied [1,2]. There is an option that the oxidation resistance of γ-TiAl might be improved by avoiding the formation of TiO$_2$ and promoting the formation of an Al$_2$O$_3$ protective dense layer. However, the Al chemical activity is reduced in Ti–Al alloys with increasing Ti-content. Therefore, this fact, in combination with the thermodynamic characteristics of the oxides, leads to a larger stability of the interfaces with TiO and TiO$_2$ rather than with Al$_2$O$_3$. Experiments [5–10] have shown that Nb, Mo, W, and Re benefit, whereas V, Mn, and Cu deteriorate the oxidation resistance of γ-TiAl. Since the experimental measurements are very sensitive to many factors, including temperature, partial pressure of oxygen, composition variation, and so forth, the obtained conclusions are often contradictory. It is believed that the alloying with impurities such as Nb, Mo, W, and so forth can increase the activity and diffusivity of aluminum and therefore enhance the formation of the alumina layer [1,11]. It was demonstrated in [12,13] that some impurities (Nb, Mo, Ta, Hf, etc.) increase the formation energy of an oxygen vacancy in TiO$_2$ and can suppress the TiO$_2$ growth.

The oxidation characteristics of γ-TiAl were studied in several studies using methods within density functional theory [14–22]. Li with coworkers [14] established that oxygen prefers to be adsorbed near the Ti atoms regardless of the concentration of oxygen. The authors of [15] emphasized also the oxidation priority of Ti atoms over Al ones. Liu et al. [16] demonstrated that the occurrence of Al self-segregation at the TiAl(111) surface can enhance the interaction between O and Al atoms and promote the growth of a pure alumina layer. It was shown in [17] that alloy doping by Nb increases the diffusion barrier of oxygen and therefore can improve the oxidation resistance. The effect of Nb on the O diffusion was also studied in [18]. Strengthening of the bonding between O and Al and weakening of the Ti–O one due to doping by Si was demonstrated in [19]. The authors conclude that the formation ability of TiO_2 can be reduced. In [20], the mechanisms of the O interaction with low-index TiAl surfaces were studied. The interaction with the Ti-rich surface was found to be much stronger than with the Al-rich one. In our previous work [21], the initial stage of γ-TiAl (001), (100), and (110) surface oxidation was also studied. The preference of the Ti-rich sites for oxygen adsorption was demonstrated, that is in agreement with experiment and earlier theoretical results. Additionally, the O diffusion migration barriers were determined in the bulk TiAl alloy and from surface into bulk [21,22]. Recently, the oxygen temperature dependent diffusion coefficient was studied in γ-TiAl using several models [23–25]. In our paper [25], a comparative study of O diffusion using a statistical approach and the Landman method was performed. We demonstrated that O diffusion along the a axis is faster than along the c axis, whereas the opposite trend was observed in Ti_3Al [26,27]. It is seen that there is much knowledge about the initial stage of TiAl surface oxidation and oxygen diffusion in the alloy.

At the same time, studies of the bonding characteristics at the TiAl–oxide interfaces remain rare. The bonding strength and stability of $TiAl/TiO_2$ and $TiAl/Al_2O_3$ interfaces were studied in [28–30]. It was shown in [28] that the O–Al bond is stronger than the O–Ti one at the $TiAl/TiO_2$ interface. Additionally, authors of [28] concluded that both Ti vacancy and Nb dopant weaken the stability of the $TiAl(110)_{Al}/TiO_2(110)_O$ interface, and Nb makes the oxygen atoms detach from TiO_2 and primarily bond with the Al atomic layer in the TiAl surface. The latter can improve the oxidation resistance of γ-TiAl. In other work [29], the strength of the O–Al and O–Ti bonds were compared at the $TiAl/Al_2O_3$ interface. It was also found that the O–Al bond is stronger, rather than the O–Ti one, owing to the O-p and Al-p hybridization, which takes part in a relatively broad energy region in comparison with the overlaps of O-p and Ti-d orbitals. In [30], the influence of alloying elements on the adhesion properties of the $TiAl(110)_{Ti}/TiO_2(110)_O$ interface was studied. It was established that among the considered impurities, Y, Nb, and Pd are suitable to improve the adhesion ability between TiO_2 and TiAl. It is necessary to point out that the surface cells $TiO_2(110)$–(1×1) and $TiAl(110)$–(2×1) were matched in papers [28] and $TiO_2(110)$–(1×2) and $TiAl(110)$–(2×2)—in [30], that leads to a misfit along one surface lattice parameter of about 22%. Therefore, a significant distortion of the atomic structure of the oxide film takes place at the interface. Furthermore, the mixed Ti–O layer second from the interface splits substantially, and the distance is comparative with the interplanar one. As a result, the interfacial oxygen atoms incorporate into the Ti interfacial layer of the alloy that leads to large work of separation (4.97 J/m^2), whereas it is only 1.53 J/m^2 if the cleavage plane locates between the shifted O atom and the oxide. Therefore, in order to better understand the bonding characteristics between oxide and alloy, a more reasonable interface with a small mismatch should be studied. In the present work, we focus on the modeling of the $TiAl/TiO_2$ interface and its adhesion properties.

2. Computational Details

The atomic and electronic structure of the $TiAl(110)/TiO_2(110)$ interface was calculated by the projector-augmented wave (PAW) method in the plane-wave basis [31,32], implemented in the Vienna Ab initio Simulation Package (VASP) code [33–35]. The generalized gradient approximation in the Perdew–Burke–Ernzerhof form (GGA–PBE) was used for the exchange-correlation functional [36]. The energy cut-off for the plane waves was set to 550 eV. For interface calculations, we adopted a k-points mesh of $2 \times 8 \times 1$ obtained by the Monkhorst–Pack method [37]. The total energies were converged up to 10^{-5} eV. The theoretical lattice parameters of the bulk TiO_2 with rutile structure ($a = 4.663$ Å, $c = 2.969$ Å) and the γ-TiAl alloy ($a = 3.977$ Å, $c = 4.081$ Å) are in good agreement with experimental ones [38,39]. More details about structural parameters of both TiAl(110) and TiO_2(110) surface and interface structures will be discussed in the corresponding section.

The work of separation (W_{sep}) or the ideal adhesion energy at the alloy–oxide interface was calculated as

$$W_{sep} = [E(TiAl) + E(TiO_2) - E(TiAl/TiO_2)]/S, \tag{1}$$

where $E(TiAl/TiO_2)$ is the total energy of a supercell containing the multilayered slabs, $E(TiAl)$ and $E(TiO_2)$ are the total energies of the same supercell containing a single slab of the alloy or the oxide, respectively, and S is the area of the interface.

In order to estimate the surface energy of stoichiometric surfaces the following formula was used:

$$\sigma = \left(E^{slab} - N \cdot E^{bulk}\right)/2S, \tag{2}$$

where E^{slab} is the total energy of the supercell containing a slab and vacuum region, E^{bulk}—the reference energy of the bulk compound per formula unit, N—the number of formula units in the slab, S—the surface area, and factor 2 corresponds to two identical surface on both slab sides.

In the case of nonstoichiometric surfaces of an A_nB_m compound, σ is a function of the chemical potential of any component and can be calculated as follows:

$$\sigma = \frac{1}{2S}\left[E^{slab} - \frac{N_B}{m}E^{bulk} - \Gamma_A\mu_A^{bulk} - \Gamma_A\Delta\mu_A\right], \tag{3}$$

where N_A (N_B) is the number of atoms of element A (B), Γ_A—the excess of element A at the surface with respect to B one:

$$\Gamma_A = N_A - \frac{n}{m}N_B, \tag{4}$$

μ_A^{bulk}—the chemical potential of element A in elementary substance under bulk conditions, $\Delta\mu_A$—he change of chemical potential of element A in A_nB_m compound with respect to the elementary substance bulk. It may change in the range:

$$-\frac{1}{n}\Delta H^f_{A_nB_m} \leq \Delta\mu_A \leq 0, \tag{5}$$

where $\Delta H^f_{A_nB_m}$ is the formation enthalpy of the compound per formula unit. More details can be found in our earlier paper [22]. Often authors draw σ as function of $\Delta\mu/\Delta H^f$ instead of $\Delta\mu$. This allows us to reach uniformity of the surface stability diagrams for different compounds.

3. Results and Discussion

3.1. Surface Energy

The surface energy of low-index surfaces of the γ-TiAl alloy as a function of the Ti chemical potential is shown in Figure 1a. The atomic structures of the γ-TiAl(001), (100), (110), and (111) surfaces were modeled by 11-layer thin films, whereas slabs of the TiO_2 with rutile structure contained 21–35 atomic layers in dependence on surface orientation. It is seen that the TiAl(100) surface with the stoichiometric composition is the most stable one in the Ti-rich region. At the same time, the TiAl(110) surface with Al termination has the lowest surface energy in the Al-rich and near-stoichiometry regions. In the limit of high aluminum concentrations, the surface energies of both TiAl(001)$_{Al}$ and TiAl(110)$_{Al}$ surfaces are almost the same. The values of σ for other considered surfaces are higher. One can see from Figure 1b that the stoichiometric TiO_2(110)$_O$ surface is stable for almost the whole interval of $\Delta\mu_{Ti}$. Only in the Ti-rich limit, the TiO_2(110)$_{TiO}$ surface with TiO termination has the smallest value of σ. The surface energy of (100)$_O$ is only higher by 0.16 J/m^2 than that of (110)$_O$.

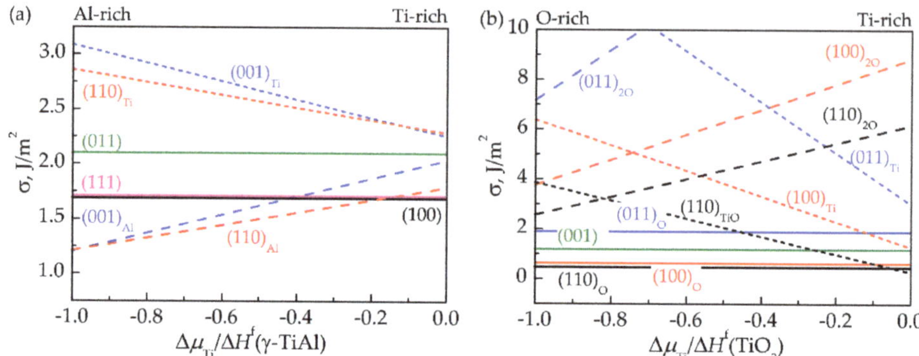

Figure 1. Surface energy of (**a**) γ-TiAl and (**b**) TiO$_2$ low-index surfaces as a function of the Ti chemical potential.

The obtained surface energies are in line with earlier data from [40–43]. For example, the surface energy of TiAl(100) is 1.70 J/m^2, that is, higher by ~5.0% than the value of 1.62 J/m^2 [40] and lower by ~11.0% than 1.91 J/m^2 obtained in [41]. It should be noted that the latter value was calculated with the local density approximation for an exchange-correlation functional (PAW-LDA). The value of 0.46 J/m^2 in the case of the TiO$_2$(110)$_O$ surface agrees well with 0.48 J/m^2 [43]. All calculated values of σ for TiAl and TiO$_2$ stoichiometric surfaces in comparison with available theoretical results are summarized in Table 1. One can see that values obtained within LDA are higher by 0.3–0.5 J/m^2 than those within GGA. Thus, based on thermodynamic findings it is reasonable to consider the interface between TiAl(110)$_{Al}$ and TiO$_2$(110)$_O$ surfaces. As the self-diffusion rate in the γ-TiAl alloy [44] is higher than in TiO$_2$ [45], a Ti segregation from the alloy toward the oxide can change the composition of interfacial layers. In this connection, we consider the TiAl(110)$_{Ti}$/TiO$_2$(110)$_O$ interface as well.

Table 1. Surface energy (in J/m^2) of stoichiometric surfaces of γ-TiAl and TiO$_2$.

Compound	TiAl			Compound	TiO$_2$			
Surface	**(100)**	**(111)**	**(011)**	**Surface**	**(110)$_O$**	**(100)$_O$**	**(001)**	**(011)$_O$**
Present result PAW-PBE	1.69	1.71	2.10	Present result PAW-PBE	0.46	0.62	1.20	1.02
[40] PAW-GGA	1.62	1.69		[28] PAW-GGA	0.46	0.56	1.44	1.05
[41] PAW-LDA	1.91			[42] PAW-GGA	0.50	0.69	1.25	1.03
				[43] US PP LDA	0.89	1.12	1.40	1.65

3.2. Work of Separation

The TiAl(110) surface cell has parameters $a_1 \times b_1 = c \times a\sqrt{2}/2 = 4.081 \times 2.812$ Å^2 while TiO$_2$(110)—$a_2 \times b_2 = a\sqrt{2} \times c = 6.594 \times 2.969$ Å^2, where a and c are lattice parameters of the corresponding compounds. It is clear that cells of $(3a_1 \times b_1)$ and $(2a_2 \times b_2)$ have enough good agreement. In this case, the misfit is $\delta_1 = 2(2a_2 - 3a_1)/(2a_2 + 3a_1)\cdot100\% = 7.4\%$ and $\delta_2 = 2(b_2 - b_1)/(b_2 + b_1)\cdot100\% = 5.4\%$. As the bulk modulus of TiO$_2$ is higher by ~62% than that of γ-TiAl, we used the parameters of the former to construct the interface structure. Therefore, the alloy structure was expanded by 7.7% and 5.6% in two directions. Atomic structures of both TiO$_2$(110)$_O$–(2 × 1) and TiAl(110)$_{Al}$–(3 × 1) surfaces are shown in Figure 2. The supercell used for the interface modelling contains twelve TiO$_2$ layers (four O–TiO–O quasi-layers) and eight TiAl layers (four Ti and four Al layers) with 72 atoms (32 O atoms, 12 Al atoms, and 28 Ti atoms) in total and a 20 Å vacuum gap in the direction perpendicular to the interface. Test calculations of surface energies demonstrated that an increase of oxide or alloy slab thickness or vacuum gap results in a change of the work of separation by less than 20 mJ/m^2. Moreover, doubling the interface cell along the [1] direction of the oxide leads to a difference in W_{sep} of 4 mJ/m^2.

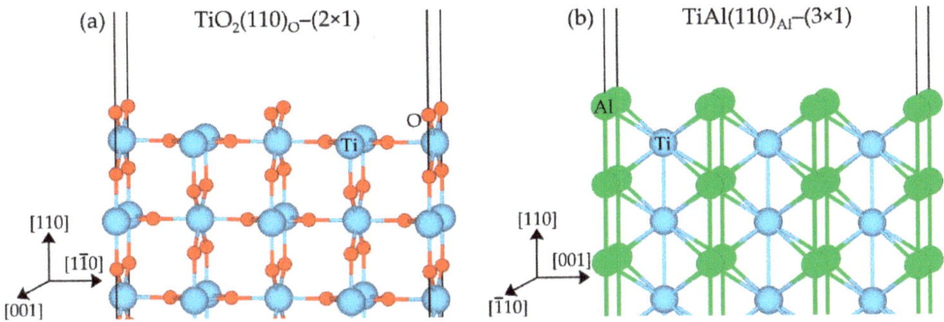

Figure 2. Atomic structure of (**a**) TiO$_2$(110)$_O$−(2 × 1) and (**b**) TiAl(110)$_{Al}$−(3 × 1) surfaces.

For the modeling of the TiAl/TiO$_2$(110) interface, two configurations (*hollow* and *top*) of contacts with the Al- and Ti-terminated γ-TiAl(110) surface were considered. Both configurations before relaxation are presented in Figure 3. In case of the *hollow* configuration, one interfacial O atom (O2$_I$ in Figure 3a) locates above the Ti$_{I-1}$ atom of the second layer. It is a so-called *H* or *hollow* site for O adsorption on the TiAl(110) surface (see, e.g., [22]) that was a reason for the configuration denotation. The other interfacial O atom (O1$_I$ in Figure 3a) occupies the *short-bridge* position between two Al1$_I$ atoms. The *top* configuration can be obtained from the *hollow* one by a shift of the oxide film along the [1] direction by $b_2/2$. As a result, the O2$_I$ atom locates now in the *long-bridge* position between two Al2$_I$ atoms and O1$_I$—in the *top* one (Figure 3b). The denotation of the *top* configuration is due to this position of the O1$_I$ atom. It should be noted that in the case of the Ti-terminated alloy surface, Al and Ti atoms of its film are swapped (Figure S1). The initial interface distance was chosen so that the interatomic distance between the nearest interfacial atoms is slightly greater than the sum of their covalent radii.

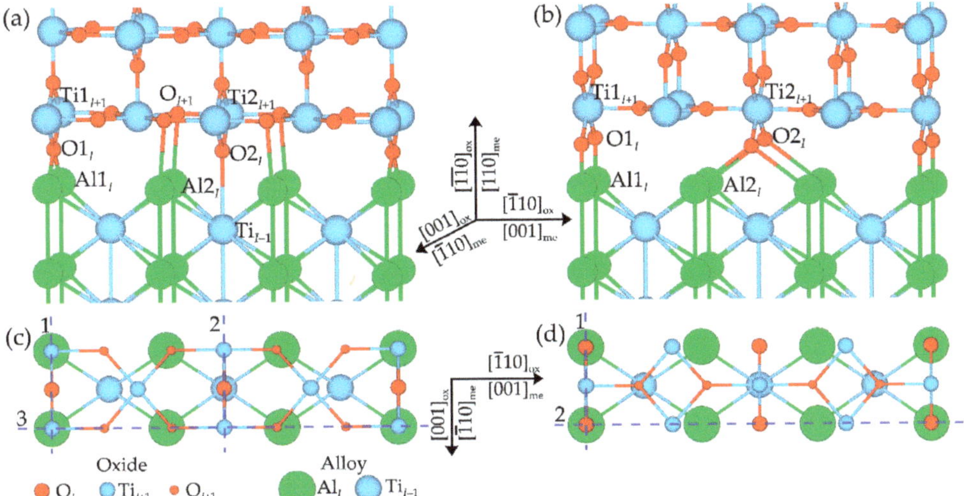

Figure 3. Atomic structures of unrelaxed interfaces with (**a**,**c**) *hollow* and (**b**,**d**) *top* configuration. Atoms that are important in interfacial bonding are marked. Symbols *I* and *I* ± 1 denote atoms of interfacial and subinterfacial layers.

The calculated values of the work of separation and the interfacial distance (d) for all considered structures are given in Table 2. Note that for the d estimation, the averaged values of positions of the interfacial atoms in the oxide and the alloy were used. One can see that for the Al-terminated alloy, the obtained values of W_{sep} are within 0.74–1.27 J/m^2 that is in line with 0.58–1.64 J/m^2 [28]. The highest value of the work of separation was calculated for the *hollow* configuration. In work [28], five interface configurations were considered. As the authors of [28] matched TiO$_2$(110)–(1 × 1) and TiAl(110)–(2 × 1) surface cells (which results in too large misfit), a direct comparison of the present results with earlier ones is impossible. Nevertheless, in the case of the most stable configuration B(Ti–Al) in [28], there are bonds between interfacial O$_I$ and subinterfacial Ti$_{I-1}$ atoms as well as between O$_{I+1}$ and Al$_I$ ones. Our *hollow* configuration demonstrates almost similar chemical bonding. The difference in W_{sep} for these configurations is 0.37 J/m^2. Some discussion will be given in Section 3.3.

Table 2. Work of separation (in J/m^2) and interface distance (d in Å) at the interface TiAl/TiO$_2$(110)$_O$.

Alloy Termination	Configuration	W_{sep}	d
	hollow	1.27	0.75
Al	*top*	0.74	1.38
	[28]	0.58–1.64	
	hollow	2.44	0.62
Ti	*top*	1.97	0.50
	[30]	4.97	

The values of the work of separation for the interface with the Ti-terminated alloy film increase to almost twice in comparison with the previous case, but they remain substantially lower than the data from [30] (Table 2). It should be noted that the authors of [30] used almost the same surface cells as in [28], they were just doubled along the shortest distance, i.e., TiO$_2$(110)–(1 × 2) and TiAl(110)–(2 × 2) cells were matched. In the present paper as well as in [30], the formation O$_I$–Al$_{I-1}$ and O$_I$–Ti$_I$ bonds is observed. At the same time, in work [30] a huge distortion of the atomic structure of TiO$_2$ near the interface takes place. The latter allows us to assume that the authors adopted the alloy lattice parameters for interface modeling. As a rule, the parameters of oxide surface cell are used in such calculations [46,47]. It should be noted that enough low values of the work of separation were calculated earlier in several papers, if oxide is terminated by one oxygen layer. For example, W_{sep} equal to 1.37 J/m^2 was obtained at the Nb(1$\bar{1}$0)/Al$_2$O$_3$(11$\bar{2}$0)$_O$ interface [48] while the value is one order more at the Nb(1$\bar{1}$0)/Al$_2$O$_3$(11$\bar{2}$0)$_{2O}$ one [49]. In the case of the U(110)/Al$_2$O$_3$(11$\bar{2}$0)$_O$ interface, the value of 1.90 J/m^2 was calculated, but 11.5 J/m^2 was obtained at the U(110)/Al$_2$O$_3$(11$\bar{2}$0)$_{2O}$ interface [50]. Analysis of both TiAl/TiO$_2$ interface characteristics will be performed in the next section.

3.3. Atomic and Electronic Factors

Figure 4 demonstrates the distribution of the charge density difference ($\Delta\rho(\mathbf{r})$) for both TiAl(110)$_{Al}$/TiO$_2$(110)$_O$ and TiAl(110)$_{Ti}$/TiO$_2$(110)$_O$ interfaces in the case of the *hollow* configuration in the planes shown in Figure 3c. We emphasize that $\Delta\rho(\mathbf{r}) = \rho_{\text{ox}} + \rho_{\text{me}} - \rho_{\text{me/ox}}$, where $\rho_{\text{me/ox}}$ is the total charge density of a supercell containing both alloy and oxide slabs; ρ_{ox} and ρ_{me} are the total charge densities of the same supercell containing a single slab of the oxide or the alloy. The section planes were chosen to give a better insight into the charge redistribution near interface bonds.

It is seen in Figure 4a that the formation of O1$_I$–Al1$_I$ bonds results in weakening of the bonds between this O1$_I$ atom and the nearest Ti$_{I+1}$ ones in the oxide. The latter can be seen by the appearance of pronounced charge depletion regions at the Ti1$_{I+1}$–O1$_I$ bonds that leads to charge redistribution around Ti1$_{I+1}$ atoms. In accordance with the analysis of charge states performed within DDEC6 method [51–53], Ti$_{I+1}$ atoms lose a smaller charge than those at the clean surface (1.99 el. instead of 2.13 el.), whereas the O1$_I$ atom gets additional 0.05 el. In comparison with that on the free oxide surface

(Table 3). Moreover, the overlap population of the Ti1$_{I+1}$–O1$_I$ bonds decreases significantly at the TiAl(110)$_{Al}$/TiO$_2$(110)$_O$ interface. We emphasize that the overlap population is used for the estimation of the bond strength. It should be noted that there is a charge depletion region between Al1$_I$ atoms in the alloy. Charge states of these atoms change from –0.24 el. on the clean surface to +0.69 el. at the interface.

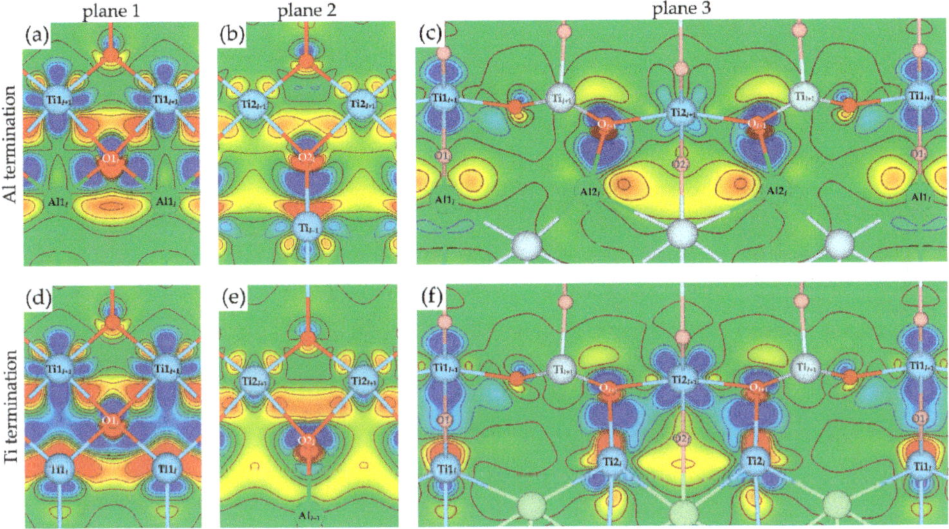

Figure 4. Distribution of the charge density difference ($\Delta\rho$) at the TiAl/TiO$_2$ interface in the case of the *hollow* configuration in planes shown in Figure 3c: (**a–c**) Al termination; (**d–f**) Ti termination. The regions of electron accumulation ($\Delta\rho < 0$) and depletion ($\Delta\rho > 0$) are given in blue and red, respectively. Contours go –0.1 to 0.1 el./Å^3 with an isoline spacing of 0.02 el./Å^3. Atoms being out of the figure plane are shown in lighter colors.

Figure 4b demonstrates that all Ti atoms surrounding the oxygen at the interface lose the charge: in particular, Ti2$_{I+1}$ atoms have a charge smaller by 0.12 el. than on the free surface; Ti$_{I-1}$ atoms lose 0.13 el., whereas the charge of O2$_I$ atoms increases by 0.04 el. It is seen that a large charge accumulation region occurs at O2$_I$–Ti$_{I-1}$ bonds, which are responsible for the interface strength. The appearance of both accumulation and depletion regions at these bonds indicates a large covalent contribution in the chemical bonding. Indeed, one can see from Table 3 that the overlap population of the O2$_I$–Ti$_{I-1}$ bond is higher than that of the O1$_I$–Al1$_I$ one by 0.11 el. At the same time, ionicity of the latter bond is higher due to the large positive charge of the Al1$_I$ atoms. It is necessary to emphasize that valence *p*-states of Al are less localized than Ti *d*-states, which allows them to be more easily involved in the interaction with oxygen. In addition, the in-plane average $<\Delta\rho(z)> = \Delta\rho(z)/S$ is given in Figure S2a, where $\Delta\rho(z)$ is a result of in-plane integration of $\Delta\rho(\mathbf{r})$ over the plane and S is the area. It illustrates where the maximum of charge accumulation/depletion is located.

Interfacial O$_{I+1}$–Al2$_I$ bonds (Figure 4c) also have an ionic character, and a charge depletion region occurs mainly near Al2$_I$ atoms. In spite of that, the O$_{I+1}$–Al2$_I$ bond length is almost the same as O1$_I$–Al1$_I$; the overlap population of the former is larger by 0.078 el. At the same time, Al2$_I$ atoms lose a smaller charge in comparison with Al1$_I$ (Table 3). All these features indicate that O$_{I+1}$–Al2$_I$ has a less ionic but more covalent character and the large charge accumulation region located near O$_{I+1}$ atoms is conditioned primarily by charge redistribution rather than by charge transfer from the alloy atoms.

All mentioned peculiarities of O–Ti and O–Al bonds are still valid in the case of the TiAl(110)$_{Ti}$/TiO$_2$(110)$_O$ interface. Formation of the O1$_I$–Ti1$_I$ interfacial bonds (Figure 4d) leads to a weakening of Ti1$_{I+1}$–O1$_I$ bonds in the oxide (overlap population is almost one and a half times less (Table 3)), to the appearance of charge redistribution around Ti1$_{I+1}$ atoms and localized charge depletion regions near $Ti1_I$ atoms. The charge state of $Ti1_I$ atoms changes from +0.16 el. on the clean alloy surface to +0.82 el. at the interface (Table 3). Formation of the $O2_I$–Al_{I-1} bond (Figure 4e) leads to a decrease of charge transfer from $Ti2_{I+1}$ atoms in the oxide to $O2_I$ by 0.24 el. As a result, the charge at Al_{I-1} atoms decreases by 0.28 el. (Table 3). Figure 4f demonstrates O_{I+1}–$Ti2_I$ interfacial bonds, which have also pronounced ionic–covalent character. Although the O_{I+1}–$Ti2_I$ bond length is larger by 0.06 Å than the $O1_I$–$Ti1_I$ one, the overlap population changes insignificantly (by 0.02 el.). Additionally, $Ti2_I$ atoms have a smaller positive charge in comparison with $Ti1_I$. In general, Figure 4 demonstrates that all formed interfacial bonds have an ionic–covalent character and the formation of the interfacial bonds leads to a decrease of the chemical bonding in both oxide and alloy. The integrated $\Delta\rho(z)$ (Figure S2b) shows that the charge depletion region associated with Al interfacial atoms is more widespread through the slab than that connected with Ti interfacial atoms. This is connected with the nature of valence electrons in Al and Ti. Additionally, it is seen that charge transfer to O interfacial atoms is larger in the case of the $TiAl(110)_{Ti}/TiO_2(110)_O$ interface.

Table 3. Charge states of some interfacial and subinterfacial atoms (Q in el.), bond length between them (d in Å) and averaged overlap population (θ in el.) for pairs of the atoms in the case of the *hollow* interface configuration. The corresponding data for free surfaces of the alloy and the oxide are given also.

Atom	Q	d	θ	Atom	Q	d	θ
$TiAl(110)_{Al}/TiO_2(110)_O$				$TiAl(110)_{Ti}/TiO_2(110)_O$			
$Ti1_{I+1}$	+1.99	2.20	0.183	$Ti1_{I+1}$	+2.00	2.07	0.253
$O1_I$	−0.93			$O1_I$	−1.00		
$Ti2_{I+1}$	+2.01	2.07	0.255	$Ti2_{I+1}$	+1.89	2.23	0.183
$O2_I$	−0.92			$O2_I$	−0.83		
$O1_I$	−0.93	1.89, 1.78 [28]	0.481	$O1_I$	−1.00	1.98, 2.22–2.31 [30]	0.483
$Al1_I$	+0.69			$Ti1_I$	+0.82		
$O2_I$	−0.92	1.92, 2.08 [28]	0.594	$O2_I$	−0.83	2.07, 1.90 [30]	0.403
Ti_{I-1}	+0.42			Al_{I-1}	−0.02		
O_{I+1}	−1.00	1.90, 1.85 [28]	0.559	O_{I+1}	−1.14	2.04, 1.92 [30]	0.505
$Al2_I$	+0.19			$Ti2_I$	+0.51		
$TiAl(110)_{Al}$ surface [1]				$TiAl(110)_{Ti}$ surface			
Al_S	−0.24	2.90	0.277	Ti_S	+0.17	2.84	0.302
Ti_{S-1}	+0.29			Al_{S-1}	−0.30		
$TiO_2(110)_O$ surface							
O_S	−0.88	1.85	0.406				
Ti_{S+1}	+2.13						
O_{S+1}	−1.13	1.95–2.03	0.279–0.340				
Ti_{S+1}	+2.13						

[1] Symbols S and $S-1$ ($S+1$) denote atoms of surface and subsurface layers in alloy (oxide).

Figure 5 demonstrates the local densities of states (DOSs) of the interfacial atoms for both interfaces in the case of the *hollow* configuration. It is seen that sharp peaks of subinterfacial O_{I+1} atoms lying at −7.2, −6.9, and −5.6 eV coincide well with similar peaks of $Al2_I$ (Figure 5a). Similarly, the thin structure of $O1_I$ DOS (position of some peaks) agrees well with that of the $Al1_I$ atom. The difference from the previous case is a wider spread of $O1_I$ and $Al1_I$ states toward negative energies. It should be noted that there is small density of states on both DOS curves for O_{I+1} and $O1_I$ atoms in the band gap, which is typical for metallic oxide due to the presence of the interface states induced by interaction with interfacial alloy atoms. It is known that the valence band of Ti atoms is almost unoccupied. The interaction of Ti_{I-1} with $O2_I$ results in the appearance of low-lying states, which spread in the region from −7.1 eV up to −3.4 eV. Additionally, the shift of the valence band of all O atoms, involved in the formation bonds through interface, toward negative energy is a consequence of charge transfer from nearest metal atoms. The increase of Ti unoccupied states in the region 0.4–0.9 eV is connected with the charge transfer from Ti to oxygen due to their interaction.

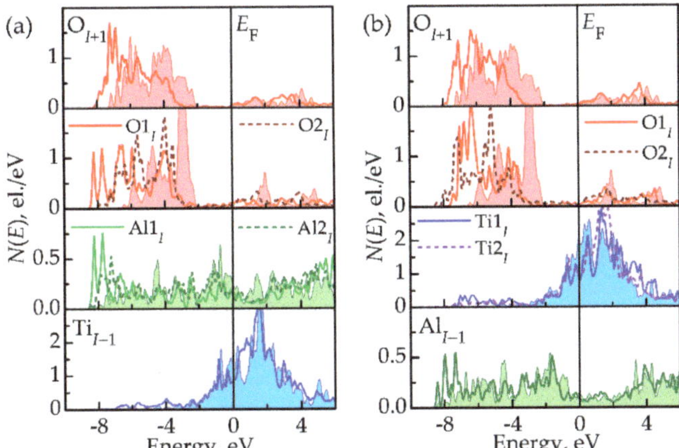

Figure 5. Local DOSs of the interfacial atoms for the *hollow* configuration at the TiAl/TiO$_2$ interface: (**a**) Al termination; (**b**) Ti termination. Shaded areas correspond to DOSs of atoms on the clean surfaces of the oxide and the alloy.

It is necessary to recall that Al *s,p*-states involved much more easily in the interaction with oxygen than Ti *d*-states, which is expressed in a stronger change of Al DOSs than that of Ti. As a result, at the interface with the Ti-terminated alloy surface (Figure 5b), the shift of occupied states of O$_{I+1}$ and O1$_I$ atoms is less pronounced than that in the previous case. However, the states of O2$_I$ atoms interacting with Al$_{I-1}$ atoms spread by 0.7 eV more toward negative energies. In general, DOS curves confirm strong interaction between interfacial atoms. Note that detailed comparison of DOS thin structures can be found in Figure S3.

Now we will discuss the difference in the structural characteristics of the present interfaces with *hollow* configuration and the most stable interfaces from papers [28,30]. The comparison of present interfacial bond lengths with those from paper [28] is given Figure 6a. Within our interface model, there are two O–Al bonds of 1.89 Å (1.78 Å [28]) and two of 1.90 Å (1.85 Å [28]), as well as one O–Ti bond of 1.92 Å (two bonds of 2.08 Å [28]), which are distributed over the area of 39.155 Å^2 (~22.95 Å^2 [28]). Thus, the total density of O–Al and O–Ti bonds (the number of bonds per unit area) is 0.102 Å^{-2} and 0.026 Å^{-2} within the present model and 0.174 Å^{-2} and 0.087 Å^{-2} according to [28], respectively. It is seen in Figure 6a that our interface model demonstrates about half of the bond density (0.128 Å^{-2} in comparison with 0.261 Å^{-2} in [28]) and weaker O–Al bonds (larger bond length than in [28]). At the same time, the O–Ti interaction is substantially stronger than that in [28]. The competition of these factors leads to a smaller value of the work of separation than that in [28], but the difference is not large.

In the case of the interface with the Ti-terminated TiAl(110) surface, the bond density is the same (only Al is replaced by Ti and vice-versa): O–Ti—0.102 Å^{-2} with bond lengths of 1.98 and 2.04 Å; O–Al—0.026 Å^{-2} with bond length of 2.07 Å. At the same time, the situation is quite different in [30]. As the TiO$_2$ atomic structure is much distorted in [30], formation of additional interfacial bonds is possible. Based on figures of atomic structure and tables with bond lengths from [30], we can conclude that there are four O–Ti bonds with length of 2.22 Å, four O–Ti bonds of 2.31 Å, two O–Ti bonds of 1.92 Å, two O–Ti bonds with unknown length (we can suppose that it is about 1.92 Å), and two O–Al bonds with a length of 1.90 Å per area ~45.9 Å^2 (Figure 6b). One can see that the bond density within the model from [30] is more than twice as high in comparison with the present model. Moreover, the lengths of some O–Ti and O–Al bonds in [30] are shorter than those in the present paper. As a result, a lower value of W_{sep} than that in [30] was obtained. It is known that compression in the interface plane results in an increase of interlayer distances and weakening of interatomic interaction inside the slab. In turn, that results in strengthening of interatomic interactions at the interface. This can explain the overestimation of the adhesion energy at the interface.

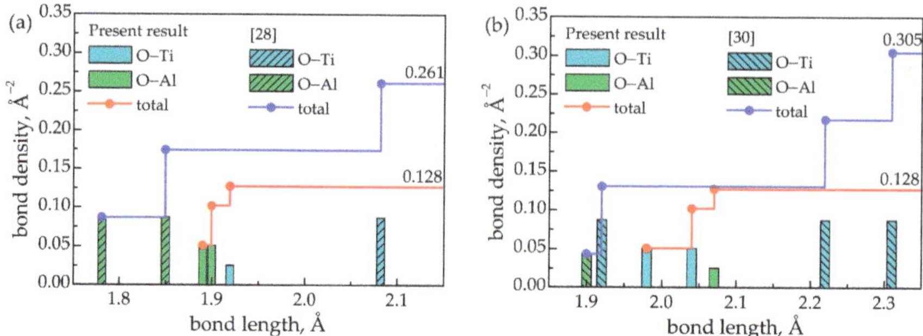

Figure 6. Bond density versus bond length at the TiAl/TiO$_2$ interface in the case of *hollow* configuration: (**a**) Al termination; (**b**) Ti termination.

It should be noted that cleavage between the oxide slab with the TiO-terminated (110) surface and TiAl(110)$_{Ti}$ with O layer (violet plane in Figure S1a) leads to a value of 1.77 J/m^2, whereas it was equal to 1.53 J/m^2 in [30]. In accordance with the Griffith fracture theory [54,55], the fracture work (G) of bulk materials along a certain lattice plane can be estimated as double surface energy. If $G < W_{sep}$, the fracture will occur in the bulk material; otherwise, the fracture will occur at the interface. Our estimation of the surface energy of TiO$_2$ (Table 1) allows to conclude that the cleavage energy along the (110) plane between two stoichiometric surfaces needs only 0.92 J/m^2, which is lower than W_{sep} obtained for both considered interfaces with the *hollow* configuration. This means that mechanical failure may be initiated in the oxide.

Finally, we will not discuss the interface with the *top* configuration. The distribution of the charge density difference shown in Figure 7 allows us to understand the peculiarities of the chemical bonding at the interfaces and together with Table S1 the lower values of W_{sep} at them in comparison with the *hollow* configuration.

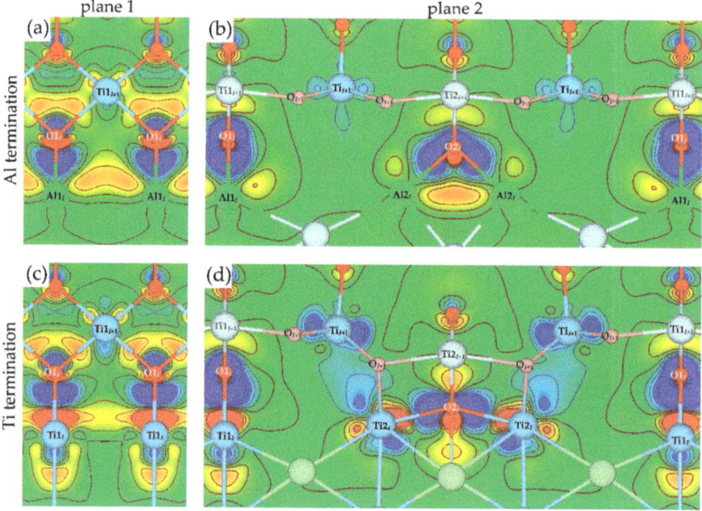

Figure 7. Distribution of the charge density difference ($\Delta\rho$) at the TiAl/TiO$_2$ interface in the case of the *top* configuration in planes shown in Figure 3d: (**a,b**) Al termination; (**c,d**) Ti termination. The regions of electron accumulation ($\Delta\rho < 0$) and depletion ($\Delta\rho > 0$) are given in blue and red, respectively. Contours go −0.1 to 0.1 el./Å^3 with an isoline spacing of 0.02 el./Å^3. Atoms being out of the figure plane are shown in lighter colors.

4. Conclusions

A comparative study of atomic and electronic structures of the TiAl/TiO$_2$(110) interface and its adhesion properties in dependence on the alloy surface termination was performed using the projector-augmented wave method. Within a more commensurable interface model than in [28,30], with a smaller misfit between the alloy and the oxide surface cells, the adhesive properties were clarified. The most preferred contact configuration was found to be the *hollow* one, in which one interfacial O atom locates above a Ti(Al) atom of the alloy subinterfacial layer and another interfacial O atom occupies the *short-bridge* position between two Al(Ti) interfacial atoms. The work of separation of 2.44 J/m^2 was obtained at the TiAl(110)$_{Ti}$/TiO$_2$(110)$_O$ interface, whereas a lower value of 1.27 J/m^2 was calculated at the TiAl(110)$_{Al}$/TiO$_2$(110)$_O$ one. The trend in the W_{sep} lowering in dependence on the alloy termination is consistent with that in [28,30]. Although the covalent contribution to chemical bonding is slightly higher at the interface with the Al-terminated TiAl(110) surface, charge transfer from metal to oxygen is higher at the interface with the Ti-terminated one. The overlap population analysis demonstrates that O–Ti bonds are stronger than O–Al ones at both interfaces. The interaction of titanium dioxide with the interfacial atoms of the alloy gives rise to the restoration of bonding in the bulk oxide. In general, the obtained small values of the work of separation at both interfaces are explained by the low bond density. In addition, it should be noted that the predicted adhesion at the TiAl/TiO$_2$ interface is considerably lower than that based on earlier obtained values of 10.43 J/m^2 at the TiAl(111)/Al$_2$O$_3$(0001)$_O$ [56] and 11.02 J/m^2 at the Ti$_3$Al(0001)/Al$_2$O$_3$(0001)$_O$ [57] interfaces. We hope that the interfaces with the oxide termination by an oxygen double layer will demonstrate much better adhesive properties as in the case of NiTi/TiO$_2$(100)$_{2O}$ [58] because of the increase of both ionic–covalent contribution to chemical bonding and bond density. In general, the obtained results allow us to get a better insight into the mechanisms of oxide scale formation on the TiAl surface. Furthermore, it can be useful for the simulation of impurity effects on the adhesion at the interface, which is a subject of our forthcoming work.

Supplementary Materials: The following are available online at http://www.mdpi.com/2075-4701/10/10/1298/s1: Figure S1: Atomic structures of unrelaxed interfaces with *hollow* and *top* configuration in the case of the Ti-terminated alloy surface, Figure S2: In-plane average charge density difference at the interface with the *hollow* configuration; Table S1: Charge states of some interfacial and subinterfacial atoms (Q in el.), bond length between them (d in Å) and averaged overlap population (θ in el.) for pairs of the atoms in the case of the *top* interface configuration.

Author Contributions: Conceptualization: A.V.B., S.E.K., and S.S.; methodology: S.E.K. and S.H.; formal analysis: A.V.B., S.S.K., and S.H.; validation and data curation: S.E.K. and S.S.; investigation: A.V.B., S.S.K., S.E.K., S.H., and S.S.; writing—original draft preparation: A.V.B., S.S.K., S.E.K., S.H., and S.S.; writing—review and editing: S.E.K. and S.S.; visualization: A.V.B., S.S.K. and S.H.; supervision: S.E.K.; project administration: A.V.B. and S.E.K.; funding acquisition: S.E.K. All authors have read and agreed to the published version of the manuscript.

Funding: The work was performed according to the Government research assignment for ISPMS SB RAS, project No. III.23.2.8. S.H. and S.S. acknowledge funding by the Deutsche Forschungsgemeinschaft (DFG).

Acknowledgments: Numerical calculations were performed on the SKIF-Cyberia supercomputer at TSU.

Conflicts of Interest: The authors declare no conflict of interest.

References

1. Li, Z.; Gao, W. High temperature corrosion of intermetallics. In *Intermetallics Research Progress*; Berdovsky, N., Ed.; Nova Science: New York, NY, USA, 2008; pp. 1–64.
2. Dai, J.; Zhu, J.; Chen, C.; Weng, F. High temperature oxidation behavior and research status of modifications on improving high temperature oxidation resistance of titanium alloys and titanium aluminides: A review. *J. Alloys Compd.* **2016**, *685*, 784–798. [CrossRef]
3. Shanabarger, M.R. Comparative study of the initial oxidation behavior of a series of titanium–aluminum alloys. *Appl. Surf. Sci.* **1998**, *134*, 179–186. [CrossRef]
4. Maurice, V.; Despert, G.; Zanna, S.; Josso, P.; Bacos, M.-P.; Marcus, P. XPS study of the initial stages of oxidation of α2-Ti$_3$Al and γ-TiAl intermetallic alloys. *Acta Mater.* **2007**, *55*, 3315–3325. [CrossRef]

5. Shida, Y.; Anada, H. Role of W, Mo, Nb and Si on oxidation of TiAl in air at high temperatures. *Mater. Trans.* **1994**, *35*, 623–631. [CrossRef]

6. Nickel, H.; Zheng, N.X.; Elschner, A.; Quadakkers, W.J. The oxidation behaviour of niobium containing γ-TiAl based intermetallics in air and argon/oxygen. *Microchim. Acta* **1995**, *119*, 23–39. [CrossRef]

7. Shida, Y.; Anada, H. The effect of various ternary additives on the oxidation behavior of TiAl in high-temperature air. *Oxid. Met.* **1996**, *45*, 197–219. [CrossRef]

8. Taniguchi, S.; Shibata, T. Influence of additional elements on the oxidation behaviour of TiAl. *Intermetallics* **1996**, *4*, S85–S93. [CrossRef]

9. Wang, F.H.; Tang, Z.L.; Wu, W.T. Effect of chromium on the oxidation resistance of TiAl intermetallics. *Oxid. Met.* **1997**, *48*, 381–390. [CrossRef]

10. Jiang, H.R.; Wang, Z.L.; Ma, W.S.; Feng, X.R.; Dong, Z.Q.; Zhang, L.; Liu, Y. Effects of Nb and Si on high temperature oxidation of TiAl. *Trans. Nonferrous Met. Soc. China* **2008**, *18*, 512–517. [CrossRef]

11. Vojtěch, D.; Popela, T.; Kubásek, J.; Maixner, J.; Novák, P. Comparison of Nb- and Ta-effectiveness for improvement of the cyclic oxidation resistance of TiAl-based intermetallics. *Intermetallics* **2011**, *19*, 493–501. [CrossRef]

12. Lin, J.P.; Zhao, L.L.; Li, G.Y.; Zhang, L.Q.; Song, X.P.; Ye, F.; Chen, G.L. Effect of Nb on oxidation behavior of high Nb containing TiAl alloys. *Intermetallics* **2011**, *19*, 131–136. [CrossRef]

13. Ping, F.-P.; Hu, Q.-M.; Bakulin, A.V.; Kulkova, S.E.; Yang, R. Alloying effects on properties of Al_2O_3 and TiO_2 in connection with oxidation resistance of TiAl. *Intermetallics* **2016**, *68*, 57–62. [CrossRef]

14. Li, H.; Liu, L.M.; Wang, S.Q.; Ye, H.Q. First-principles study of oxygen atom adsorption on γ-TiAl(111) surface. *Acta Metall. Sin.* **2006**, *42*, 897–902.

15. Li, H.; Wang, S.Q.; Ye, H.Q. Initial oxidation of γ-TiAl(111) surface: Density functional theory study. *J. Mater. Sci. Technol.* **2009**, *25*, 569–576.

16. Liu, S.Y.; Shang, J.X.; Wang, F.H.; Zhang, Y. *Ab initio* study of surface self-segregation effect on the adsorption of oxygen on the γ-TiAl(111) surface. *Phys. Rev. B* **2009**, *79*, 075419. [CrossRef]

17. Li, H.; Wang, S.Q.; Ye, H.Q. Influence of Nb doping on oxidation resistance of γ-TiAl: A first principles study. *Acta Phys. Sin.* **2009**, *58*, S224–S229.

18. Zhao, C.Y.; Wang, X.; Wang, F.H. First-principles study of Nb doping effect on the diffusion of oxygen atom in γ-TiAl. *Adv. Mater. Res.* **2011**, *304*, 148–153. [CrossRef]

19. Liu, S.Y.; Shang, J.X.; Wang, F.H.; Zhang, Y. Surface segregation of Si and its effect on oxygen adsorption on a γ-TiAl(111) surface from first principles. *J. Phys. Condens. Matter* **2009**, *21*, 225005–225007. [CrossRef]

20. Song, Y.; Dai, J.H. Mechanism of oxygen adsorption on surfaces of γ-TiAl. *Surf. Sci.* **2012**, *606*, 852–857. [CrossRef]

21. Bakulin, A.V.; Kulkova, S.E.; Hu, Q.M.; Yang, R. Adsorption and diffusion of oxygen on (001) and (100) TiAl surface. *Comput. Mater. Sci.* **2015**, *97*, 55–63. [CrossRef]

22. Bakulin, A.V.; Kulkova, S.E.; Hu, Q.M.; Yang, R. Theoretical study of oxygen sorption and diffusion in the volume and on the surface of a γ-TiAl alloy. *J. Exp. Theor. Phys.* **2015**, *120*, 257–267. [CrossRef]

23. Epifano, E.; Hug, G. First-principle study of the solubility and diffusion of oxygen and boron in γ-TiAl. *Comput. Mater. Sci.* **2020**, *174*, 107475. [CrossRef]

24. Connétable, D.; Prillieux, A.; Thenot, C.; Monchoux, J.-P. Theoretical study of oxygen insertion and diffusivity in the γ-TiAl $L1_0$ system. *J. Phys. Condens. Matter* **2020**, *32*, 175702. [CrossRef] [PubMed]

25. Bakulin, A.V.; Kulkov, S.S.; Kulkova, S.E. Diffusion properties of oxygen in the γ-TiAl alloy. *J. Exp. Theor. Phys.* **2020**, *130*, 579–590. [CrossRef]

26. Koizumi, Y.; Kishimoto, M.; Minamino, Y.; Nakajima, H. Oxygen diffusion in Ti_3Al single crystals. *Philos. Mag.* **2008**, *88*, 2991–3010. [CrossRef]

27. Bakulin, A.V.; Latyshev, A.M.; Kulkova, S.E. Absorption and diffusion of oxygen in the Ti_3Al alloy. *J. Exp. Theor. Phys.* **2017**, *125*, 138–147. [CrossRef]

28. Song, Y.; Xing, F.J.; Dai, J.H.; Yang, R. First-principles study of influence of Ti vacancy and Nb dopant on the bonding of $TiAl/TiO_2$ interface. *Intermetallics* **2014**, *49*, 1–6. [CrossRef]

29. Wang, B.D.; Dai, J.H.; Wu, X.; Song, Y.; Yang, R. First-principles study of the bonding characteristics of $TiAl(111)/Al_2O_3(0001)$ interface. *Intermetallics* **2015**, *60*, 58–65. [CrossRef]

30. Dai, J.H.; Song, Y.; Yang, R. Influence of alloying elements on stability and adhesion ability of $TiAl/TiO_2$ interface by first-principles calculations. *Intermetallics* **2017**, *85*, 80–89. [CrossRef]

31. Blöchl, P.E. Projector augmented-wave method. *Phys. Rev. B* **1994**, *50*, 17953–17979. [CrossRef]
32. Kresse, G.; Joubert, J. From ultrasoft pseudopotentials to the projector augmented-wave method. *Phys. Rev. B* **1999**, *59*, 1758–1775. [CrossRef]
33. Kresse, G.; Hafner, J. Ab initio molecular dynamics for open-shell transition metals. *Phys. Rev. B* **1993**, *48*, 13115–13118. [CrossRef]
34. Kresse, G.; Furthmüller, J. Efficient iterative schemes for *ab initio* total-energy calculations using a plane-wave basis set. *Phys. Rev. B* **1996**, *54*, 11169–11186. [CrossRef] [PubMed]
35. Kresse, G.; Furthmüller, J. Efficiency of ab-initio total energy calculations for metals and semiconductors using a plane-wave basis set. *Comput. Mater. Sci.* **1996**, *6*, 15–50. [CrossRef]
36. Perdew, J.P.; Burke, K.; Ernzerhof, M. Generalized gradient approximation made simple. *Phys. Rev. Lett.* **1996**, *77*, 3865–3868. [CrossRef]
37. Monkhorst, H.J.; Pack, J.D. Special points for Brillouin-zone integrations. *Phys. Rev. B* **1976**, *13*, 5188–5192. [CrossRef]
38. Beltrán, A.; Andrés, J.; Sambrano, J.R.; Longo, E. Density functional theory study on the structural and electronic properties of low index rutile surfaces for $TiO_2/SnO_2/TiO_2$ and $SnO_2/TiO_2/SnO_2$ composite systems. *J. Phys. Chem. A* **2008**, *112*, 8943–8952. [CrossRef]
39. Tanaka, K.; Ichitsubo, T.; Inui, H.; Yamaguchi, M.; Koiwa, M. Single-crystal elastic constants of γ-TiAl. *Philos. Mag. Lett.* **1996**, *73*, 71–78. [CrossRef]
40. Wang, L.; Shang, J.-X.; Wang, F.-H.; Zhang, Y.; Chroneos, A. Unexpected relationship between interlayer distances and surface/cleavage energies in γ-TiAl: Density functional study. *J. Phys. Condens. Matter* **2011**, *23*, 265009. [CrossRef]
41. Wang, J.W.; Gong, H.R. Adsorption and diffusion of hydrogen on Ti, Al, and TiAl surfaces. *Int. J. Hydrogen Energy* **2014**, *39*, 6068–6075. [CrossRef]
42. Ramamoorthy, M.; Vanderbilt, D.; King-Smith, R.D. First-principles calculations of the energetics of stoichiometric TiO_2 surfaces. *Phys. Rev. B* **1994**, *49*, 16721–16727. [CrossRef] [PubMed]
43. Perron, H.; Domain, C.; Roques, J.; Drot, R.; Simoni, E.; Catalette, H. Optimisation of accurate rutile TiO_2 (110), (100), (101) and (001) surface models from periodic DFT calculations. *Theor. Chem. Acc.* **2007**, *117*, 565–574. [CrossRef]
44. Moore, D.K.; Cherniak, D.J.; Watson, E.B. Oxygen diffusion in rutile from 750 to 1000 °C and 0.1 to 1000 MPa. *Am. Mineral.* **1998**, *83*, 700–711. [CrossRef]
45. Herzig, C.; Przeorski, T.; Mishin, Y. Self-diffusion in γ-TiAl: An experimental study and atomistic calculations. *Intermetallics* **1999**, *7*, 389–404. [CrossRef]
46. Zhang, W.; Smith, J.R.; Evans, A.G. The connection between *ab initio* calculations and interface adhesion measurements on metal/oxide systems: Ni/Al_2O_3 and Cu/Al_2O_3. *Acta Mater.* **2002**, *50*, 3803–3816. [CrossRef]
47. Feng, J.; Zhang, W.; Jiang, W. *Ab initio* study of Ag/Al_2O_3 and Au/Al_2O_3 interfaces. *Phys. Rev. B* **2005**, *72*, 115423. [CrossRef]
48. Melnikov, V.V.; Eremeev, S.V.; Kulkova, S.E. Adhesion of niobium films to differently oriented α-Al_2O_3 surfaces. *Tech. Phys.* **2011**, *56*, 1494–1500. [CrossRef]
49. Du, J.L.; Fang, Y.; Fu, E.G.; Ding, X.; Yu, K.Y.; Wang, Y.G.; Wang, Y.Q.; Baldwin, J.K.; Wang, P.P.; Bai, Q. What determines the interfacial configuration of Nb/Al_2O_3 and Nb/MgO interface. *Sci. Rep.* **2006**, *6*, 33931. [CrossRef]
50. Mei, Z.-G.; Bhattacharya, S.; Yacout, A.M. Adhesion of ZrN and Al_2O_3 coatings on U metal from first-principles. *Appl. Surf. Sci.* **2019**, *473*, 121–126. [CrossRef]
51. Manz, T.A.; Limas, N.G. Introducing DDEC6 atomic population analysis: Part 1. Charge partitioning theory and methodology. *RSC Adv.* **2016**, *6*, 47771–47801. [CrossRef]
52. Limas, N.G.; Manz, T.A. Introducing DDEC6 atomic population analysis: Part 2. Computed results for a wide range of periodic and nonperiodic materials. *RSC Adv.* **2016**, *6*, 45727–45747. [CrossRef]
53. Manz, T.A. Introducing DDEC6 atomic population analysis: Part 3. Comprehensive method to compute bond orders. *RSC Adv.* **2017**, *7*, 45552–45581. [CrossRef]
54. Griffith, A.A. The phenomena of rupture and ow in solids. *Trans. R. Soc. Lond.* **1921**, *A221*, 163–198. [CrossRef]
55. Zehnder, A. Griffith theory of fracture. In *Encyclopedia of Tribology*; Wang, Q.J., Chung, Y.-W., Eds.; Springer: New York, NY, USA, 2013; pp. 1570–1573. [CrossRef]

56. Bakulin, A.V.; Kulkov, S.S.; Kulkova, S.E. Adhesive properties of the TiAl/Al$_2$O$_3$ interface. *Russ. Phys. J.* **2020**, in press. [CrossRef]
57. Bakulin, A.V.; Kulkov, S.S.; Kulkova, S.E. Adhesion properties of clean and doped Ti$_3$Al/Al$_2$O$_3$ interface. *Appl. Surf. Sci.* **2020**, *536*, 147639. [CrossRef]
58. Bakulin, A.V.; Kulkova, S.E. Adhesion at the Ta(Mo)/NiTi Interface. *Tech. Phys. Lett.* **2019**, *45*, 620–624. [CrossRef]

MDPI

St. Alban-Anlage 66

4052 Basel

Switzerland

Tel. +41 61 683 77 34

Fax +41 61 302 89 18

www.mdpi.com

Metals Editorial Office

E-mail: metals@mdpi.com

www.mdpi.com/journal/metals